Sequences and Series

Calculus Practice Workbook

$$\sum_{n=1}^{\infty} \frac{1}{n(n+2)} = \frac{3}{4}$$

$$\frac{1}{1\cdot 3} + \frac{1}{2\cdot 4} + \frac{1}{3\cdot 5} + \frac{1}{4\cdot 6} + \cdots = \frac{3}{4}$$

Chris McMullen, Ph.D.

Sequences and Series
Calculus Practice Workbook
with Full Solutions
Chris McMullen, Ph.D.

www.improveyourmathfluency.com
www.monkeyphysicsblog.wordpress.com
www.chrismcmullen.com

Zishka Publishing
ISBN: 978-1-941691-48-9

Mathematics > Calculus

Contents

Introduction

This book is designed to help students learn or review essential concepts relating to the part of a calculus course that covers sequences and series. Sample topics include:

- arithmetic series
- geometric series
- infinite series
- limit theorems
- convergence and divergence
- comparison, integral, ratio, and root tests
- absolute convergence
- alternating series
- calculating or estimating the sum for a series
- power series
- Taylor series
- Maclaurin series
- binomial expansions
- convergence of power series

1 Sequences

A **sequence** is a set of elements in a particular order. In mathematics, we are primarily concerned with sequences of numbers, such as 1, 2, 4, 8, 16, 32, ..., though a sequence could consist of other kinds of elements like letters (a, b, c, d, e, ...). A sequence may be finite (like the set of 2-digit numbers) or infinite (like the set of odd numbers). In calculus, we are primarily interested in infinite sequences.

Another way to define **sequence** is a **function** where the domain consists of positive integers (1, 2, 3, 4, 5, ...). If we let $a(n)$ represent the function for positive integer n, the sequence is $a(1), a(2), a(3), a(4), a(5), \ldots$ The nth value of the sequence is $a(n)$. For example, the function $a(n) = n(n + 1)$ makes the sequence 2, 6, 12, 20, 30, 42,... Since n starts at 1, the first value is $a(1) = 1(2) = 2$, the second value is $a(2) = 2(3) = 6$, the third value is $a(3) = 3(4) = 12$, and so on. As another example, the function $b(k) = \frac{1}{k^2}$ makes the sequence $1, \frac{1}{4}, \frac{1}{9}, \frac{1}{16}, \frac{1}{25}, \frac{1}{36}, \frac{1}{49}, \ldots$

The elements of a sequence are called **terms**. For example, 2, 6, 12, and 20 are the first four terms of the sequence $a(n) = n(n + 1)$. The symbol used for the domain of the function is called the **index**. For example, n is the index for $a(n) = n(n + 1)$, whereas k is the index for $b(k) = \frac{1}{k^2}$. Letters from near the middle of the English alphabet (like i, j, k, ℓ, m, and n) are often used as index symbols. In contrast, letters from the end of the alphabet (like t, u, v, w, x, y, and z) are often used as continuous variables. Although f, g, and h are common for functions, we often use letters in the beginning of the alphabet (especially a and b) for sequences (since their domains are restricted to the set of positive integers). A sequence can also be represented by using **curly braces**. For example, $\{n(n + 1)\}$ means the entire sequence, whereas $n(n + 1)$ means the nth term of the sequence. A **subscript** can also represent the nth term, as in $a_n = n(n + 1)$.

Another way to specify a sequence is to express how consecutive terms are related. This is called a **recursion relation**. For example, the recursion relation $a_{n+1} = \frac{a_n}{2}$ along with the initial value $a_1 = \frac{1}{2}$ makes the sequence $\frac{1}{2}, \frac{1}{4}, \frac{1}{8}, \frac{1}{16}, \frac{1}{32}, \ldots$ As a second example, the recursion relation $b_{n+1} = b_n + 2$ along with the initial value $b_1 = 0$ makes the sequence $0, 2, 4, 6, 8, 10, \ldots$ The first example is equivalent to $\{a_n\} = \left\{\frac{1}{2^n}\right\}$ while the second example is equivalent to $\{b_n\} = \{2(n-1)\}$. Although in these cases it happens to be possible to specify the sequence either using a formula in terms of n or in terms of a recursion relation, sometimes only the recursion relation is known for a sequence. Other times, a formula is known, but the recursion relation is more practical. That's the case with factorials. For example, for the sequence $\{a_n\} = \{n!\} = 1, 2, 6, 24, 120, 720, \ldots$ it's simpler to find $a_7 = 7a_6 = 7(720) = 5040$ using the previous term than it is to multiply out $a_7 = 7! = 7(6)(5)(4)(3)(2)(1)$. Also, note that a recursion relation can involve more than just two consecutive terms. For example, the recursion relation $c_{n+2} = \frac{c_{n+1}}{c_n}$ where $c_1 = 2$ and $c_2 = 3$ makes the sequence $2, 3, \frac{3}{2}, \frac{1}{2}, \frac{1}{3}, \frac{2}{3}, 2, 3, \ldots$ (which is a repeating sequence). To find any given term, divide the previous term by the term prior to that. For example, we get the 5th term when $n = 3$ by dividing the 4th term by the 3rd term: $c_{3+2} = c_5 = \frac{c_{3+1}}{c_3} = \frac{c_4}{c_3} = \frac{1/2}{3/2} = \frac{1}{2} \div \frac{3}{2} = \frac{1}{2} \times \frac{2}{3} = \frac{1}{3}$. (Recall that the way to divide by a fraction is to multiply by its reciprocal.)

An important question in calculus is, "Which value, if any, does the sequence approach as the index grows to infinity?" Some sequences **converge** to a finite value, whereas other sequences don't – those are said to **diverge**. The sequence $\left\{\frac{1}{n}\right\} = 1, \frac{1}{2}, \frac{1}{3}, \frac{1}{4}, \frac{1}{5}, \ldots$ is an example of a converging sequence; it converges to zero as n grows to infinity. In contrast, the sequence $\{n^2\} = 1, 4, 9, 16, 25, 36, \ldots$ diverges; it becomes infinite as n grows to infinity. We will explore the convergence and divergence of **sequences** in Chapter 2. (This is not to be confused with the convergence and divergence of **series**, which will be explored in Chapters 4-10.)

For an **alternating** sequence, the sign of each term changes. The way to show this in a formula for a sequence is to include $(-1)^n$ or $(-1)^{n+1}$. For example, the sequence $\left\{\frac{(-1)^n}{n}\right\}$ is $-1, \frac{1}{2}, -\frac{1}{3}, \frac{1}{4}, -\frac{1}{5}, \ldots$ whereas the sequence $\left\{\frac{(-1)^{n+1}}{n}\right\}$ is $1, -\frac{1}{2}, \frac{1}{3}, -\frac{1}{4}, \frac{1}{5}, \ldots$ The reason for this is that $(-1)^1 = -1$, $(-1)^3 = -1$, $(-1)^5 = -1$, etc. In the second case, when $n = 2, 4, 6, 8, \ldots$ then $(n+1)$ is an odd number. For example, $(-1)^{4+1} = -1$.

A few common sequences worth noting include:

- The **prime number** sequence is 2, 3, 5, 7, 11, 13, 17, 19, 23, ...
- The **Fibonacci sequence** begins with ones and adds the two previous terms together to make the next term. It has the recursion relation $a_{n+1} = a_n + a_{n-1}$ with $a_1 = a_2 = 1$. This sequence is 1, 1, 2, 3, 5, 8, 13, 21, 34, ...
- The **harmonic** sequence is $\left\{\frac{1}{n}\right\} = 1, \frac{1}{2}, \frac{1}{3}, \frac{1}{4}, \frac{1}{5}, \ldots$ As we will see in future chapters, the corresponding **series** is particularly important in calculus.

Example 1. Write out the first five terms for the sequence $\{5(n-1)\}$.
Solution: Beginning with $n = 1$, plug successive values of n into the formula $5(n-1)$.

- $a_1 = 5(1-1) = 5(0) = 0$
- $a_2 = 5(2-1) = 5(1) = 5$
- $a_3 = 5(3-1) = 5(2) = 10$
- $a_4 = 5(4-1) = 5(3) = 15$
- $a_5 = 5(5-1) = 5(4) = 20$

Example 2. Write out the first five terms for the sequence $\left\{\frac{2}{n^3}\right\}$.

Solution: Beginning with $n = 1$, plug successive values of n into the formula $\frac{2}{n^3}$.

- $a_1 = \frac{2}{1^3} = \frac{2}{1} = 2$
- $a_2 = \frac{2}{2^3} = \frac{2}{8} = \frac{1}{4}$
- $a_3 = \frac{2}{3^3} = \frac{2}{27}$
- $a_4 = \frac{2}{4^3} = \frac{2}{64} = \frac{1}{32}$
- $a_5 = \frac{2}{5^3} = \frac{2}{125}$

Example 3. Write out the first five terms for the sequence starting with $a_1 = 3$ given by the recursion formula $a_{n+1} = a_n^2 - 2$.
Solution: Begin with $a_1 = 3$. Use the recursion formula to get the remaining terms: $a_2 = 3^2 - 2 = 9 - 2 = 7$, $a_3 = 7^2 - 2 = 49 - 2 = 47$, $a_4 = 47^2 - 2 = 2209 - 2 = 2207$, and $a_5 = 2207^2 - 2 = 4{,}870{,}849 - 2 = 4{,}870{,}847$.

Example 4. Write out the first five terms for the sequence starting with $a_1 = \frac{1}{4}$ and $a_2 = \frac{1}{2}$ given by the recursion formula $a_{n+2} = a_n a_{n+1}$.

Solution: Begin with $a_1 = \frac{1}{4}$ and $a_2 = \frac{1}{2}$. Use the recursion formula to get the remaining terms: $a_3 = \frac{1}{4}\left(\frac{1}{2}\right) = \frac{1}{8}$, $a_4 = \frac{1}{2}\left(\frac{1}{8}\right) = \frac{1}{16}$, and $a_5 = \frac{1}{8}\left(\frac{1}{16}\right) = \frac{1}{128}$.

Example 5. For the sequence below, (A) give a formula for the nth term in terms of n and (B) predict the next two terms of the sequence.

$$5, 10, 17, 26, 37, 50, 65, \ldots$$

Solution: (A) The challenge is to identify some kind of pattern that you know about. Some **common patterns** include multiples (like 7, 14, 21, 28, 35, ...), prime numbers, powers (like squares or cubes), reciprocals (like $1, \frac{1}{2}, \frac{1}{3}, \frac{1}{4}, \frac{1}{5}, \ldots$), the Fibonacci sequence (1, 1, 2, 3, 5, 8, 13, ...), functions (like $\sin 30°$, $\sin 60°$, $\sin 90°$, $\sin 120°$, ...), factorials (like 1!, 2!, 3!, 4!, 5!, ...), or variations or combinations of these. Now examine this list of common patterns and look at the sequence above. Does this sequence look similar to multiples, primes, powers, reciprocals, the Fibonacci sequence, a known function, or a factorial? If you don't see it yet, it's because this sequence has been shifted. Try subtracting one from every value and see what we get: 4, 9, 16, 25, 36, 49, 64, ... Aha! These are perfect squares. If we take the perfect squares, starting with 2^2, and add 1, we can make the sequence. But there is a problem. The way that a sequence has been defined, we need to begin the sequence with $n = 1$, not 2. That's easy to fix; just add 1 to each value of n. We'll add 1 to n, square this sum, and then add 1 again. This gives us the formula $a_n = (n+1)^2 + 1$. Try plugging in different values of n, starting with $n = 1$, and you'll make the sequence 5, 10, 17, 26, 37, 50, 65, ...

(B) When $n = 7$, we get $(7+1)^2 + 1 = 8^2 + 1 = 65$. Use $n = 8$ and 9 to get the next two terms: $(8+1)^2 + 1 = 9^2 + 1 = 82$ and $(9+1)^2 + 1 = 10^2 + 1 = 101$.

Example 6. For the sequence below, (A) give a formula for the nth term in terms of n and (B) predict the next two terms of the sequence.

$$\frac{\sqrt{3}}{2}, \frac{1}{2}, 0, -\frac{1}{2}, -\frac{\sqrt{3}}{2}, \ldots$$

Solution: (A) Review the list of common kinds of sequences from Example 4. Does this sequence resemble anything you know? You should recognize that these values are $\cos 30°, \cos 60°, \cos 90°, \cos 120°, \cos 150°, \ldots$ To start with $n = 1$, we need to write the angle as the product of 30° and n. The formula is $a_n = \cos(30°n)$. If you prefer to work with radians, the answer is $a_n = \cos\left(\frac{n\pi}{6}\right)$, since $\frac{\pi}{6}$ radians corresponds to 30°.

(B) When $n = 5$, we get $a_n = \cos(150°)$. Use $n = 6$ and 7 to get the next two terms: $\cos(180°) = -1$ and $\cos(210°) = -\frac{\sqrt{3}}{2}$.

Example 7. For the sequence below, (A) give a recursion formula for a_{n+1} in terms of the previous term, and (B) predict the next two terms of the sequence.

$$5, 7, 11, 19, 35, 67, 131, \ldots$$

Solution: (A) To figure this question out, think about how each term is related to the previous term. Notice that for the bigger numbers, each number is nearly double the number before it. For example, 35 times 2 equals 70, which is 3 more than 67. Also, 19 times 2 equals 38, which is 3 more than 35. This works for all of the numbers. The recursion formula is $a_{n+1} = 2a_n - 3$. For example, $a_4 = 2a_3 - 3 = 2(11) - 3 = 19$.

(B) Use the recursion formula: $2(131) - 3 = 259$ and $2(259) - 3 = 515$.

Chapter 1 Problems

Directions: Write out the first five terms for each sequence.

(1) $\{3n-4\}$

(2) $\left\{\frac{m}{m+1}\right\}$

(3) $a_{n+1} = 5a_n$ with $a_1 = 4$

(4) $\{(-1)^n(2n)\}$

(5) $\{k(k-2)\}$

(6) $b_{i+2} = 5b_{i+1} - 3b_i$ with $b_1 = b_2 = 1$

(7) $\left\{\frac{24}{n!}\right\}$

(8) $a_{k+1} = \frac{a_k}{k}$ with $a_1 = 2$

(9) $c_{n+1} = c_n + 2n$ with $c_1 = 3$

(10) $\{3^\ell\}$

(11) $\left\{\frac{3m^2-2}{m^2+5}\right\}$

(12) $\left\{\frac{2^k}{k^3}\right\}$

(13) $\{1 + (-1)^i\}$

(14) $a_{j+2} = a_j a_{j+1}$ with $a_1 = 2$ and $a_2 = 3$

(15) $B_{n+1} = 2B_n^2 - 12$ with $B_1 = 2$

(16) $\left\{\frac{(-1)^{n+1}}{n+1}\right\}$

(17) $\{(-5)^n\}$

(18) $c_{n+1} = c_n(c_n - 4)$ with $c_1 = 6$

(19) $\{0.2^{\ell-1}\}$

(20) $\frac{1}{A_{k+2}} = \frac{1}{A_{k+1}} + \frac{1}{A_k}$ with $A_1 = 2$ and $A_2 = 3$

Directions: (A) For each sequence below, either give a formula for the nth term or give a recursion formula. (B) Predict the next two terms of each sequence.

(21) $-2, 1, 4, 7, 10, 13, \ldots$

(22) $\frac{3}{5}, \frac{2}{3}, \frac{5}{7}, \frac{3}{4}, \frac{7}{9}, \frac{4}{5}, \ldots$

(23) $3, 4, 7, 11, 18, 29, \ldots$

(24) $7, 26, 63, 124, 215, 342, \ldots$

(25) $1, -\frac{1}{3}, \frac{1}{9}, -\frac{1}{27}, \frac{1}{81}, -\frac{1}{243}, \ldots$

(26) $0.9, 0.99, 0.999, 0.9999, 0.99999, \ldots$

(27) $-4, 22, -130, 778, -4666, 27{,}994$

(28) $\frac{1}{9}, 1, 1, \frac{1}{9}, \frac{1}{25}, \frac{1}{49}, \frac{1}{81}, \frac{1}{121}, \ldots$

(29) $2\sqrt{3}, 4, 2\sqrt{5}, 2\sqrt{6}, 2\sqrt{7}, 4\sqrt{2}, \ldots$

(30) $-7, -11, -16, -22, -29, -37, \ldots$

2 Limiting Values

Some sequences approach a unique, finite **limiting value** as the index grows to infinity; such sequences are said to **converge** to that limiting value. As examples, the harmonic sequence $\left\{\frac{1}{n}\right\} = 1, \frac{1}{2}, \frac{1}{3}, \frac{1}{4}, \frac{1}{5}, \frac{1}{6}, \frac{1}{7}, \ldots$ approaches the limiting value of zero as n grows to infinity and the sequence $\left\{\frac{2k}{k+1}\right\} = 1, \frac{4}{3}, \frac{3}{2}, \frac{8}{5}, \frac{5}{3}, \frac{12}{7}, \frac{7}{4}, \frac{16}{9}, \ldots$ approaches the limiting value of 2 as k grows to infinity. Other sequences **diverge**. They either become infinite as the index grows to infinity, like the sequence $\{m\} = 1, 2, 3, 4, 5, 6, 7, \ldots$, or they remain finite without approaching a unique value, like the sequence $\left\{\sin\frac{i\pi}{6}\right\} = \frac{1}{2}, \frac{\sqrt{3}}{2}, 1, \frac{\sqrt{3}}{2}, \frac{1}{2}, 0, -\frac{1}{2}, -\frac{\sqrt{3}}{2}, -1, -\frac{\sqrt{3}}{2}, -\frac{1}{2}, 0, \frac{1}{2}, \frac{\sqrt{3}}{2}, \ldots$ which repeats forever without approaching any one specific value. (The angle was given in radians.)

A **limiting value** is defined more formally in calculus as follows. Let $\{a_n\}$ represent a sequence and let n be the index. If $\{a_n\}$ **converges** to the limiting value L, then for any positive value ϵ, there exists a value N where for every value of the index for which $n > N$ it is true that $|a_n - L| < \epsilon$. Otherwise, $\{a_n\}$ **diverges**. Consider these examples:

- For $\left\{\frac{1}{n}\right\}$, pick any positive number ϵ. If $N > \frac{1}{\epsilon}$, then for any $n > N$, we will have $n > \frac{1}{\epsilon}$, which means that $\frac{1}{n} < \epsilon$. (For example, if $\epsilon = 0.1$, then $\frac{1}{\epsilon} = 10$. One value of n that satisfies $n > \frac{1}{\epsilon}$ is 11. Observe that $\frac{1}{11} < 0.1$.) Since $\frac{1}{n} < \epsilon$, the condition $|a_n - L| < \epsilon$ will be satisfied if $|a_n - L| = \frac{1}{n}$, which is true if $L = 0$. This shows that the harmonic sequence $\{a_n\} = \left\{\frac{1}{n}\right\}$ converges to the limiting value $L = 0$.
- Although the sequence $\left\{\sin\frac{n\pi}{6}\right\}$ is range-bound between -1 and 1, it diverges because it doesn't converge to any specific value. For example, we will show that it doesn't converge to $L = 0$. Let $\epsilon = 0.01$. In this case, we would need $|a_n - L| = \left|\sin\frac{n\pi}{6} - 0\right| = \sin\frac{n\pi}{6}$ to be less than 0.01. For this repeating sequence, for any N, you can find an $n > N$ for which $\sin\frac{n\pi}{6} = \frac{1}{2}$, which is greater than 0.01.

Fortunately, many problems about the convergence or divergence of sequences can be solved without applying the formal definition directly. We will learn some more **practical** limit theorems after discussing the uniqueness of limiting values.

If the sequence $\{a_n\}$ **converges** to the limiting value L, then L must be unique. The way to see this is to assume that $\{a_n\}$ converges to two limiting values L_1 and L_2 where $L_2 \neq L_1$. If so, there exists some N such that for every $n > N$, $|a_n - L_1| < \epsilon$ for any positive value of ϵ. If we choose any positive ϵ smaller than one-half of the difference between L_1 and L_2, that is $\epsilon < \frac{L_2 - L_1}{2}$, then you won't be able to find an M such that for every $n > M$, $|a_n - L_2|$ is less than ϵ. This is easier to see with numbers. Suppose that $L_1 = 1$ and $L_2 = 2$. Then $\frac{L_2 - L_1}{2} = \frac{2-1}{2} = \frac{1}{2} = 0.5$. Consider $\epsilon = 0.4$. If there is some N such that for every $n > N$, then $|a_n - 1| < 0.4$ if $\{a_n\}$ converges to $L_1 = 1$. In order for the sequence to also converge to $L_2 = 2$, we need to find some M such that for every $n > M$, then $|a_n - 2| < 0.4$. This is an obvious contradiction, which shows that the same sequence can't converge to both $L_1 = 1$ and $L_2 = 2$. (Why? a_n must be less than 1.4 if $|a_n - 1| < 0.4$, but for any value of a_n less than 1.4, $|a_n - 2|$ will be at least 0.6). A sequence can converge to L_1 or L_2, but not both (if $L_2 \neq L_1$).

Now we will explore some **practical** limit theorems. These will be useful for solving problems concerning the convergence or divergence of **sequences**. (If instead you want to test the convergence or divergence of a **series**, see Chapters 4-10.)

If $\{a_n\}$ **converges** to the **limiting value** L, we can express this as a limit as $\lim_{n\to\infty} a_n = L$. If $\lim_{n\to\infty} a_n = \infty$, if $\lim_{n\to\infty} a_n = -\infty$, or if $\lim_{n\to\infty} a_n$ does not exist then $\{a_n\}$ **diverges**. For example, for $\{a_n\} = \{n\}$, $\lim_{n\to\infty} a_n = \infty$, for $\{a_n\} = \{-n\}$, $\lim_{n\to\infty} a_n = -\infty$, and for $\{a_n\} = \left\{\sin\frac{n\pi}{6}\right\}$, $\lim_{n\to\infty} a_n$ does not exist (because this function doesn't converge to any specific value, as discussed on the previous page). If we are concerned with the convergence or divergence of a sequence, we wish to investigate the limit $\lim_{n\to\infty} a_n$. The properties of limits that are taught in first-semester calculus apply to limits of the form $\lim_{n\to\infty} a_n$. In particular, if $\lim_{n\to\infty} a_n = L_a$ and $\lim_{n\to\infty} b_n = L_b$, then:

- $\lim_{n\to\infty}(a_n + b_n) = L_a + L_b$. For example, $\left\{n + \frac{1}{n}\right\}$ diverges because $\lim_{n\to\infty}\left(n + \frac{1}{n}\right)$ $= \lim_{n\to\infty} n + \lim_{n\to\infty}\frac{1}{n} = \infty + 0 = \infty$.
- $\lim_{n\to\infty}(ca_n) = cL_a$ where c doesn't depend on the index n. For example, $\left\{\frac{42}{n}\right\}$ converges to zero because $\lim_{n\to\infty}\frac{42}{n} = 42\lim_{n\to\infty}\frac{1}{n} = 42(0) = 0$.
- $\lim_{n\to\infty}\frac{a_n}{b_n} = \frac{L_a}{L_b}$ (provided that the b_n's and $L_b \neq 0$). **Tip**: If the numerator and denominator are polynomials, divide each by the highest power of n. For example, for $\left\{\frac{2n^2}{n^2+n}\right\}$, divide by n^2 in the numerator and denominator: $\lim_{n\to\infty}\frac{2n^2}{n^2+n}$ $= \lim_{n\to\infty}\frac{2}{1+\frac{1}{n}}$. Now apply the rule for dividing limits: $\lim_{n\to\infty}\frac{2}{1+\frac{1}{n}} = \frac{\lim_{n\to\infty} 2}{\lim_{n\to\infty}\left(1+\frac{1}{n}\right)} = \frac{2}{1} = 1$. In the last step, we also applied the rule for adding limits.
- Similar to the previous rule, $\lim_{n\to\infty} a_n b_n = L_a L_b$ and $\lim_{n\to\infty} a_n^c = L_a^c$ (if $c, a_n > 0$).

The **sandwich theorem** takes the following form. If there exists a value N where for every value of the index for which $n > N$ it is true that $a_n \leq b_n \leq c_n$, if $\lim_{n\to\infty} a_n = L$ and $\lim_{n\to\infty} c_n = L$, it follows that $\lim_{n\to\infty} b_n = L$. For example, consider $\left\{\frac{\sin n}{n}\right\}$. Let $\{a_n\} = \{0\}$, $\{b_n\} = \left\{\frac{\sin n}{n}\right\}$, and $\{c_n\} = \frac{1}{n}$. Observe that $a_n \leq b_n \leq c_n$ for all n. Also note that $\lim_{n\to\infty} a_n = \lim_{n\to\infty} 0 = 0$ and $\lim_{n\to\infty} c_n = \lim_{n\to\infty}\frac{1}{n} = 0$. It follows that $\lim_{n\to\infty} b_n = \lim_{n\to\infty}\frac{\sin n}{n} = 0$. (Note here that $n \to \infty$. You're probably familiar with the important result from first-semester calculus that $\lim_{x\to 0}\frac{\sin x}{x} = 1$, where $x \to 0$, whereas in the sandwich theorem, we instead have $n \to \infty$. You don't want to confuse these two different limits.)

The sandwich theorem is helpful with **alternating** sequences of the form $\{(-1)^n c_n\}$ or $\{(-1)^{n+1} c_n\}$ when $\lim_{n\to\infty} c_n = 0$. Let $\{a_n\} = \{-c_n\}$ and $\{b_n\} = \{(-1)^n c_n\}$ or $\{(-1)^{n+1} c_n\}$. Then $a_n \leq b_n \leq c_n = -c_n \leq b_n \leq c_n$ since $b_n = \pm c_n$ for all n. In such cases, $\lim_{n\to\infty} a_n = \lim_{n\to\infty} c_n = 0$, such that $\lim_{n\to\infty} b_n = 0$. For example, $\left\{\frac{(-1)^n}{n^2}\right\}$ converges to zero since $\left\{\frac{1}{n^2}\right\}$ converges to zero. Before using this result, make sure that $\lim_{n\to\infty} c_n = 0$.

If the sequence $\{a_n\}$ converges to the limiting value L, meaning that $\lim_{n\to\infty} a_n = L$, then the sequence $\{f(a_n)\}$ converges to the limiting value $f(L)$, meaning that $\lim_{n\to\infty} f(a_n) = f(L)$, provided that the **function** f is defined at the a_n's and is continuous at L. For example, consider $\left\{\cos\left(\frac{1}{n}\right)\right\}$. Let $f(a_n) = \cos a_n$ and $a_n = \frac{1}{n}$. Since $\{a_n\} = \left\{\frac{1}{n}\right\}$ converges to $L = 0$, $f(a_n)$ is defined for all a_n, and $f(a_n) = \cos a_n$ is continuous at $f(L) = \cos 0$, the sequence $\{f(a_n)\} = \{\cos a_n\} = \left\{\cos\left(\frac{1}{n}\right)\right\}$ converges to $f(L) = \cos 0 = 1$. (Unlike the example with sine that we saw earlier in this chapter, here the index n appears in the denominator of the angle. As n grows to infinity, the angle in radians approaches zero, and the cosine of zero is one.)

You'll want to distinguish the previous theorem concerning **functions** with the one that follows. In the previous theorem, we know that the sequence $\{a_n\}$ converges and use it to find the limiting value of $\{f(a_n)\}$. In the following theorem, we know that the function $f(x)$ converges and use it to find the limiting value of $\{a_n\} = \{f(n)\}$. Another important distinction is that the previous theorem involves the function $f(a_n)$, while the following theorem involves the function $f(n)$, which has a different argument.

If the **function** $f(x)$ approaches L as x grows to infinity, meaning that $\lim_{x\to\infty} f(x) = L$ (this is a standard first-semester calculus limit that has nothing to do with sequences), then the sequence $\{a_n\} = \{f(n)\}$ converges to the limiting value L, meaning that $\lim_{n\to\infty} a_n = L$, provided that $f(x)$ is defined for all $x \geq N$. Here, $a_n = f(n)$ for $x \geq N$. This theorem is often applied using **l'Hôpital's rule**. You should have learned l'Hôpital's rule in first-semester calculus, but we will review it now just in case.

When finding the limit of a ratio of two functions, if both functions approach zero or if both functions grow to $\pm\infty$ (these are called indeterminant forms), apply **l'Hôpital's rule** as follows: Take the derivative of the numerator and the denominator separately and evaluate each at the limit. **Don't** try to apply l'Hôpital's rule to every limit that has a ratio of two functions; only use it when both functions approach zero or when both functions approach $\pm\infty$ in the limit. For example, you may apply l'Hôpital's rule to $\lim_{x\to 0} \frac{\sin x}{x}$ ($\sin x$ and x both $\to 0$), but **not** to $\lim_{x\to\infty} \frac{\sin x}{x}$ (x grows to ∞, but $\sin x$ doesn't).

As an example using the previous limit theorem with l'Hôpital's rule, consider $\left\{\frac{\ln n}{n^2}\right\}$. The corresponding function $f(x) = \frac{\ln x}{x^2}$ is defined for all $x \geq 1$. (It's actually defined for $x > 0$, but since n starts at 1 for a sequence, we don't need to check below 1. We don't even need it to be defined at 1; we just need any N for which the function is defined for all $x \geq N$. In this example, we may let $N = 1$, since that works.) Now we need to check if the limit $\lim_{x\to\infty} f(x) = \lim_{x\to\infty} \frac{\ln x}{x^2}$ exists. Observe that $\ln x$ and x^2 both grow to ∞ as $x \to \infty$, which is one of the indeterminate forms. Thus, we may apply l'Hôpital's rule. Take a derivative of the numerator with respect to x, take a derivative of the denominator with respect to x, and evaluate each as $x \to \infty$.

$$\lim_{x\to\infty} \frac{\ln x}{x^2} = \frac{\left.\frac{d}{dx}\ln x\right|_{x\to\infty}}{\left.\frac{d}{dx}x^2\right|_{x\to\infty}} = \frac{\lim_{x\to\infty}\frac{1}{x}}{\lim_{x\to\infty} 2x} = \frac{1/\infty}{2(\infty)} = \frac{0}{\infty} = 0$$

Since $\lim_{x\to\infty} \frac{\ln x}{x^2} = 0$ and $\frac{\ln x}{x^2}$ is defined for all $x \geq 1$, it follows that $\lim_{n\to\infty} \frac{\ln n}{n^2} = 0$.

Since l'Hôpital's rule involves derivatives and since expressions of the form c^n show up in a variety of sequences, it's worth remembering the following derivative (that you should have encountered in first-semester calculus) where c is a constant:

$$\frac{d}{dx}c^x = c^x \ln c$$

For the special case where the base is Euler's number ($e \approx 2.71828$), this simplifies to

$$\frac{d}{dx}e^x = e^x$$

according to the identity $\ln e = 1$.

If $\{a_n\}$ converges to the limiting value $\lim_{n\to\infty} a_n = L$ and $L \neq 0$ (and $a_n \neq 0$), then $\left\{\frac{1}{a_n}\right\}$ converges to the limiting value $\lim_{n\to\infty} \frac{1}{a_n} = \frac{1}{L}$. If $\lim_{n\to\infty} a_n = 0$, then $\left\{\frac{1}{a_n}\right\}$ diverges.

What really matters for the convergence of a sequence is the **tail** of the sequence. The **tail** includes all of the terms greater than some value N. The beginning of a sequence doesn't matter; the behavior of the tail determines whether or not it converges. You can see this if you review the formal definition of a limiting value (specifically, $n > N$).

A few special cases are worth remembering because they show up often.

- $\left\{\frac{1}{n^c}\right\}$ converges to $\lim_{n\to\infty} \frac{1}{n^c} = 0$ provided that $c > 0$. (What if $c < 0$? That's equivalent to the next bullet point. What if $c = 0$? Then you get $\left\{\frac{1}{n^c}\right\} = \left\{\frac{1}{n^0}\right\} = \{1\}$.) For example, $\left\{\frac{1}{n^2}\right\}$ converges to zero.
- $\{n^c\}$ diverges if $c > 0$. (Note that if $|c| < 1$, that's equivalent to the previous bullet point.) For example, $\{n^2\}$ diverges.
- $\left\{\frac{1}{c^n}\right\}$ converges to $\lim_{n\to\infty} \frac{1}{c^n} = 0$ provided that $|c| > 1$. (What if $|c| < 1$? That's equivalent to the next bullet point. What if $|c| = 1$? Then you get $\lim_{n\to\infty} 1^n = 1$.) For example, $\left\{\frac{1}{2^n}\right\}$ converges to zero.
- $\{c^n\}$ diverges if $|c| > 1$. (Note that if $|c| < 1$, that's equivalent to the previous bullet point.) For example, $\{2^n\}$ diverges.
- $\left\{\frac{c^n}{n!}\right\}$ converges to $\lim_{n\to\infty} \frac{c^n}{n!} = 0$ for any value of c. For example, $\left\{\frac{7^n}{n!}\right\}$ converges to 0. The reason behind this is that the factorial function grows much more rapidly than the power function.
- $\left\{\frac{n!}{c^n}\right\}$ diverges. Let $a_n = \frac{c^n}{n!}$. Since $\lim_{n\to\infty} a_n = 0$, $\left\{\frac{1}{a_n}\right\} = \left\{\frac{n!}{c^n}\right\}$ diverges. For example, $\left\{\frac{n!}{7^n}\right\}$ diverges.
- A **repeating** pattern like 1, 3, 5, 7, 9, 1, 3, 5, 7, 9, 1, 3, ... diverges because it doesn't converge to any one value. (However, 2, 2, 2, 2, 2, 2, ... converges to 2.)

Example 1. Does $\left\{\frac{6}{n^4}\right\}$ converge or diverge? If it converges, give the limiting value.

Solution: $\left\{\frac{1}{n^4}\right\}$ converges to zero because $\left\{\frac{1}{n^c}\right\}$ converges to zero if $c > 0$. Then $\left\{\frac{6}{n^4}\right\}$ **converges** to zero according to the rule $\lim_{n\to\infty}(ca_n) = cL_a$. Here, $a_n = \frac{1}{n^4}$ and $L_a = 0$. (The c's appearing in these two different limit theorems aren't the same, but this point doesn't really matter here. If you prefer, you may call one c_1 and the other c_2.)

Example 2. Does $\{-3k^4\}$ converge or diverge? If it converges, give the limiting value.
Solution: $\{k^4\}$ **diverges** because $\{k^c\}$ diverges if $c > 0$. If you multiply the sequence $\{k^4\}$ by -3, it will still diverge.

Example 3. Does $\left\{\sin\frac{n}{2}\right\}$ converge or diverge? If it converges, give the limiting value.

Solution: $\left\{\sin\frac{n}{2}\right\}$ **diverges** because sine is periodic; this repeating pattern diverges as it never converges to any single value.

Example 4. Does $\left\{\sin\frac{2}{n}\right\}$ converge or diverge? If it converges, give the limiting value.

Solution: Unlike the previous example, this isn't a repeating pattern. With n in the denominator of the angle and starting at 1, this sequence is in Quadrant I for $n \geq 2$. We'll use the first limit theorem pertaining to functions. Let $f(a_n) = \sin a_n$ and $a_n = \frac{2}{n}$. We know that $\{a_n\} = \left\{\frac{2}{n}\right\}$ converges to $L = 0$ since $\left\{\frac{1}{n}\right\}$ converges to zero, according to the rule $\lim_{n\to\infty}(ca_n) = cL_a$. Since $f(a_n) = \sin a_n$ is continuous at $f(L) = \sin 0$, $\{f(a_n)\}$ $= \{\sin a_n\} = \left\{\sin\left(\frac{2}{n}\right)\right\}$ **converges** to $f(L) = \sin 0 = 0$.

Example 5. Does $\left\{\frac{1}{2^\ell}\right\}$ converge or diverge? If it converges, give the limiting value.

Solution: $\left\{\frac{1}{2^\ell}\right\}$ **converges** to zero because $\left\{\frac{1}{c^\ell}\right\}$ converges to $\lim_{\ell\to\infty}\frac{1}{c^\ell} = 0$ if $|c| > 1$. We could prove this directly using the sandwich theorem, since $0 \leq \frac{1}{2^\ell} \leq \frac{1}{\ell}$. (When $\ell = 1$, $\frac{1}{2^1} < \frac{1}{1}$, when $\ell = 2$, $\frac{1}{2^2} < \frac{1}{2}$, when $\ell = 3$, $\frac{1}{2^3} < \frac{1}{3}$, when $\ell = 4$, $\frac{1}{2^4} < \frac{1}{4}$, and so on.)

Example 6. Does $\{(-1)^m(m+2)\}$ converge or diverge? If it converges, give the limiting value.

Solution: $\{m\}$ diverges, which means that $\{m+2\}$ also diverges; multiplying by $(-1)^m$ will make the sequence alternate, but won't keep it from **diverging**.

Example 7. Does $\left\{\frac{(-1)^i}{i+2}\right\}$ converge or diverge? If it converges, give the limiting value.

Solution: $\left\{\frac{1}{i}\right\}$ converges to zero; $\left\{\frac{1}{i+2}\right\}$ also converges to zero because all it does is shift the index by two. If $\{c_i\} = \left\{\frac{1}{i+2}\right\}$ converges to zero, $\{(-1)^i c_i\} = \left\{\frac{(-1)^i}{i+2}\right\}$ **converges** to zero according to the sandwich theorem. See the paragraph in the chapter following the sandwich theorem regarding alternating sequences of the form $\{(-1)^i c_i\}$ for which $\lim_{i\to\infty} c_i = 0$.

Example 8. Does $a_n = \frac{a_{n-1}}{3}$ with $a_1 = \frac{1}{3}$ converge or diverge? If it converges, give the limiting value.

Solution: This problem gave the recursion formula instead of a formula for n. First use the recursion formula to see what the sequence looks like: $a_1 = \frac{1}{3}$, $a_2 = \frac{1}{9}$, $a_3 = \frac{1}{27}$, $a_4 = \frac{1}{81}$, $a_5 = \frac{1}{243}$, ... This sequence is $\frac{1}{3}, \frac{1}{9}, \frac{1}{27}, \frac{1}{81}, \frac{1}{243}, \ldots$ These are the reciprocals of the powers of 3. This sequence is the same as $\left\{\frac{1}{3^n}\right\}$, which **converges** because $\left\{\frac{1}{n^c}\right\}$ converges to zero if $c > 0$. Here, $c = 3$.

Example 9. Does $\left\{\frac{2j}{j-\frac{3}{2}}\right\}$ converge or diverge? If it converges, give the limiting value.

Solution: When the sequence is a ratio of polynomials, there are two ways to solve the problem. One way is to apply l'Hôpital's rule to the function $f(x) = \frac{2x}{x-\frac{3}{2}}$:

$$\lim_{x\to\infty} \frac{2x}{x-\frac{3}{2}} = \frac{\left.\frac{d}{dx}2x\right|_{x\to\infty}}{\left.\frac{d}{dx}\left(x-\frac{3}{2}\right)\right|_{x\to\infty}} = \frac{\lim\limits_{x\to\infty} 2}{\lim\limits_{x\to\infty} 1} = \frac{2}{1} = 2$$

Since $\lim\limits_{x\to\infty} \frac{2x}{x-\frac{3}{2}} = 2$ and $\frac{2x}{x-\frac{3}{2}}$ is defined for all $x \geq 2$ (observe that $x = \frac{3}{2}$ is a problem, as it involves division by zero), it follows that $\lim\limits_{j\to\infty} \frac{2j}{j-\frac{3}{2}} = 2$. The sequence **converges**.

Another way to solve the problem is to divide the numerator and denominator each by the highest power of the index, which in this case is simply j.

$$\lim_{j\to\infty} \frac{2j}{j-\frac{3}{2}} = \lim_{j\to\infty} \frac{2}{1-\frac{3}{2j}} = \frac{2}{1-0} = \frac{2}{1} = 2$$

Example 10. Does $\left\{\frac{j^2}{2j+1}\right\}$ converge or diverge? If it converges, give the limiting value.

Solution: As in Example 9, you may either apply l'Hôpital's rule or divide by j^2 in the numerator and denominator. (Here, j^2 is the highest power of the index.)

$$\lim_{x\to\infty} \frac{x^2}{2x+1} = \frac{\left.\frac{d}{dx}x^2\right|_{x\to\infty}}{\left.\frac{d}{dx}(2x+1)\right|_{x\to\infty}} = \frac{\lim\limits_{x\to\infty} 2x}{\lim\limits_{x\to\infty} 2} = \frac{\infty}{2} = \infty$$

Since $\lim_{x\to\infty} \frac{x^2}{2x+1} = \infty$, this sequence **diverges**.

Another way to solve the problem is to divide the numerator and denominator each by the highest power of the index, which in this case is j^2.

$$\lim_{j\to\infty} \frac{j^2}{2j+1} = \lim_{j\to\infty} \frac{1}{\frac{2}{j}+\frac{1}{j^2}} = \frac{1}{0+0} = \frac{2}{0} = \infty$$

Tip: If the formula for a sequence involves a ratio of polynomials, like Examples 9-10, look at the **degree** of each polynomial. The **degree** is the highest power.

- If the degree of the numerator is less than the degree of the denominator, like $\left\{\frac{4n^3-5n}{8n^4+1}\right\}$, the sequence converges to zero. Here, the 3 of n^3 is less than the 4 of n^4.
- If the degree of the numerator is higher than the degree of the denominator, like $\left\{\frac{3n^6+5n^3-7}{n^5-n^2}\right\}$ or like Example 10, the sequence diverges. Here, the 6 of n^6 is higher than the 5 of n^5.
- If the degree of the numerator and denominator are the same, like $\left\{\frac{6n^2-n}{2n^2+9}\right\}$ or like Example 9, the sequence converges to the ratio of the coefficients of the leading powers. Here, the sequence converges to $\frac{6}{2} = 3$ since $6n^2$ and $2n^2$ are the leading powers of the polynomials.

Example 11. Does $\left\{\sqrt{\frac{1}{m}}\right\}$ converge or diverge? If it converges, give the limiting value.

Solution: Use the first limit theorem pertaining to functions. Let $f(a_m) = \sqrt{a_m}$ and $a_m = \frac{1}{m}$. Since $\left\{\frac{1}{m}\right\}$ converges to $L = 0$ and since $f(a_m) = \sqrt{a_m}$ is continuous at $f(L) = \sqrt{0} = 0$, $\{f(a_m)\} = \sqrt{a_m}$ **converges** to $f(L) = \sqrt{0} = 0$.

Example 12. Does $\left\{\frac{2^n}{n!}\right\}$ converge or diverge? If it converges, give the limiting value.

Solution: $\left\{\frac{2^n}{n!}\right\}$ **converges** to $\lim_{n\to\infty} \frac{2^n}{n!} = 0$ because $\left\{\frac{c^n}{n!}\right\}$ converges to zero for all c. Here, $c = 2$.

Example 13. Does $\left\{\frac{n!}{2^n}\right\}$ converge or diverge? If it converges, give the limiting value.

Solution: $\left\{\frac{n!}{2^n}\right\}$ **diverges** because $\left\{\frac{n!}{c^n}\right\}$ diverges for all c. Here, $c = 2$.

Example 14. Does $\left\{\frac{\ln n}{2^n}\right\}$ converge or diverge? If it converges, give the limiting value.

Solution: Apply l'Hôpital's rule to the function $f(x) = \frac{\ln x}{2^x}$. Recall from calculus that $\frac{d}{dx}\ln x = \frac{1}{x}$ and $\frac{d}{dx}2^x = 2^x \ln 2$. The sequence **converges**.

$$\lim_{x\to\infty}\frac{\ln x}{2^x} = \frac{\left.\frac{d}{dx}\ln x\right|_{x\to\infty}}{\left.\frac{d}{dx}2^x\right|_{x\to\infty}} = \frac{\lim\limits_{x\to\infty}\frac{1}{x}}{\lim\limits_{x\to\infty}2^x \ln 2} = \frac{0}{\infty} = 0$$

Since $\lim\limits_{x\to\infty}\frac{\ln x}{2^x} = 0$ and $\frac{\ln x}{2^x}$ is defined for all $x \geq 1$, it follows that $\lim\limits_{n\to\infty}\frac{\ln n}{2^n} = 0$. (This function is actually defined for $x > 0$, but since n starts at 1 for a sequence, we don't need to check below 1. We don't even need it to be defined at 1; we just need any N for which the function is defined for all $x \geq N$. In this example, we may let $N = 1$, since that works.)

Example 15. Does $\{2^{1/n}\}$ converge or diverge? If it converges, give the limiting value.

Solution: The 'trick' to some problems with exponents is to take a natural logarithm. Let $a_n = 2^{1/n}$. Now take the natural logarithm of both sides: $\ln a_n = \ln 2^{1/n}$. Recall the rule $\ln p^q = q \ln p$ to get $\ln a_n = \frac{1}{n}\ln 2$. This shows that $\{\ln a_n\} = \left\{\frac{\ln 2}{n}\right\}$ converges to zero (since $\ln 2$ is just a constant). Now recall the property of logarithms that $e^{\ln a_n} = a_n$, which equals $2^{1/n}$. Use the first limit theorem that relates to functions with $f(b_n) = e^{b_n}$ and $b_n = \ln a_n$. We already know that $\{b_n\} = \{\ln a_n\}$ converges to $L = 0$. Since $f(b_n) = e^{b_n}$ is continuous at $f(L) = e^0 = 1$, $\{f(b_n)\} = \{e^{b_n}\} = \{e^{\ln a_n}\} = \{a_n\}$ $= \{2^{1/n}\}$ **converges** to $f(L) = e^0 = 1$. The main idea here is easy to see: $\frac{1}{n}$ converges to zero, so the exponent of $2^{1/n}$ approaches zero, and $2^0 = 1$.

Tip: When a sequence involves an exponent and none of the usual limit theorems seem to be helpful, try taking the natural logarithm of both sides as we did in Example 15.

Chapter 2 Problems

Directions: Determine whether each sequence converges or diverges. If the sequence converges, give the limiting value.

(1) $\left\{\frac{500}{n^5}\right\}$

(2) $\left\{-\frac{m^{10}}{1,000,000}\right\}$

(3) $\left\{\frac{(-1)^k}{k}\right\}$

(4) $\left\{\frac{(-1)^{i+1}}{2i+3}\right\}$

(5) $a_n = 5$

(6) $\left\{\frac{\ell-4}{\ell}\right\}$

(7) $\left\{\frac{j-4}{j}(-1)^j\right\}$

(8) $a_{k+1} = \frac{a_k}{4}$ with $a_1 = 8$

(9) $\left\{\frac{9}{3^m}\right\}$

(10) $\left\{\frac{3^j}{2^j}\right\}$

(11) $\left\{\frac{2\ell-5}{8\ell+3}\right\}$

(12) $\left\{\frac{n-3n^2}{6-n}\right\}$

(13) $\left\{\frac{m}{m^2+2m+4}\right\}$

(14) $\left\{\frac{5+9i-2i^2}{4-3i+6i^2}\right\}$

(15) $\left\{7-\frac{2}{k}\right\}$

(16) $\left\{\left(3+\frac{1}{j}\right)\left(2-\frac{1}{j}\right)\right\}$

(17) $\{\sqrt{n}\}$

(18) $\left\{\frac{1}{\sqrt{m+6}}\right\}$

(19) $\left\{\sqrt{\frac{4i^2+3}{9i^2-8}}\right\}$

(20) $\left\{\sqrt{\frac{\ell^3-1}{\ell^2+5}}\right\}$

(21) $\left\{\frac{\sqrt{j}}{j^2}\right\}$

(22) $\{\sqrt{\cos n}\}$

(23) $a_{m+2} = \frac{a_{m+1}}{a_m}$ with $a_1 = 2$ and $a_2 = 8$

(24) $\{\tan(2k)\}$

(25) $\left\{\tan\frac{1}{2\ell}\right\}$

(26) $\left\{\sin\frac{1}{n}+\cos\frac{1}{n}\right\}$

(27) $\left\{\frac{120}{k!}\right\}$

(28) $\left\{\frac{1}{(j+2)!}\right\}$

(29) $\left\{-\frac{n!}{3^n}\right\}$

(30) $\left\{\frac{3^m}{m!}\right\}$

(31) $\{e^k\}$

(32) $\{e^{-n}\}$

(33) $\left\{\frac{m^2}{2^m}\right\}$

(34) $\left\{\frac{\ell}{\ln \ell}\right\}$

(35) $\left\{\frac{\ln(3k)}{\ln(2k)}\right\}$

(36) $\left\{\frac{n^2}{e^n}\right\}$

(37) $\left\{\sqrt{\frac{1}{j!}}\right\}$

(38) $\left\{\frac{1}{3^{1/i}}\right\}$

(39) $\left\{n^{1/n}\right\}$

(40) $\left\{\left(1+\frac{3}{k}\right)^k\right\}$

3 Series

The distinction between a series and a sequence is that the terms of a **series** are added together. For example, the series $1 + 2 + 3 + 4 + 5 + \cdots$ corresponds to the sequence 1, 2, 3, 4, 5, ... The only difference is that we **add** the numbers together in the series.

Since a series may involve a very large number of terms (or may even be infinite), we use special notation for a series. The uppercase Greek letter **sigma** (Σ) represents a series; this is called the **summation symbol**. Below the summation symbol, we write the name of the index along with its starting value; above the summation symbol, we write the final value of the index (which is ∞ for an infinite series). To the right of the summation symbol, we write the formula for each term in terms of the index. The first example below represents an infinite series beginning with $n = 1$; the formula for n for the nth term of this series is $\frac{2}{n^2}$; the ellipsis (...) indicates that the series continues indefinitely. The second example below represents a finite series beginning with $i = 3$ and ending with $i = 7$; the formula for the ith term of this series is $i(i + 1)$; this series includes exactly 5 terms. The last example below represents an infinite series beginning with $k = 1$ where each term of the series is the kth term of the sequence $\{a_k\}$; the ellipsis (...) indicates that the series continues indefinitely.

$$\sum_{n=1}^{\infty} \frac{2}{n^2} = \frac{2}{1^2} + \frac{2}{2^2} + \frac{2}{3^2} + \frac{2}{4^2} + \frac{2}{5^2} + \frac{2}{6^2} + \frac{2}{7^2} + \frac{2}{8^2} + \cdots$$

$$= 2 + \frac{1}{2} + \frac{2}{9} + \frac{1}{8} + \frac{2}{25} + \frac{1}{18} + \frac{2}{49} + \frac{1}{32} + \cdots$$

$$\sum_{i=3}^{7} i(i+1) = 3(3+1) + 4(4+1) + 5(5+1) + 6(6+1) + 7(7+1)$$

$$= 12 + 20 + 30 + 42 + 56 = 160$$

$$\sum_{k=1}^{\infty} a_k = a_1 + a_2 + a_3 + a_4 + a_5 + a_6 + a_7 + a_8 + \cdots$$

The first sum is read as, "The sum as n goes from 1 to infinity of 2 over n squared." The second sum is read as, "The sum as i goes from 3 to 7 of i times (i plus 1)."

In this chapter, we will practice using series notation and get acquainted with a few common **finite** series. (Chapters 4 and onward focus on **infinite** series, including how to determine whether an infinite series converges or diverges.)

Perhaps the simplest series is the series consisting of consecutive integers:

$$\sum_{n=1}^{N} n = 1 + 2 + 3 + 4 + 5 + \cdots + N$$

Here, lowercase n is the index whereas uppercase N is the total number of terms in the series. This series forms **triangular numbers**:

- If $N = 1$, there is just one term in the series, so the sum is 1.
- If $N = 2$, the sum is $1 + 2 = 3$.
- If $N = 3$, the sum is $1 + 2 + 3 = 6$.
- If $N = 4$, the sum is $1 + 2 + 3 + 4 = 10$.
- If $N = 5$, the sum is $1 + 2 + 3 + 4 + 5 = 15$.
- If $N = 6$, the sum is $1 + 2 + 3 + 4 + 5 + 6 = 21$.

We call these sums triangular numbers because they can be formed by arranging objects in rows in the shapes of triangles as depicted below.

$N = 1$	$N = 2$	$N = 3$	$N = 4$	$N = 5$	$N = 6$
sum $= 1$	sum $= 3$	sum $= 6$	sum $= 10$	sum $= 15$	sum $= 21$

The formula for a finite series of triangular numbers beginning at $n = 1$ is:

$$\sum_{n=1}^{N} n = 1 + 2 + 3 + 4 + 5 + \cdots + N = \frac{N(N+1)}{2}$$

For example, when $N = 4$ the sum equals $\frac{4(4+1)}{2} = \frac{4(5)}{2} = \frac{20}{2} = 10$, and when $N = 6$ the sum equals $\frac{6(6+1)}{2} = \frac{6(7)}{2} = \frac{42}{2} = 21$. One way to understand the formula $\frac{N(N+1)}{2}$ is to divide the series into pairs. For example, if $N = 10$, there are 5 pairs of numbers that add up to 11. These pairs are $1 + 10$, $2 + 9$, $3 + 8$, $4 + 7$, and $5 + 6$. Here, $\frac{N}{2} = \frac{10}{2} = 5$ is the number of pairs and each pair has a sum equal to $N + 1 = 10 + 1 = 11$. When we multiply $\frac{N}{2} = 5$ and $N + 1 = 11$ together, we get $\frac{N(N+1)}{2} = 5(11) = 55$. If needed, you can add $1 + 2 + 3 + 4 + 5 + 6 + 7 + 8 + 9 + 10$ together to verify that the sum equals 55. This same reasoning works even if there are an odd number of terms. For example, when $N = 7$, there are $\frac{N}{2} = \frac{7}{2} = 3.5$ pairs where each pair has a sum of $N + 1 = 8$. These pairs are $1 + 7$, $2 + 6$, $3 + 5$, and then 4 is one-half of a pair (so there are 3.5 pairs), which is also one-half of 8. When we multiply $\frac{N}{2} = \frac{7}{2} = 3.5$ and $N + 1 = 7 + 1 = 8$ together, we get $\frac{N(N+1)}{2} = (3.5)(8) = 28$, which agrees with $1 + 2 + 3 + 4 + 5 + 6 + 7 = 28$.

Triangular numbers are an example of an arithmetic series. In an **arithmetic series**, any two consecutive terms always have the same difference. For example, for the triangular numbers $1 + 2 + 3 + 4 + 5 + 6 + \cdots + N$, the difference is 1, whereas for the arithmetic series $7 + 11 + 15 + 19 + 23 + \cdots + N$, the difference is 4. (Note that the word "arithmetic" is pronounced differently in "arithmetic series"; as an adjective in "arithmetic series" it sounds more like **air**-ith-**met**-ick, whereas as a noun by itself as "arithmetic" it sounds like the more usual uh-**rith**-muh-tick, with emphasis on the **rith**.) A general arithmetic series has the following form, where a is the first term, d is the difference between any two consecutive terms, and N is the number of terms. The two sums below, which are equivalent, offer an important lesson about series. **It's possible to rewrite a series by shifting the index.** Here, we defined $k = n - 1$. When $n = 1, k = 0$ and when $n = N, k = N - 1$, so the second sum varies from $k = 0$ to $N - 1$. If you shift an index as we have done here (which is sometimes useful), remember to shift the first and last values of the index, too.

$$\sum_{n=1}^{N} [a + (n-1)d] = a + (a + d) + (a + 2d) + (a + 3d) + \cdots + [a + (N-1)d]$$

$$= \sum_{k=0}^{N-1} (a + kd) = a + (a + d) + (a + 2d) + (a + 3d) + \cdots + [a + (N-1)d]$$

The series above can be separated into two parts as follows:

$$\sum_{n=1}^{N} [a + (n-1)d] = (a + a + a + \cdots + a) + [d + 2d + 3d + \cdots + (N-1)d]$$

$$\sum_{n=1}^{N}[a+(n-1)d] = Na + [1+2+3+\cdots+(N-1)]d$$

$$\sum_{n=1}^{N}[a+(n-1)d] = Na + \frac{(N-1)(N-1+1)d}{2} = Na + \frac{(N-1)Nd}{2} = Na + \frac{N(N-1)d}{2}$$

The first part has one a for each term, so the first part adds up to Na. The second part has the series of triangular numbers times d, so it has a sum of $\frac{N(N-1)}{2}d$. For example, the series $7+11+15+19+23+27$ consists of $N=6$ terms, a first term of $a=7$, and a constant difference of $d=11-7=4$. The formula for the sum gives $Na+\frac{N(N-1)d}{2} = (6)(7)+\frac{6(6-1)(4)}{2} = 42+\frac{120}{2} = 102$. There is a simpler way to write the formula for the sum for an arithmetic series. Let b represent the last term in the series. Looking back at our beginning formula for the series, $b=a+(N-1)d$. In terms of the last term b, we can rewrite the above formula as

$$\sum_{n=1}^{N}[a+(n-1)d] = \frac{2Na+N(N-1)d}{2} = \frac{Na+N[a+(N-1)d]}{2} = \frac{N(a+b)}{2}$$

For example, the series $7+11+15+19+23+27$ has $N=6$ terms, a first term of $a=7$, and a last term of $b=27$. The above formula gives $\frac{N(a+b)}{2} = \frac{6(7+27)}{2} = 102$. An interesting special case occurs with the series of **odd numbers beginning with one**, $1+3+5+7+9+11+13+\cdots+[1+(N-1)2]$. In this case, the first term is $a=1$ and the last term is $b=1+(N-1)2$, such that the series has a sum equal to

$$\sum_{n=1}^{N}[1+(n-1)2] = \frac{N[1+1+(N-1)2]}{2} = \frac{N(2+2N-2)}{2} = \frac{N(2N)}{2} = N^2$$

For example, the series $1+3+5+7+9+11+13+15+17$ includes $N=9$ terms. You can verify that the last term is $b=1+(9-1)2=1+16=17$ and that the sum equals $N^2=9^2=81$.

An arithmetic series is one way to generalize the series of triangular numbers $1+2+3+4+5+6+7+\cdots$. We saw the triangular numbers show up when we derived the formula $\frac{N(a+b)}{2}$ for an arithmetic series; that's where the $\frac{N(N-1)}{2}d$ came from. Another way to generalize an arithmetic series is to raise integers to **powers**, like this:

$$\sum_{n=1}^{N} n^p = 1^p + 2^p + 3^p + 4^p + 5^p + \cdots + N^p$$

For $p = 2, 3$, or 4, the formulas for such series are:

$$\sum_{n=1}^{N} n^2 = 1^2 + 2^2 + 3^2 + 4^2 + 5^2 + \cdots + N^2 = 1 + 4 + 9 + 16 + 25 + \cdots + N^2$$

$$\sum_{n=1}^{N} n^2 = \frac{N(N+1)(2N+1)}{6}$$

$$\sum_{n=1}^{N} n^3 = 1^3 + 2^3 + 3^3 + 4^3 + 5^3 + \cdots + N^3 = 1 + 8 + 27 + 64 + 125 + \cdots + N^3$$

$$\sum_{n=1}^{N} n^3 = \frac{N^2(N+1)^2}{4}$$

$$\sum_{n=1}^{N} n^4 = 1^4 + 2^4 + 3^4 + 4^4 + 5^4 + \cdots + N^4 = 1 + 16 + 81 + 256 + 625 + \cdots + N^4$$

$$\sum_{n=1}^{N} n^4 = \frac{N(N+1)(2N+1)(3N^2+3N-1)}{30}$$

For example, $1 + 4 + 9 + 16 + 25 + 36$ has $N = 6$ terms, such that the sum is equal to $\frac{N(N+1)(2N+1)}{6} = \frac{6(6+1)[2(6)+1]}{6} = 7(13) = 91$.

Another common type of finite series is the geometric series. In a **geometric series**, any two consecutive terms always have the same ratio. (Whereas for an arithmetic series the difference was constant, in a geometric series the ratio is constant.) For example, in the series $4 + 12 + 36 + 108 + 324 + 972 + \cdots$, each term is 3 times the previous term. The recursion formula for the $(n+1)^{\text{th}}$ term of the corresponding sequence is $b_{n+1} = 3b_n$. For example, in the series just given, $b_4 = b_{3+1} = 108 = 3b_3 = 3(36)$. A general geometric series has the following form, where a is the first term, r is the ratio of consecutive terms (meaning that $b_{n+1} = rb_n$), and N is the number of terms.

$$\sum_{n=1}^{N} ar^{n-1} = a + ar + ar^2 + ar^3 + ar^4 + ar^5 + \cdots + ar^{N-1}$$

$$= \sum_{k=0}^{N-1} ar^k = a + ar + ar^2 + ar^3 + ar^4 + ar^5 + \cdots + ar^{N-1}$$

Just as we had done for the general arithmetic series, we demonstrated how to rewrite the sum in terms of the index $k = n - 1$. The two series above are identical; the only difference is how the index is defined in the summation notation. In what follows, we will focus on the first form,

$$\sum_{n=1}^{N} ar^{n-1} = a + ar + ar^2 + ar^3 + ar^4 + ar^5 + \cdots + ar^{N-1}$$

Consider what happens if we multiply the entire series by r:

$$r \sum_{n=1}^{N} ar^{n-1} = r(a + ar + ar^2 + ar^3 + ar^4 + ar^5 + \cdots + ar^{N-1})$$

$$ar + ar^2 + ar^3 + ar^4 + ar^5 + ar^6 + \cdots + ar^N = \sum_{n=1}^{N} ar^n$$

Compare the original series to the new series where we multiplied by r. All of the terms are identical except for the first term of the original series (which is a) and the last term of the new series (which is ar^N). If we subtract these two series, everything else will cancel out:

$$\sum_{n=1}^{N} ar^{n-1} - r \sum_{n=1}^{N} ar^{n-1} = a - ar^N$$

We will factor out the summations on the left-hand side and factor out a on the right.

$$(1 - r) \sum_{n=1}^{N} ar^{n-1} = a(1 - r^N)$$

If $r \neq 1$, we may divide by $(1 - r)$ on both sides to get a formula for the original sum.

$$\sum_{n=1}^{N} ar^{n-1} = \frac{a(1 - r^N)}{1 - r}$$

Since the last term of the original series is $b = ar^{N-1}$ and since $br = ar^N$, we may alternatively express the above formula as

$$\sum_{n=1}^{N} ar^{n-1} = \frac{a(1 - r^N)}{1 - r} = \frac{a - ar^N}{1 - r} = \frac{a - br}{1 - r}$$

The previous formulas are only valid if $r \neq 1$. (The case $r = 1$ trivially makes the series $a + a + a + a + a + \cdots + a = Na$, where each term equals a.) As an example, the series $4 + 12 + 36 + 108 + 324 + 972$ has $N = 6$ terms, a ratio of $r = \frac{b_n+1}{b_n} = \frac{12}{4}$ $= 3$, a first term of $a = 4$, and a last term of $b = ar^{N-1} = 4(3)^{6-1} = 4(3^5) = 4(243)$ $= 972$. The sum equals $\frac{a(1-r^N)}{1-r} = \frac{4(1-3^6)}{1-3} = \frac{4(1-729)}{-2} = -2(-728) = 1456$. This is the same as $\frac{a-br}{1-r} = \frac{4-972(3)}{1-3} = \frac{4-2916}{-2} = \frac{-2912}{-2} = 1456$. You can verify that if you add all 6 terms together, you get this same answer. If $r > 1$, you might prefer to write the two formulas as $\frac{a(r^N-1)}{r-1}$ and $\frac{br-a}{r-1}$; the reason we prefer the form with $1 - r$ is that usually we work with infinite series in calculus, and $|r| < 1$ if an infinite geometric series converges (see Chapter 4).

What if we combine the two kinds of series that we've discussed so far? In that case, you get an **arithmetic-geometric series** (provided that $r \neq 1$):

$$\sum_{n=1}^{N} [a + (n-1)d]r^{n-1}$$

$$= a + (a + d)r + (a + 2d)r^2 + (a + 3d)r^3 + \cdots + [a + (n-1)d]r^{N-1}$$

$$= \frac{a(1 - r^N)}{1 - r} + \frac{rd[1 - Nr^{N-1} + (N-1)r^N]}{(1-r)^2} = \frac{a + (d - a)r + br^2 - (a + Nd)r^N}{(1-r)^2}$$

where N is the number of terms, a is the first term, and $b = [a + (N-1)d]r^{N-1}$ is the last term. Note that we put the d **after** r because dr would like look differentiation in the context of calculus; rd avoids such confusion. As an example, the series

$$1 + (1 + 2)3 + [1 + 2(2)]3^2 + [1 + 3(2)]3^3 + [1 + 4(2)]3^4$$

$$= 1 + 3(3) + 5(9) + 7(27) + 9(81) = 1 + 9 + 45 + 189 + 729 = 973$$

has $N = 5$ terms, $d = 2$ is the difference in the coefficients $(1, 3, 5, 7, 9)$, $r = 3$ is the ratio of the powers $(3, 9, 27, 81)$, $a = 1$ is the first term, and $b = [1 + (5-1)(2)]3^{5-1}$ $= 9(3^4) = 9(81) = 729$ is the last term. The formulas for the sums give:

$$\frac{a(1 - r^N)}{1 - r} + \frac{rd[1 - Nr^{N-1} + (N-1)r^N]}{(1-r)^2} = \frac{1(1 - 3^5)}{1 - 3} + \frac{3(2)[1 - 5(3^{5-1}) + (5-1)3^5]}{(1-3)^2}$$

$$= \frac{-242}{-2} + \frac{6[1 - 5(81) + 4(243)]}{4} = 121 + \frac{6(568)}{4} = 973$$

$$\frac{a + (d - a)r + br^2 - (a + Nd)r^N}{(1 - r)^2} = \frac{1 + (2 - 1)(3) + 729(3^2) - [1 + 5(2)]3^5}{(1 - 3)^2}$$

$$= \frac{1 + 3 + 6561 - 2673}{4} = \frac{3892}{4} = 973$$

Sometimes a sum can be evaluated by rearranging the terms to produce cancellation. We will illustrate this with the following series.

$$\sum_{n=1}^{N} \frac{1}{n(n+1)} = \frac{1}{1(2)} + \frac{1}{2(3)} + \frac{1}{3(4)} + \frac{1}{4(5)} + \frac{1}{5(6)} + \cdots + \frac{1}{N(N+1)}$$

The 'trick' here is to use the method of **partial fractions** that is taught when you learn techniques of integration. If you don't know this method, don't worry; it just means to write the fraction in the following form and then solve for the numerators.

$$\frac{1}{n(n+1)} = \frac{f}{n} + \frac{g}{n+1}$$

To determine the values of f and g, multiply both sides by $n(n + 1)$.

$$1 = f(n + 1) + gn = fn + f + gn = f + (f + g)n$$

We want 1 to equal $f + (f + g)n$ for any value of n, yet the symbol n doesn't appear on the left-hand side. How can $f + (f + g)n$ equal 1 for any value of n? This happens if $f + g = 0$. This means that $f = 1$ and $1 + g = 0$ such that $g = -1$. The equation with fractions above becomes

$$\frac{1}{n(n+1)} = \frac{1}{n} - \frac{1}{n+1}$$

When we use this formula to write the previous sum, we get an interesting result.

$$\sum_{n=1}^{N} \frac{1}{n(n+1)} = \sum_{n=1}^{N} \left(\frac{1}{n} - \frac{1}{n+1}\right)$$

$$= \left(\frac{1}{1} - \frac{1}{2}\right) + \left(\frac{1}{2} - \frac{1}{3}\right) + \left(\frac{1}{3} - \frac{1}{4}\right) + \left(\frac{1}{4} - \frac{1}{5}\right) + \cdots + \left(\frac{1}{N} - \frac{1}{N+1}\right)$$

Observe that all of the terms cancel out except for $\frac{1}{1}$ and $-\frac{1}{N+1}$. For example, $-\frac{1}{2} + \frac{1}{2}$ equals zero, $-\frac{1}{3} + \frac{1}{3}$ equals zero, and so on.

$$\sum_{n=1}^{N} \frac{1}{n(n+1)} = \frac{1}{1} - \frac{1}{N+1} = \frac{N+1}{N+1} - \frac{1}{N+1} = \frac{N+1-1}{N+1} = \frac{N}{N+1}$$

For example, $\frac{1}{1(2)} + \frac{1}{2(3)} + \frac{1}{3(4)} + \frac{1}{4(5)} + \frac{1}{5(6)} = \frac{1}{2} + \frac{1}{6} + \frac{1}{12} + \frac{1}{20} + \frac{1}{30}$ has $N = 5$ terms such that this sum equals $\frac{5}{5+1} = \frac{5}{6}$. Let's check this by making a common denominator of 60 to add these five fractions directly: $\frac{30}{60} + \frac{10}{60} + \frac{5}{60} + \frac{3}{60} + \frac{2}{60} = \frac{50}{60} = \frac{5}{6}$.

Note that it's possible to have **double sums**, triple sums, etc. For example, consider the double sum below, where m and n each vary from 1 to 3.

$$\sum_{m=1}^{3}\sum_{n=1}^{3} mn = 1(1) + 1(2) + 1(3) + 2(1) + 2(2) + 2(3) + 3(1) + 3(2) + 3(3)$$

$$= 1 + 2 + 3 + 2 + 4 + 6 + 3 + 6 + 9 = 36$$

Example 1. Write out the series below longhand and evaluate the sum.

$$\sum_{k=4}^{9} \frac{1}{k}$$

Solution: Each term has the form $\frac{1}{k}$. Begin by plugging $k = 4$ into $\frac{1}{k}$. Then plug in $k = 5$, 6, and so on until reaching the last value, $k = 9$. Add these terms together.

$$\sum_{k=4}^{9} \frac{1}{k} = \frac{1}{4} + \frac{1}{5} + \frac{1}{6} + \frac{1}{7} + \frac{1}{8} + \frac{1}{9}$$

This isn't one of the series that we discussed in this chapter, so we don't have a special formula for it. Just make a common denominator to add the fractions. If you want to make the lowest common denominator, observe that every denominator can be made using three 2's, two 3's, one 5, and one 7. Multiply this out to make the lowest common denominator of $2(2)(2)(3)(3)(5)(7) = 2520$. For example, since $\frac{2520}{7} = 360$, we will rewrite $\frac{1}{7}$ as $\frac{1}{7} = \frac{1}{7}\frac{360}{360} = \frac{360}{2520}$.

$$\sum_{k=4}^{9} \frac{1}{k} = \frac{630}{2520} + \frac{504}{2520} + \frac{420}{2520} + \frac{360}{2520} + \frac{315}{2520} + \frac{280}{2520} = \frac{2509}{2520}$$

Example 2. Rewrite the series below using summation notation.

$$\frac{4}{9}+\frac{5}{16}+\frac{6}{25}+\frac{7}{36}+\frac{8}{49}$$

Solution: Figure out the pattern. It may help to recall Example 5 from Chapter 1, which mentioned a variety of common patterns. You should recognize that the denominators are perfect squares: $3^2 = 9, 4^2 = 16, 5^2 = 25, 6^2 = 36$, and $7^2 = 49$. Also, observe that the numerator is one more than the number that is squared in the denominator. For example, in $\frac{6}{25} = \frac{6}{5^2}$, the numerator is 6 while the denominator is 5 squared. Let's use n to make the denominator of $\frac{1}{n^2}$. The numerator is then $n + 1$. If we do this, n begins at 3 and increases by 1 until it reaches 7. It turns out that there isn't just one way to express this series. All of the sums below represent the same series. For example, in the second sum, $k = n + 1$ varies from 4 to 8. For example, when $n = 3$, the left sum gives the term $\frac{3+1}{3^2} = \frac{4}{9}$ and when $k = 3 + 1 = 4$, the middle sum gives $\frac{4}{(4-1)^2} = \frac{4}{9}$.

$$\sum_{n=3}^{7}\frac{n+1}{n^2} = \sum_{k=4}^{8}\frac{k}{(k-1)^2} = \sum_{i=1}^{5}\frac{i+3}{(i+2)^2}$$

Example 3. Rewrite the series below using summation notation.

$$35 - 63 + 99 - 143 + 195 - 255 + \cdots$$

Solution: The ellipsis (...) indicates that the series continues indefinitely, so the upper limit is infinity. Can you figure out this pattern? Note that each term is the product of consecutive odd numbers: $5(7) = 35, 7(9) = 63, 9(11) = 99, 11(13) = 143, 13(15) = 195, 15(17) = 255, \ldots$ Now we have a problem; in summation notation, the index includes consecutive integers, but we need to skip the even numbers. To address that, use $2n + 1$. For example, when $n = 4$, this gives $2(4) + 1 = 9$, and when $n = 5$, this gives $2(5) + 1 = 11$. The formula $2n + 1$ makes consecutive odd numbers. We need the first odd number to be 5, so we start the series at $n = 2$, since $2(2) + 1 = 5$. In each term, consecutive odd numbers are multiplied together. The first odd number is $2n + 1$. For the same value of n, the next odd number is $2n + 3$. Note that this is an **alternating series**. Note: $(-1)^n$ makes terms negative for odd n, whereas $(-1)^{n+1}$ makes terms negative for even n. We need $(-1)^n$ since the $n = 3$ term is negative (recall that we're starting n at 2). Each term is $(-1)^n(2n + 1)(2n + 3)$.

$$\sum_{n=2}^{\infty}(-1)^n(2n+1)(2n+3) = \sum_{n=2}^{\infty}(-1)^n(4n^2+8n+3)$$

For example, when $n = 2$, the first term is $(-1)^2[2(2)+1][2(2)+3] = (1)5(7) = 35$, which agrees with $(-1)^2[4(2)^2+8(2)+3] = 16+16+3 = 35$. As with the previous example, we can make an equivalent series by shifting the index. It turns out that we will obtain a compact expression if we let $k = n+1$, such that $n = k-1$. Note that k begins from 3 instead of 2. Writing the product in terms of k, we get

$$(-1)^n(2n+1)(2n+3) = (-1)^{k-1}[2(k-1)+1][2(k-1)+3]$$

$$= (-1)^{k-1}(2k-2+1)(2k-2+3) = (-1)^{k-1}(2k-1)(2k+1) = (-1)^{k-1}(4k^2-1)$$

In terms of k, the series can be expressed either of the following ways.

$$\sum_{k=3}^{\infty}(-1)^{k-1}(2k-1)(2k+1) = \sum_{k=3}^{\infty}(-1)^{k-1}(4k^2-1)$$

For example, when $k = 3$, we get $(-1)^2[2(3)-1][2(3)+1] = 5(7) = 35$, which agrees with $(-1)^2[4(3)^2-1] = 4(9)-1 = 36-1 = 35$. Any of the four answers that we've given are correct, and there are an infinite number of ways to express the answer. The point is that if you express your index a little differently (especially, if you write your sum to have a different starting value), you might get a somewhat different looking answer than the answer given at the back of the book. If you get a different answer and want to check your answer, write out the series longhand like we did in Example 1 to see if it matches the given series.

Example 4. Rewrite the series below with the index k where $k = n+2$.

$$\sum_{n=5}^{12}\frac{n+1}{n+4}$$

Solution: Use $k = n+2$ to find the lower and upper limits of the sum. When $n = 5$, $k = 5+2 = 7$ and when $n = 12$, $k = 12+2 = 14$. Solve for n in the equation given: $n = k-2$. Plug this into the expressions for the numerator and denominator.

$$\sum_{k=7}^{14}\frac{(k-2)+1}{(k-2)+4} = \sum_{k=7}^{14}\frac{k-1}{k+2}$$

To check the answer, write out each series longhand and compare.

$$\sum_{n=5}^{12} \frac{n+1}{n+4} = \frac{5+1}{5+4} + \frac{6+1}{6+4} + \frac{7+1}{7+4} + \frac{8+1}{8+4} + \frac{9+1}{9+4} + \frac{10+1}{10+4} + \frac{11+1}{11+4} + \frac{12+1}{12+4}$$

$$= \frac{6}{9} + \frac{7}{10} + \frac{8}{11} + \frac{9}{12} + \frac{10}{13} + \frac{11}{14} + \frac{12}{15} + \frac{13}{16}$$

$$\sum_{k=7}^{14} \frac{k-1}{k+2} = \frac{7-1}{7+2} + \frac{8-1}{8+2} + \frac{9-1}{9+2} + \frac{10-1}{10+2} + \frac{11-1}{11+2} + \frac{12-1}{12+2} + \frac{13-1}{13+2} + \frac{14-1}{14+2}$$

$$= \frac{6}{9} + \frac{7}{10} + \frac{8}{11} + \frac{9}{12} + \frac{10}{13} + \frac{11}{14} + \frac{12}{15} + \frac{13}{16}$$

Example 5. Use a formula to evaluate the series below.

$$1 + 2 + 3 + 4 + 5 + 6 + 7 + 8 + 9 + 10$$

Solution: Use the triangular number formula with $N = 10$. (To use this formula, the terms must be consecutive integers beginning with 1.)

$$\frac{N(N+1)}{2} = \frac{10(11)}{2} = 55$$

Example 6. Use a formula to evaluate the series below.

$$15 + 26 + 37 + 48 + 59 + 70 + 81$$

Solution: Since the difference between any two consecutive terms is 11, this is an arithmetic series with $N = 7$ terms, a difference of $d = 11$, a first term of $a = 15$, and a last term of $b = 81$. Use either formula for an arithmetic series.

$$Na + \frac{N(N-1)d}{2} = 7(15) + \frac{7(7-1)(11)}{2} = 105 + \frac{7(6)11}{2} = 105 + 231 = 336$$

$$\frac{N(a+b)}{2} = \frac{7(15+81)}{2} = \frac{7(96)}{2} = 336$$

Example 7. Use a formula to evaluate the series below.

$$24 + 12 + 6 + 3 + \frac{3}{2} + \frac{3}{4} + \frac{3}{8} + \frac{3}{16}$$

Solution: Since each term is one-half of the preceding term, this is a geometric series with $N = 8$ terms, a ratio of $r = \frac{b_{n+1}}{b_n} = \frac{1}{2}$, a first term of $a = 24$, and a last term of $b = \frac{3}{16}$. Use either formula for a geometric series.

$$\frac{a(1-r^N)}{1-r} = \frac{24\left[1-\left(\frac{1}{2}\right)^8\right]}{1-\frac{1}{2}} = \frac{24\left(1-\frac{1}{256}\right)}{1/2} = \frac{24-\frac{3}{32}}{1/2} = \frac{32\left(24-\frac{3}{32}\right)}{32\left(\frac{1}{2}\right)} = \frac{768-3}{16} = \frac{765}{16}$$

$$\frac{a-br}{1-r} = \frac{24-\left(\frac{3}{16}\right)\left(\frac{1}{2}\right)}{1-\frac{1}{2}} = \frac{24-\frac{3}{32}}{1/2} = \frac{32\left(24-\frac{3}{32}\right)}{32\left(\frac{1}{2}\right)} = \frac{768-3}{16} = \frac{765}{16}$$

Note: In each case, we multiplied the numerator and denominator each by 32, which removed the fractions from the numerator and denominator.

Example 8. Use a formula to evaluate the series below.

$$\sum_{n=1}^{6}(2+3n)2^{n-1}$$

Solution: This is an arithmetic-geometric series. Before we can identify the constants, we need to write it in the standard form $[a+(n-1)d]r^{n-1}$. It's nearly there, but not quite. The problem is that $2+3n$ doesn't quit have the same form as $a+(n-1)d$. Note that $a+(n-1)d = a+nd-d$. If we identify $d=3$, this becomes $a+3n-3$. This equals $2+3n$ provided that $a-3=2$, which tells us that $a=5$. To find r, just compare 2^{n-1} to r^{n-1}; this tells us that $r=2$. There are $N=6$ terms. Plug $N=6, d=3, r=2$, and $a=5$ into the formula for an arithmetic-geometric series.

$$\frac{a(1-r^N)}{1-r} + \frac{rd[1-Nr^{N-1}+(N-1)r^N]}{(1-r)^2}$$

$$= \frac{5(1-2^6)}{1-2} + \frac{2(3)[1-6(2^{6-1})+(6-1)2^6]}{(1-2)^2} = \frac{5(1-64)}{-1} + \frac{6[1-6(2^5)+(5)64]}{(-1)^2}$$

$$= \frac{5(-63)}{-1} + \frac{6(1-192+320)}{1} = 315+774 = 1089$$

Alternatively, first find the last term and use the formula involving the last term.

$$b = [a+(N-1)d]r^{N-1} = [5+(6-1)3]2^{6-1} = [5+5(3)]2^5 = 20(32) = 640$$

$$\frac{a+(d-a)r+br^2-(a+Nd)r^N}{(1-r)^2} = \frac{5+(3-5)2+640(2)^2-[5+6(3)]2^6}{(1-2)^2}$$

$$= \frac{5-4+2560-1472}{(-1)^2} = 1089$$

We can check the answer by writing out the series longhand.

$$\sum_{n=1}^{6}(2+3n)2^{n-1}$$

$$= [2+3(1)]1+[2+3(2)]2+[2+3(3)]4+[2+3(4)]8 + [2+3(5)]16+[2+3(6)]32$$

$$= 5(1)+8(2)+11(4)+14(8)+17(16)+20(32)$$

$$= 5+16+44+112+272+640 = 1089$$

Compare with the standard arithmetic-geometric form:

$$\sum_{n=1}^{6}[a+(n-1)d]r^{n-1}$$

$$= 5(1)+(5+3)(2)+[5+2(3)](4)+[5+3(3)](8) + [5+4(3)](16)+[5+5(3)](32)$$

$$= 5(1)+8(2)+11(4)+14(8)+17(16)+20(32)$$

$$= 5+16+44+112+272+640 = 1089$$

Example 9. Use a formula to evaluate the series below.

$1000+1331+1728+2197+2744+3375+4096+4913+5832+6859+8000$

Solution: These are cubes: $10^3+11^3+12^3+13^3+14^3+15^3+16^3+17^3+18^3+19^3+20^3$. We have a formula for a sum of cubes starting with 1^3, but this series starts with 10^3. What shall we do? The 'trick' is to find $S_1 = 1^3+2^3+3^3+\cdots+19^3+20^3$ and then subtract $S_2 = 1^3+2^3+3^3+\cdots+8^3+9^3$ to get $S_3 = 10^3+11^3+12^3+\cdots+19^3+20^3$. Note that the series S_1 has $N = 20$ terms while the series S_2 has $N_2 = 9$ terms.

$$S_1 = \frac{N_1^2(N_1+1)^2}{4} = \frac{20^2 21^2}{4} = \frac{400(441)}{4} = 44{,}100$$

$$S_2 = \frac{N_2^2(N_2+1)^2}{4} = \frac{9^2 10^2}{4} = \frac{81(100)}{4} = 2025$$

$$S_3 = S_1 - S_2 = 44{,}100 - 2025 = 42{,}075$$

Example 10. Use the method of partial fractions and then cancel pairs of terms in order to evaluate the series below.

$$\sum_{n=1}^{7} \frac{6}{n(n+2)}$$

Solution: Split the expression into two separate fractions as follows.

$$\frac{6}{n(n+2)} = \frac{f}{n} + \frac{g}{n+2}$$

Multiply by $n(n+2)$ on both sides.

$$6 = f(n+2) + gn = fn + 2f + gn = 2f + (f+g)n$$

This will be true for any value of n if $f + g = 0$ and $6 = 2f$, which leads to $f = 3$ and $g = -3$. Now we can rewrite the above series as

$$\sum_{n=1}^{7} \frac{6}{n(n+2)} = \sum_{n=1}^{7} \left(\frac{3}{n} - \frac{3}{n+2}\right)$$

$$= \frac{3}{1} - \frac{3}{1+2} + \frac{3}{2} - \frac{3}{2+2} + \frac{3}{3} - \frac{3}{3+2} + \frac{3}{4} - \frac{3}{4+2} + \frac{3}{5} - \frac{3}{5+2} + \frac{3}{6} - \frac{3}{6+2} + \frac{3}{7} - \frac{3}{7+2}$$

$$= \frac{3}{1} - \frac{3}{3} + \frac{3}{2} - \frac{3}{4} + \frac{3}{3} - \frac{3}{5} + \frac{3}{4} - \frac{3}{6} + \frac{3}{5} - \frac{3}{7} + \frac{3}{6} - \frac{3}{8} + \frac{3}{7} - \frac{3}{9}$$

Many pairs of terms cancel. The ones that remain are

$$\frac{3}{1} + \frac{3}{2} - \frac{3}{8} - \frac{3}{9} = 3 + \frac{3}{2} - \frac{3}{8} - \frac{1}{3}$$

Make a common denominator of 24.

$$\frac{72}{24} + \frac{36}{24} - \frac{9}{24} - \frac{8}{24} = \frac{91}{24}$$

Example 11. Evaluate the double sum below.

$$\sum_{m=1}^{3} \sum_{n=1}^{3} \frac{m}{n}$$

Solution: Here, m and n each vary from 1 to 3, creating 9 terms.

$$\sum_{m=1}^{3} \sum_{n=1}^{3} \frac{m}{n} = \frac{1}{1} + \frac{2}{1} + \frac{3}{1} + \frac{1}{2} + \frac{2}{2} + \frac{3}{2} + \frac{1}{3} + \frac{2}{3} + \frac{3}{3}$$

$$= 1 + 2 + 3 + \frac{1}{2} + 1 + \frac{3}{2} + \frac{1}{3} + \frac{2}{3} + 1 = 6 + 3 + 2 = 11$$

Chapter 3 Problems

Directions: Write out each series longhand and evaluate the sum.

(1)

$$\sum_{n=1}^{4} (-1)^n n^4$$

(2)

$$\sum_{k=1}^{5} \frac{(-1)^{k+1}}{k(k+1)}$$

(3)

$$\sum_{j=4}^{10} (3j^2 - 12)$$

(4)

$$\sum_{m=14}^{23} 7$$

(5) Note: The argument is in radians.

$$\sum_{\ell=3}^{9} \cos\left(\frac{\ell\pi}{3}\right)$$

(6)

$$\sum_{i=5}^{11} \frac{6i(-1)^i}{(i-2)^2}$$

(7)

$$\sum_{m=1}^{3} \sum_{n=1}^{3} \frac{m^2 n}{2}$$

(8)

$$\sum_{i=1}^{3}\sum_{j=1}^{3}(3i - j)$$

(9)

$$\sum_{k=1}^{3}\sum_{\ell=k}^{3}\frac{6\ell}{k}$$

(10)

$$\sum_{i=1}^{3}\sum_{j=1}^{3}\sum_{k=1}^{3}(k^2 - ij)$$

Directions: Rewrite each series using summation notation.

(11)

$$\frac{2}{11}+\frac{1}{6}+\frac{2}{13}+\frac{1}{7}+\frac{2}{15}+\frac{1}{8}$$

(12)

$$14-16+18-20+22-24+26-28+30$$

(13)

$$-2+6-10+14-18+22-26+30$$

(14)

$$\frac{3}{16}+\frac{4}{25}+\frac{5}{36}+\frac{6}{49}+\frac{7}{64}+\frac{8}{81}+\frac{9}{100}$$

(15)

$$7 + 5 + 3 + 1 - 1 - 3 - 5 - 7$$

(16)

$$1 + 3 + 1 + 3 + 1 + 3 + 1 + 3 + 1 + 3 + \cdots$$

(17)

$$7 - 21 + 63 - 189 + 567$$

(18)

$$175 + 200 + 225 + 250 + 275 + 300 + 325$$

(19)

$$\frac{1}{6}+\frac{1}{3}+\frac{1}{2}+\frac{2}{3}+\frac{5}{6}$$

(20)

$$32+8+2+\frac{1}{2}+\frac{1}{8}+\frac{1}{32}+\frac{1}{128}+\frac{1}{512}$$

(21)

$$\frac{9}{4}+\frac{11}{10}+\frac{13}{18}+\frac{15}{28}+\frac{17}{40}+\frac{19}{54}+\frac{21}{70}$$

(22)

$$4+\frac{8}{5}+\frac{2}{9}+\frac{64}{7}+\frac{8}{125}+24+\frac{10}{343}+\frac{512}{11}+\frac{4}{243}+\frac{1000}{13}+\frac{14}{1331}$$

Directions: Rewrite each series below using the indicated index.

(23) $k = n - 2$

$$\sum_{n=3}^{11} (4n - 2)$$

(24) $j = k + 1$

$$\sum_{k=0}^{8} \frac{k - 2}{k + 2}$$

(25) $\ell = -m$

$$\sum_{m=0}^{4} \frac{m^2}{m + 1}$$

(26) $n = i + 3$

$$\sum_{i=1}^{7} \frac{i^2 + 6i + 3}{i + 3}$$

Directions: Evaluate each series using a formula from this chapter.

(27)

$1 + 2 + 3 + 4 + 5 + 6 + 7 + 8 + 9 + 10 + 11 + 12 + 13 + 14 + 15 + 16 + 17 + 18$

(28)

$$\sum_{k=1}^{144} k$$

(29)

$$\sum_{i=25}^{75} 4i$$

(30)

$8 + 16 + 24 + 32 + 40 + 48 + 56 + 64 + 72 + 80 + 88 + 96 + 104 + 112 + 120$

(31)

$$18 + 25 + 32 + 39 + 46 + 53 + 60 + 67 + 74 + 81 + 88 + 95$$

(32)

$$\sum_{m=1}^{50} [6 + 4(m - 1)]$$

(33)

$$\sum_{\ell=1}^{15} (5\ell - 2)$$

(34)

$$-18 - 7 + 4 + 15 + 26 + 37 + 48 + 59 + 70$$

(35)

$$43 + 33 + 23 + 13 + 3 - 7 - 17 - 27 - 37 - 47 - 57 - 67$$

(36)

$$1 + 3 + 9 + 27 + 81 + 243 + 729 + 2187 + 6561 + 19{,}683 + 59{,}049 + 177{,}147$$

(37)

$$12 + 48 + 192 + 768 + 3072 + 12{,}288 + 49{,}152 + 196{,}608 + 786{,}432$$

(38)

$$\sum_{j=1}^{12} 5^{j-1}$$

(39)

$$\sum_{n=5}^{10} 4^n$$

(40)

$$5 - 30 + 180 - 1080 + 6480 - 38{,}880 + 233{,}280 - 1{,}399{,}680 + 8{,}398{,}080$$

(41)

$$\frac{1}{2} + \frac{2}{3} + \frac{5}{6} + 1 + \frac{7}{6} + \frac{4}{3} + \frac{3}{2} + \frac{5}{3} + \frac{11}{6} + 2 + \frac{13}{6} + \frac{7}{3} + \frac{5}{2} + \frac{8}{3} + \frac{17}{6} + 3$$

(42)

$$\frac{16}{243} + \frac{8}{81} + \frac{4}{27} + \frac{2}{9} + \frac{1}{3} + \frac{1}{2} + \frac{3}{4} + \frac{9}{8} + \frac{27}{16} + \frac{81}{32} + \frac{243}{64}$$

(43)

$$\sum_{k=1}^{20} \frac{9}{2^k}$$

(44)

$$6 + 32 + 160 + 768 + 3584 + 16{,}384 + 73{,}728 + 327{,}680 + 1{,}441{,}792$$

(45)

$$\sum_{m=1}^{12} [5 + 2(m-1)]3^{m-1}$$

(46)

$$\sum_{k=1}^{9} \frac{5k-3}{4^k}$$

(47)

$$-1 - 16 - 81 - 256 - 625 - 1296 - 2401 - 4096 - 6561 - 10{,}000 - 14{,}641$$

(48)

$$\sum_{n=30}^{40} \frac{n^2}{5}$$

(49)

$$3 + 24 + 81 + 192 + 375 + 648 + 1029 + 1536$$

(50)

$$\sum_{i=1}^{12} (i^3 - 2i^2 + i - 4)$$

Directions: Cancel pairs of terms in order to evaluate each series.

(51)

$$\sum_{k=2}^{30} \frac{3}{k^2 - k}$$

(52)

$$\sum_{n=1}^{18} \frac{2}{4n^2 - 1}$$

4 Series Convergence

An infinite series has an infinite number of terms added together. Is it possible for the sum to be finite? Yes, it is possible. We will illustrate this with an example. Suppose that a robot will travel in a straight line from point A to point B as follows. First, the robot will travel halfway from A to B; call this point P_1; the distance AP_1 is $1/2$ of the full trip. Next, the robot will travel halfway from P_1 to B; call this point P_2; the distance P_1P_2 is $1/4$ of the full trip. Next, the robot will travel halfway from P_2 to B; call this point P_3; the distance P_2P_3 is $1/8$ of the full trip. Next, the robot will travel halfway from P_3 to B; call this point P_4; the distance P_3P_4 is $1/16$ of the full trip. Next, the robot will travel halfway from P_4 to B; call this point P_5; the distance P_4P_5 is $1/32$ of the full trip. The robot continues to travel halfway from the previous position to point B each time. The total distance traveled can be expressed by the following series. We know that this sum is finite because we can visualize the trip as shown below. If we add all of the trips together, they will make the total distance AB. Since we represented AP_1 with the fraction $1/2$, the sum of these fractions must be equal to one. In this example, an infinite number of fractions add up to the finite value of one.

$$\sum_{n=1}^{\infty} \frac{1}{2^n} = \frac{1}{2} + \frac{1}{4} + \frac{1}{8} + \frac{1}{16} + \frac{1}{32} + \cdots = 1$$

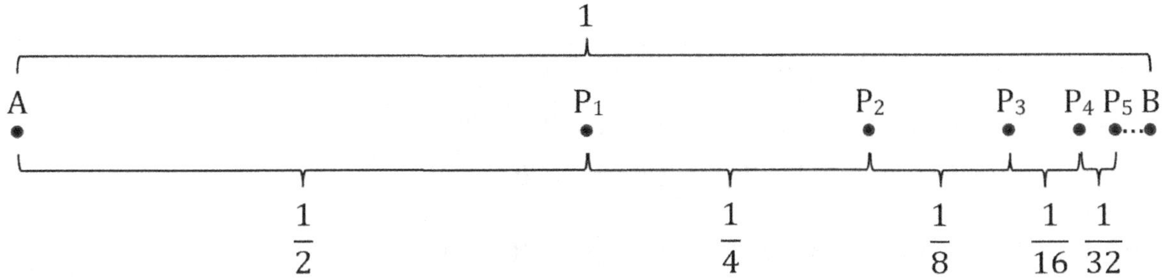

Another question one might wonder is, "Can you actually add an infinite number of terms together?" That's a good question. What we really have in mind is the limit as the number of terms grows extremely large. Let $S_n = \frac{1}{2} + \frac{1}{4} + \frac{1}{8} + \frac{1}{16} + \frac{1}{32} + \cdots + \frac{1}{2^n}$ be the sum of the first n terms. What we're really saying is $\lim_{n\to\infty} S_n = 1$. As n gets larger and larger, S_n approaches 1.

The question we'll address now and thru Chapter 11 is how to determine whether or not an infinite series converges or diverges. As we'll see, there are a variety of tests that we can apply to see whether or not an infinite series converges.

The first step is to define what we mean by convergence and divergence as it relates to a series. An infinite series has the following form. The initial value is often 1, but it could be 0, -1, 3, 8, or any other finite integer; the same concept applies provided that the index begins from a finite initial value (but see a note later in the chapter).[1]

$$\sum_{n=1}^{\infty} a_n = a_1 + a_2 + a_3 + \cdots$$

For a finite value of n, let S_n be the **partial sum** below. The distinction between the sum above and the sum below is that the above sum is infinite whereas the sum below has exactly n terms.

$$S_n = \sum_{k=1}^{n} a_k = a_1 + a_2 + a_3 + \cdots + a_n$$

The partial sums $S_1 = a_1$, $S_2 = a_1 + a_2$, $S_3 = a_1 + a_2 + a_3$, $S_4 = a_1 + a_2 + a_3 + a_4$, etc. form a sequence $\{S_n\}$ called the **sequence of partial sums**. Whether or not a series converges depends on this sequence of partial sums:

- If the sequence of partial sums, $\{S_n\}$, converges to a (finite) limiting value L, then the corresponding infinite series **converges** and its sum is equal to L:

$$\sum_{n=1}^{\infty} a_n = a_1 + a_2 + a_3 + \cdots = L$$

- If the sequence of partial sums, $\{S_n\}$, diverges (as discussed in Chapter 2), then the corresponding infinite series **diverges**.

Consider the definitions above with regard to the series below.

$$\sum_{n=1}^{\infty} \frac{1}{2^n} = \frac{1}{2} + \frac{1}{4} + \frac{1}{8} + \frac{1}{16} + \frac{1}{32} + \cdots$$

---

[1] An infinite series could also go from $-\infty$ to a finite value. If so, the same basic idea will apply, except that we have to think in terms of the index becoming very negative rather than positive. An infinite series could also go from $-\infty$ to $+\infty$; if so, it can be split into two separate series (one from $-\infty$ to 0 and one from 1 to ∞).

The partial sum S_n for this series is:

$$S_n = \sum_{k=1}^{n} \frac{1}{2^k} = \frac{1}{2} + \frac{1}{4} + \frac{1}{8} + \frac{1}{16} + \frac{1}{32} + \cdots + \frac{1}{2^n}$$

The sequence of partial sums $\{S_n\}$ for this series is:

- $S_1 = \frac{1}{2}$ just has one term.
- $S_2 = \frac{1}{2} + \frac{1}{4} = \frac{3}{4}$.
- $S_3 = \frac{1}{2} + \frac{1}{4} + \frac{1}{8} = \frac{7}{8}$.
- $S_4 = \frac{1}{2} + \frac{1}{4} + \frac{1}{8} + \frac{1}{16} = \frac{15}{16}$.
- Continuing the pattern, $S_n = \frac{2^n - 1}{2^n}$. For example, when $n = 5$, $S_5 = \frac{2^5 - 1}{2^5} = \frac{31}{32}$.

The sequence of partial sums is $\{S_n\} = \left\{\frac{2^n - 1}{2^n}\right\} = \left\{1 - \frac{1}{2^n}\right\}$. Applying what we learned in Chapter 2 about limiting values, the sequence $\left\{1 - \frac{1}{2^n}\right\}$ converges to 1. Since the sequence of partial sums converges to the limiting value $L = 1$, the corresponding infinite series converges to $L = 1$.

$$\sum_{n=1}^{\infty} \frac{1}{2^n} = 1$$

Notice that the lower limit is important. For example, if we change the lower limit to zero, we get one more term and the sum changes to 2 (since $\frac{1}{2^0} = \frac{1}{1} = 1$).

$$\sum_{n=0}^{\infty} \frac{1}{2^n} = 2$$

It's important to distinguish between sequences and series. Notice that we're looking at a **<u>sequence</u>** of partial sums to determine whether or not a **<u>series</u>** converges. When we apply the convergence properties of sequences that we learned in Chapter 2, we must apply this to the **<u>sequence</u>** of partial sums; it's a common mistake for students to accidentally apply what they know about the convergence of sequences to the individual terms of the series (rather than to the sequence of partial sums). You don't want to look at the series $\sum a_n$ and think about the sequence $\{a_n\}$; that's a big mistake. You want to look at the partial sums $S_n = a_1 + a_2 + \cdots + a_n$ and think about $\{S_n\}$.

Now let's look at an example of an infinite series that **diverges**. Consider the series below.

$$\sum_{n=1}^{\infty} n = 1 + 2 + 3 + 4 + 5 + \cdots$$

The sequence of partial sums $\{S_n\}$ for this series is:

- $S_1 = 1$ just has one term.
- $S_2 = 1 + 2 = 3$.
- $S_3 = 1 + 2 + 3 = 6$.
- $S_4 = 1 + 2 + 3 + 4 = 10$.
- $S_5 = 1 + 2 + 3 + 4 + 5 = 15$.
- The partial sums are **triangular numbers** (Chapter 3): $S_n = \frac{n(n+1)}{2}$. For example, $S_6 = \frac{6(7)}{2} = 21$.

The sequence of partial sums is $\{S_n\} = \left\{\frac{n(n+1)}{2}\right\}$. Applying what we learned in Chapter 2 about limiting values, the sequence $\left\{\frac{n(n+1)}{2}\right\}$ **diverges**.

In Chapter 3, we learned about arithmetic and geometric series. Let's explore whether or not infinite arithmetic or geometric series converge. We'll begin with arithmetic series. The general form of an infinite **arithmetic series** is

$$\sum_{n=1}^{\infty} [a + (n-1)d] = a + (a + d) + (a + 2d) + \cdots$$

First note that we can split this up into three series:

$$\sum_{n=1}^{\infty} [a + (n-1)d] = \sum_{n=1}^{\infty} a + \sum_{n=1}^{\infty} nd - \sum_{n=1}^{\infty} d$$

The first series diverges if $a \neq 0$ because its partial sum is $a + a + a + \cdots = na$; as n grows to infinity, this sequence of partial sums diverges. The last series diverges if $d \neq 0$ for the same reason. In the special case $a = d$, the first and third series cancel out, but that won't matter because of the middle series. The middle sum equals $d(1 + 2 + 3 + \cdots)$, which we just saw as an example of a diverging series in a previous paragraph. Therefore, any infinite **arithmetic** series **diverges**.

Now let's consider a geometric series. The general form of an infinite **geometric series** is:

$$\sum_{n=1}^{\infty} ar^{n-1} = a + ar + ar^2 + \cdots$$

The partial sum S_n for this series is:

$$S_n = \sum_{k=1}^{n} ar^{k-1} = a + ar + ar^2 + \cdots + ar^{n-1}$$

In Chapter 3, we learned that this partial sum (which is a finite geometric series) is:

$$S_n = \frac{a(1 - r^n)}{1 - r} \text{ (if } r \neq 1) \quad , \quad S_n = na \text{ (if } r = 1)$$

The convergence or divergence of an infinite geometric series depends on the value of r. We will look at the sequence of partial sums $\{S_n\}$, where S_n is given above.

- If $r = 0$, the series trivially **converges** (but it's not really a geometric series).
- If $0 < |r| < 1$, $\{S_n\} = \left\{\frac{a(1-r^n)}{1-r}\right\}$ converges to the limiting value $L = \frac{a}{1-r}$ because r^n approaches zero as $n \to \infty$ if $|r| < 1$ (recall Chapter 2). An infinite geometric series **converges** to $L = \frac{a}{1-r}$ in this case.
- If $r = 1$, $\{S_n\} = \{na\}$ diverges if $a \neq 0$. (In the trivial case $a = 0$, the series is simply $0 + 0 + 0 + \cdots = 0$, which converges, but since that isn't a geometric series, we don't need to consider it.) An infinite geometric series **diverges** in this case.
- If $r = -1$, $\{S_n\} = \left\{\frac{a[1-(-1)^n]}{1-(-1)}\right\} = \left\{\frac{a[1-(-1)^n]}{2}\right\}$. In this case, $S_1 = a, S_2 = 0, S_3 = a$, $S_4 = 0$, $S_5 = a$, etc. This sequence of partial sums diverges because it doesn't converge to any one value (recall Chapter 2). An infinite geometric series **diverges** in this case.
- If $|r| > 1$, $\{S_n\} = \left\{\frac{a(1-r^n)}{1-r}\right\}$ diverges because r^n becomes infinite as $n \to \infty$ if $|r| > 1$ (recall Chapter 2). An infinite geometric series **diverges** in this case.

We can summarize the above points concisely as follows:

- If $|r| < 1$, an infinite geometric series **converges** to $L = \frac{a}{1-r}$.
- If $|r| \geq 1$, an infinite geometric series **diverges** (assuming $a \neq 0$).

As examples, consider the two different infinite **geometric series** below.

$$\sum_{n=1}^{\infty} \frac{6}{2^{n-1}} = \sum_{n=1}^{\infty} 6\left(\frac{1}{2}\right)^{n-1} = 6 + 3 + \frac{3}{2} + \frac{3}{4} + \frac{3}{8} + \cdots$$

- The geometric series above has $a = 6$ and $r = \frac{1}{2}$. Since $|r| < 1$, it converges to $\frac{a}{1-r} = \frac{6}{1-\frac{1}{2}} = \frac{6}{1/2} = 6 \div \frac{1}{2} = 6 \times 2 = 12$. Observe that $S_5 = \frac{93}{8} = 11.625$. If you try higher values of n, you'll see that S_n gets closer to 12.

$$\sum_{n=1}^{\infty} 3(2)^{n-1} = 3 + 6 + 12 + 24 + 48 + \cdots$$

- The geometric series above has $a = 3$ and $r = 2$. Since $|r| \geq 1$, it diverges.

Recall from Chapter 3 that for some finite series, we can calculate the sum by making pairs of terms that cancel out; we saw examples and exercises of this type that we were able to solve using the method of **partial fractions**. This approach also works even for some infinite series, as we'll see now. Consider the series below.

$$\sum_{n=2}^{\infty} \frac{1}{n(n-1)}$$

We will split the fraction into two fractions by introducing the unknowns x and y. (If you want more practice with this concept, review the technique of partial fractions from the part of calculus that covers techniques of integration.)

$$\frac{1}{n(n-1)} = \frac{x}{n} + \frac{y}{n-1}$$

Multiply both sides by $n(n-1)$.

$$1 = x(n-1) + ny = nx - x + ny = n(x+y) - x$$

In order for this equation to be true for any value of n, we need $1 = -x$ and $x + y = 0$ such that $y = -x = -(-1) = 1$. The above series can therefore be expressed as

$$\sum_{n=2}^{\infty} \frac{1}{n(n-1)} = \sum_{n=2}^{\infty} \left(-\frac{1}{n} + \frac{1}{n-1}\right) = -\frac{1}{2} + \frac{1}{1} - \frac{1}{3} + \frac{1}{2} - \frac{1}{4} + \frac{1}{3} - \frac{1}{5} + \frac{1}{4} + \cdots$$

The partial sum S_n for this series is:

$$S_n = \sum_{k=2}^{n} \left(-\frac{1}{k} + \frac{1}{k-1}\right) = -\frac{1}{2} + \frac{1}{1} - \frac{1}{3} + \frac{1}{2} - \frac{1}{4} + \frac{1}{3} + \cdots - \frac{1}{n-1} + \frac{1}{n-2} - \frac{1}{n} + \frac{1}{n-1}$$

Every pair cancels out except for two terms:

$$S_n = \sum_{k=2}^{n} \left(-\frac{1}{k} + \frac{1}{k-1}\right) = 1 - \frac{1}{n} = \frac{n}{n} - \frac{1}{n} = \frac{n-1}{n}$$

The sequence of partial sums is $\{S_n\} = \left\{1 - \frac{1}{n}\right\}$. Applying what we learned in Chapter 2 about limiting values, the sequence $\left\{1 - \frac{1}{n}\right\}$ converges to 1. Since the sequence of partial sums converges to the limiting value $L = 1$, the corresponding infinite series converges to $L = 1$.

There are several tests and theorems concerning the convergence and divergence of series. We'll discuss a simple test now and explore others in Chapters 5-11.

The **simple divergence test** can be useful for showing that a variety of series diverge, but beware that in many cases this test is inconclusive. We'll show how to use the simple divergence test for the general series below.

$$\sum_{n=1}^{\infty} a_n = a_1 + a_2 + a_3 + \cdots$$

- If the sequence $\{a_n\}$ doesn't converge to zero, meaning that $\lim_{n\to\infty} a_n \neq 0$ (or if the sequence $\{a_n\}$ diverges), then the corresponding infinite series **diverges**. Notice how this rule applies to the **individual terms** a_n (rather than the partial sums S_n that we considered previously) of the series. The basic idea is that if a_n doesn't approach zero as $n \to \infty$, then the sum of the infinite series can't be finite (but this isn't a proof; to see how this is proven, see the next paragraph).
- If the sequence $\{a_n\}$ converges to zero, meaning that $\lim_{n\to\infty} a_n = 0$, **you need to use a different test** to determine whether or not the infinite series converges. In this case, **you can't draw any conclusions from this test**.

We now offer a quick proof of the simple divergence test. Consider the partial sum $S_n = a_1 + a_2 + a_3 + \cdots + a_n$. If the infinite series converges to L, this means that the sequence of partial sums $\{S_n\}$ converges to the limiting value L. That is, $S_n \to L$ as $n \to \infty$. Note that $S_n - S_{n-1} = a_n$. As $n \to \infty$, S_n and S_{n-1} both $\to L$, such that $S_n - S_{n-1} \to L - L = 0$. Since $S_n - S_{n-1} = a_n$, this shows that $a_n \to 0$ as $n \to \infty$ if $\{S_n\}$ converges.

Following are three examples involving the simple divergence test.

- The series below **diverges** since the sequence $\left\{\frac{n}{n+2}\right\}$ converges to 1 as $n \to \infty$.

$$\sum_{n=1}^{\infty} \frac{n}{n+2}$$

- The series below **diverges** because the sequence $\{1 + (-1)^n\} = 0, 1, 0, 1, 0, \ldots$ diverges; this sequence alternates between two values without ever converging to a single value. Technically, the similar series $\sum(-1)^n$ also diverges, even though $\{(-1)^n\} = -1, 1, -1, 1, -1, 1, \ldots$ appears to have an infinite number of pairs that cancel out. Again, the problem is that the sequence $\{(-1)^n\}$ diverges because it doesn't converge to a single value. If you feel strongly that $\sum(-1)^n$ is finite, you have a point, but now consider this; you can't determine which value the sum equals. Does it add up to -1, 0, or 1? You can't know because the series never ends; it's infinite.

$$\sum_{n=1}^{\infty} [1 + (-1)^n]$$

- The simple divergence test is **inconclusive** for the harmonic series below because the sequence $\left\{\frac{1}{n}\right\}$ converges to 0 as $n \to \infty$. When $\lim_{n\to\infty} a_n = 0$, **you need to use a different test**. (As we'll learn in future chapters, the harmonic series actually diverges even though $\lim_{n\to\infty} \frac{1}{n} = 0$.)

$$\sum_{n=1}^{\infty} \frac{1}{n}$$

Beware: If a series converges, then $\lim_{n\to\infty} a_n = 0$, but the converse isn't necessarily true. If $\lim_{n\to\infty} a_n = 0$, **the series may converge or diverge**. The problem is that some series that diverge, like the harmonic series above, have $\lim_{n\to\infty} a_n = 0$.

On the other hand, if the sequence of **partial sums** converges, that is $\lim_{n\to\infty} S_n = L$, then the series does converge. Students who don't have these important distinctions firmly in mind tend to make more mistakes trying to apply these concepts.

What if the index doesn't start at 1? What if it starts at 0, 2, 15, or -6? The same basic concepts still apply. Consider the two sums below where n_1 and n_2 are two different finite **starting values** and where $n_2 > n_1$. The right sum simply removes one or more terms from the left sum.

$$\sum_{n=n_1}^{\infty} a_n \quad , \quad \sum_{n=n_2}^{\infty} a_n$$

Removing a finite number of terms doesn't alter the convergence or divergence of a series; if the left sum converges, the right sum also converges, and if the left sum diverges, the right sum also diverges. The same is true if we add a finite number of terms, provided that each term that is added is finite. If the right sum converges, the left sum also converges, **provided that the added terms are all finite**; if the right sum diverges, the left sum also diverges. As an example of an exception, consider the series below. Suppose that we already know that the right sum converges. The problem with the left sum is that $\frac{1}{n^2}$ is undefined at $n = 0$; it isn't finite at $n = 0$. (However, if the right sum diverged, the left sum would also diverge. But in this example, the right sum converges, as we'll learn in a future chapter.)

$$\sum_{n=0}^{\infty} \frac{1}{n^2} \quad , \quad \sum_{n=1}^{\infty} \frac{1}{n^2}$$

While the starting value of the index doesn't impact whether a series converges or diverges (apart from the exception noted above), if the series converges, the starting value can affect the value of the sum. For example, consider the three sums below. All three series converge, but the value of the sum is different in each case. The middle sum includes the $n = 1$ term, which is $\frac{1}{2^1} = \frac{1}{2}$, which the right sum is missing, which is why the middle sum is $\frac{1}{2}$ larger than the right sum. The left sum includes the $n = 0$ term, which is $\frac{1}{2^0} = \frac{1}{1} = 1$, which is why the left sum is 1 larger than the middle sum.

$$\sum_{n=0}^{\infty} \frac{1}{2^n} = 2 \quad , \quad \sum_{n=1}^{\infty} \frac{1}{2^n} = 1 \quad , \quad \sum_{n=2}^{\infty} \frac{1}{2^n} = \frac{1}{2}$$

If two infinite series are identical apart from the starting value, the two sums are related as follows, where $n_2 > n_1$. Basically, we're just adding the missing terms to the series with the higher starting value.

$$\sum_{n=n_1}^{\infty} a_n = a_{n_1} + a_{n_1+1} + a_{n_1+2} + \cdots + a_{n_2-1} + \sum_{n=n_2}^{\infty} a_n$$

The example below illustrates this for a specific case.

$$\sum_{n=1}^{\infty} \frac{1}{3^n} = \frac{1}{3^1} + \frac{1}{3^2} + \frac{1}{3^3} + \frac{1}{3^4} + \sum_{n=5}^{\infty} \frac{1}{3^n}$$

Similarly, **<u>shifting the index</u>** (which we learned how to do in Chapter 3) of a series doesn't alter its convergence or divergence. For example, the two series below are equivalent. Note that $n = k + 1$.

$$\sum_{k=0}^{\infty} \frac{1}{(k+1)^3} \quad , \quad \sum_{n=1}^{\infty} \frac{1}{n^3}$$

Written out longhand, both series above are:

$$\frac{1}{1^3} + \frac{1}{2^3} + \frac{1}{3^3} + \frac{1}{4^3} + \frac{1}{5^3} + \cdots$$

Multiplying an entire series by the same nonzero, finite constant doesn't affect whether or not it converges. For example, if the series on the left below converges to L, then the series on the right converges to cL; if instead the series on the left diverges, then the series on the right also diverges.

$$\sum_{n=1}^{\infty} a_n = L \quad \to \quad c\sum_{n=1}^{\infty} a_n = \sum_{n=1}^{\infty} ca_n = cL$$

If one series converges to L_a and another series converges to L_b, then

$$\sum_{n=1}^{\infty} a_n = L_a \quad , \quad \sum_{n=1}^{\infty} b_n = L_b \quad \to \quad \sum_{n=1}^{\infty} (a_n + b_n) = L_a + L_b$$

<u>Beware</u> that multiplication like $a_n b_n$ or division like $\frac{a_n}{b_n}$ **<u>doesn't work the same way</u>**. If the series $\sum a_n$ converges to L_a and if the series $\sum b_n$ converges to L_b, it's possible for the series $\sum a_n b_n$ to converge to something different from $L_a L_b$, and it's possible for the series $\sum \frac{a_n}{b_n}$ to diverge (even if $L_b \neq 0$). Chapter 12 shows how to multiply series.

Also, beware it's possible for the sum of two diverging series to be converging. This can happen if there is cancellation. We saw an example of this when we used partial fractions; $\sum -\frac{1}{n}$ and $\sum \frac{1}{n+1}$ each diverge (as we'll learn in a future chapter), yet the sum $\sum\left(-\frac{1}{n}+\frac{1}{n+1}\right)$ converges (if they don't include $n = 0$) due to cancellation.

If the series $\sum a_n$ converges to L and all of the a_n are nonzero, then the series made of **reciprocals**, $\sum \frac{1}{a_n}$, diverges. For example, $\sum \frac{1}{2^n}$ converges whereas $\sum 2^n$ diverges. However, knowing that a series diverges **doesn't** guarantee that the series made of reciprocals converges. For example, $\sum \frac{2n-1}{2n+1}$ and $\sum \frac{2n+1}{2n-1}$ both diverge according to the simple divergence test (because both of the corresponding sequences converge to 1).

$$\sum_{n=1}^{\infty} a_n = L \quad \rightarrow \quad \sum_{n=1}^{\infty} \frac{1}{a_n} \text{ diverges}$$

Example 1. What does the simple divergence test reveal about the series below?

$$\sum_{n=1}^{\infty}(n-4)$$

Solution: The sequence $\{n-4\}$ doesn't converge to zero (it actually diverges) as $n \rightarrow \infty$. Therefore, the corresponding series **diverges** according to the simple divergence test.

Example 2. What does the simple divergence test say about the series below?

$$\sum_{k=3}^{\infty} \frac{2k}{k^2-3k+5}$$

Solution: Divide the numerator and denominator each by the highest power of the index (which is k^2 in this example): $\frac{2k}{k^2-3k+5} = \frac{\frac{2}{k}}{1-\frac{3}{k}+\frac{5}{k^2}}$. The numerator converges to zero while the denominator converges to 1 as $k \rightarrow \infty$, which shows that the sequence $\left\{\frac{2k}{k^2-3k+5}\right\}$ converges to zero as $k \rightarrow \infty$. When the sequence converges to zero, the simple divergence test is **inconclusive**. It doesn't reveal anything about the series.

Example 3. What does the simple divergence test say about the series below?

$$1 + 4 + 7 + 1 + 4 + 7 + 1 + 4 + 7 + \cdots$$

Solution: The sequence 1, 4, 7, 1, 4, 7, 1, 4, 7, ... repeats forever without converging to any one value; this sequence diverges. Therefore, the corresponding series **diverges** according to the simple divergence test.

Example 4. Does the series below converge or diverge? If it converges, find the sum.

$$\sum_{i=3}^{\infty} \frac{5}{2^i}$$

Solution: This is a geometric series. Before we can find the sum, we need to rewrite the series using the standard form

$$\sum_{i=1}^{\infty} ar^{i-1}$$

Note that $\left(\frac{1}{2}\right)^i = \frac{1^i}{2^i} = \frac{1}{2^i}$. With this, we can rewrite the series as

$$\sum_{i=3}^{\infty} \frac{5}{2^i} = \sum_{i=3}^{\infty} 5\left(\frac{1}{2}\right)^i$$

The series still isn't in standard form because the exponent needs to be $i - 1$ and the starting value isn't 1 (we need the correct starting value in order to use the sum formula for a geometric series). Let's shift the index to $i = k - 1$, such that $i + 1 = k$, like we learned how to do in Chapter 3. When $i = 3$, $k = 3 + 1 = 4$.

$$\sum_{i=3}^{\infty} 5\left(\frac{1}{2}\right)^i = \sum_{k=4}^{\infty} 5\left(\frac{1}{2}\right)^{k-1}$$

Now we need to fix the starting value. Let's calculate what the first three terms would be if the sum started at $k = 1$.

$$5\left(\frac{1}{2}\right)^{1-1} + 5\left(\frac{1}{2}\right)^{2-1} + 5\left(\frac{1}{2}\right)^{3-1} = 5\left(\frac{1}{2}\right)^0 + 5\left(\frac{1}{2}\right)^1 + 5\left(\frac{1}{2}\right)^2 = 5(1) + \frac{5}{2} + \frac{5}{4}$$

$$= \frac{20}{4} + \frac{10}{4} + \frac{5}{4} = \frac{35}{4}$$

We can start the sum at $k = 1$ if we also subtract $\frac{35}{4}$.

$$\sum_{k=4}^{\infty} 5\left(\frac{1}{2}\right)^{k-1} = -\frac{35}{4} + \sum_{k=1}^{\infty} 5\left(\frac{1}{2}\right)^{k-1}$$

Now the sum is in standard form. Compare $5\left(\frac{1}{2}\right)^{k-1}$ with ar^{k-1} to see that $a = 5$ and $r = \frac{1}{2}$. Since $|r| < 1$, this sum **converges** to $L = \frac{a}{1-r}$. Remember to include the $-\frac{35}{4}$.

$$\sum_{i=3}^{\infty} \frac{5}{2^i} = -\frac{35}{4} + \frac{a}{1-r} = -\frac{35}{4} + \frac{5}{1-\frac{1}{2}} = -\frac{35}{4} + \frac{5}{1/2} = -\frac{35}{4} + 5 \div \frac{1}{2}$$

$$= -\frac{35}{4} + 5 \times \frac{2}{1} = -\frac{35}{4} + \frac{10}{1} = -\frac{35}{4} + \frac{40}{4} = \frac{5}{4}$$

Example 5. Does the series below converge or diverge? If it converges, find the sum.

$$\frac{1}{100} - \frac{7}{100} + \frac{49}{100} - \frac{343}{100} + \frac{2401}{100} - \frac{16{,}807}{100} + \cdots$$

Solution: Since each term is -7 times the previous term, this is a geometric series with $a = \frac{1}{100}$ and $r = -7$. Written using summation notation, this series is

$$\sum_{n=1}^{\infty} \frac{(-7)^{n-1}}{100}$$

Since $|r| \geq 1$ (and $a \neq 0$), this geometric series **diverges**.

Example 6. Does the series below converge or diverge? If it converges, find the sum.

$$\sum_{m=2}^{\infty} \frac{3m-2}{m}$$

Solution: Apply the simple divergence test. Divide the numerator and denominator each by m to see that the sequence $\left\{\frac{3m-2}{m}\right\} = \left\{\frac{3-\frac{2}{m}}{1}\right\}$ converges to $L = \frac{3}{1} = 3$. Since the sequence doesn't converge to zero, the corresponding series **diverges**.

Example 7. Does the series below converge or diverge? If it converges, find the sum.

$$\frac{1}{2}+\frac{1}{2}+\frac{1}{3}+\frac{1}{3}+\frac{1}{3}+\frac{1}{4}+\frac{1}{4}+\frac{1}{4}+\frac{1}{4}+\frac{1}{5}+\frac{1}{5}+\frac{1}{5}+\frac{1}{5}+\frac{1}{5}+\cdots$$

Solution: There are two $\frac{1}{2}$'s, which add up to one. There are three $\frac{1}{3}$'s, which also add up to one. There are four $\frac{1}{4}$'s, which also add up to one. There are five $\frac{1}{5}$'s, which again add up to one. Each set of fractions adds up to one, and there are an infinite number of sets, so this series diverges. (Note that the simple divergence test is inconclusive for this series since the individual terms of this sequence converge to zero.)

Example 8. Does the series below converge or diverge? If it converges, find the sum.

$$\sum_{j=0}^{\infty}\frac{2}{4j^2-1}$$

Solution: Try the method of partial fractions from Chapter 3. First factor $4j^2-1$ as $(2j-1)(2j+1)$.

$$\frac{2}{4j^2-1}=\frac{2}{(2j-1)(2j+1)}=\frac{x}{2j-1}+\frac{y}{2j+1}$$

Multiply both sides by $(2j-1)(2j+1)$.

$$2=x(2j+1)+y(2j-1)=2jx+x+2jy-y=x-y+2j(x+y)$$

For this equation to be true for any value of j, we need $2=x-y$ and $x+y=0$, for which $y=-x$. Plug this into the first equation to get $2=x-(-x)=2x$, for which $x=1$. With $x=1$ and $y=-x=-1$, we can rewrite the series as

$$\sum_{j=0}^{\infty}\frac{2}{4j^2-1}=\sum_{j=0}^{\infty}\left(\frac{1}{2j-1}-\frac{1}{2j+1}\right)$$

The partial sum S_j for this series is:

$$S_j=\sum_{k=0}^{j}\left(\frac{1}{2k-1}-\frac{1}{2k+1}\right)=\frac{1}{-1}-\frac{1}{1}+\frac{1}{1}-\frac{1}{3}+\frac{1}{3}-\frac{1}{5}+\frac{1}{5}-\frac{1}{7}+\cdots+\frac{1}{2j-1}-\frac{1}{2j+1}$$

All of the terms cancel except for the first and last.

$$S_j=-1-\frac{1}{2j+1}$$

The sequence of partial sums is $\{S_j\} = \left\{-1 - \frac{1}{2j+1}\right\}$. Applying what we learned in Chapter 2 about limiting values, the sequence $\left\{-1 - \frac{1}{2j+1}\right\}$ converges to -1 (since $\left\{\frac{1}{2j+1}\right\}$ converges to zero). Since the sequence of partial sums converges to the limiting value $L = -1$, the corresponding infinite series converges to $L = -1$.

$$\sum_{j=0}^{\infty} \frac{2}{4j^2 - 1} = -1$$

Note: In this example, we looked at the sequence of partial sums $S_j = -1 - \frac{1}{2j+1}$ (**not** the sequence of terms $a_j = \frac{2}{4j^2-1}$). Here, we applied the definition of the convergence of a series, which uses partial sums. Contrast this with Example 6, where we used the simple divergence test, which uses terms rather than partial sums. Another important distinction is that here the series converges to the same value that the sequence of partial sums converges to, whereas when we apply the simple divergence test we're only concerned with whether or not the sequence of terms converges to zero. It's very important to distinguish between these two methods.

Example 9. Does the series below converge or diverge? If it converges, find the sum.

$$\sum_{\ell=1}^{\infty} \left(\frac{6}{3^{\ell-1}} + \frac{8}{4^{\ell-1}}\right)$$

Solution: Here we have two geometric series added together. Both series converge, so we may separate the two series and add their sums together. As with Example 4, first we need to write each series in the standard form $ar^{\ell-1}$.

$$\sum_{\ell=1}^{\infty} \left(\frac{6}{3^{\ell-1}} + \frac{8}{4^{\ell-1}}\right) = \sum_{\ell=1}^{\infty} 6\left(\frac{1}{3}\right)^{\ell-1} + \sum_{\ell=1}^{\infty} 8\left(\frac{1}{4}\right)^{\ell-1} = \frac{a_1}{1-r_1} + \frac{a_2}{1-r_2}$$

$$= \frac{6}{1-\frac{1}{3}} + \frac{8}{1-\frac{1}{4}} = \frac{6}{2/3} + \frac{8}{3/4} = 6 \div \frac{2}{3} + 8 \div \frac{3}{4} = 6 \times \frac{3}{2} + 8 \times \frac{4}{3}$$

$$= \frac{18}{2} + \frac{32}{3} = 9 + \frac{32}{3} = \frac{27}{3} + \frac{32}{3} = \frac{59}{3}$$

Chapter 4 Problems

Directions: What, if anything, does the simple divergence test reveal about each series?

(1)

$$\sum_{k=1}^{\infty} \frac{k^2}{9}$$

(2)

$$\sum_{j=2}^{\infty} \frac{9}{(j+1)^2}$$

(3)

$$\sum_{n=0}^{\infty} \cos\left(\frac{n\pi}{4}\right)$$

(4)

$$\sum_{m=4}^{\infty} \cos\left(\frac{4}{m}\right)$$

(5)

$$97 - 85 + 73 - 61 + 49 - 37 + 25 - 13 + \cdots$$

(6)

$$12 + 6 + 2 + \frac{1}{2} + \frac{1}{10} + \frac{1}{60} + \frac{1}{420} + \cdots$$

(7)

$$\sum_{i=-1}^{\infty} \frac{5i - 2}{4i + 3}$$

(8)

$$\sum_{\ell=1}^{\infty} \frac{3\ell - 6}{7 - \ell^2}$$

(9)

$$\sum_{n=2}^{\infty} \frac{5n^3}{8n^2 - 5n + 2}$$

(10)

$$0.1 + 0.2 + 0.3 + 0.4 + 0.5 + 0.6 + \cdots$$

(11)

$$10 - 5 + \frac{5}{2} - \frac{5}{4} + \frac{5}{8} - \frac{5}{16} + \frac{5}{32} - \frac{5}{64} + \cdots$$

(12)

$$\sum_{i=1}^{\infty} \left(1 - \frac{1}{i}\right)$$

(13)

$$\sum_{m=0}^{\infty}(m^2 - 4m + 11)$$

(14)

$$\sum_{k=3}^{\infty}\sqrt{\frac{k+1}{k-1}}$$

(15)

$$\sum_{j=-1}^{\infty}(-1)^j\frac{2j+1}{4j-3}$$

(16)

$$\sum_{\ell=1}^{\infty}\frac{\ell^2}{\ell!}$$

Directions: Does the series below converge or diverge? If it converges, find the sum.

(17)

$$\sum_{n=5}^{\infty} \frac{2^{n-1}}{5}$$

(18)

$$\sum_{i=2}^{\infty} 2\left(\frac{1}{5}\right)^{i-1}$$

(19)

$$\sum_{\ell=500}^{\infty} \frac{\ell}{1{,}000{,}000}$$

(20)

$$\sum_{k=-1}^{\infty} \frac{12}{4^k}$$

(21)

$$\sum_{j=1}^{\infty} \frac{9}{3^{j+2}}$$

(22)

$$\sum_{m=2}^{\infty} \frac{3}{m(m+3)}$$

(23)

$$\sum_{i=4}^{\infty} \frac{i+3}{i}$$

(24)

$$\sum_{n=2}^{\infty} \frac{4}{n^2-1}$$

(25)

$$5 - 20 + 80 - 320 + 1280 - 5120 + 20{,}480 + \cdots$$

(26)

$$0.2 + 0.04 + 0.008 + 0.0016 + 0.00032 + 0.000064 + 0.0000128 + \cdots$$

(27)

$$\sum_{k=1}^{\infty} \frac{e^k}{k^2}$$

(28)

$$\sum_{m=2}^{\infty} \left[\frac{1}{m^2} - \frac{1}{(m+4)^2}\right]$$

(29)

$$\frac{1}{2}+\frac{1}{8}+\frac{1}{8}+\frac{1}{32}+\frac{1}{32}+\frac{1}{32}+\frac{1}{32}+\frac{1}{128}+\frac{1}{128}+\frac{1}{128}+\frac{1}{128}+\frac{1}{128}+\frac{1}{128}+\frac{1}{128}+\frac{1}{128}+\cdots$$

(30)

$$\frac{1}{2}+\frac{1}{3}+\frac{1}{3}+\frac{1}{4}+\frac{1}{4}+\frac{1}{4}+\frac{1}{5}+\frac{1}{5}+\frac{1}{5}+\frac{1}{5}+\frac{1}{6}+\frac{1}{6}+\frac{1}{6}+\frac{1}{6}+\frac{1}{6}+\cdots$$

(31)

$$\sum_{\ell=1}^{\infty}\left(\frac{4}{1.5^{\ell-1}}-\frac{2}{1.25^{\ell-1}}\right)$$

(32)

$$\sum_{j=3}^{\infty}\left[5(0.3)^{j-1}-3(1.1)^{j-1}\right]$$

(33)

$$\sum_{k=5}^{\infty} \frac{1}{2k+2} - \sum_{m=3}^{\infty} \frac{1}{2m}$$

(34) Note: Explore the convergence as i approaches $-\infty$.

$$\sum_{i=-\infty}^{0} \frac{1}{2^{i-4}}$$

(35) Note: Explore the convergence as m approaches both ∞ and $-\infty$.

$$\sum_{m=-\infty}^{\infty} \frac{2}{3^{|m|}}$$

5 Comparison Test

If we know some series that converge and some series that diverge, we can sometimes use this knowledge to help determine whether other series converge or diverge. We can do this by using a comparison test. Before we learn about the comparison test, it will be useful to list a few common series that are known to converge or diverge.

Following are examples of common infinite series that are known to **converge**.

- Recall from Chapter 4 that a **geometric** series with terms of the form ar^{n-1} converges if $|r| < 1$. If we let $q = \frac{1}{r}$, we can write the series with terms of the form $\frac{a}{q^{n-1}}$; in this case, the series converges if $|q| > 1$. We can also shift the index to $k = n - 1$ to express the terms in the form $\frac{a}{q^k}$.

$$\sum_{n=1}^{\infty} ar^{n-1} = \sum_{n=1}^{\infty} \frac{a}{q^{n-1}} = \sum_{k=0}^{\infty} \frac{a}{q^k} \quad \text{converges if } |r| < 1, |q| > 1$$

- As we saw in Chapter 4, some series can be shown to converge using **partial fractions**. For example, the series with terms of the form $\frac{1}{n(n-1)}$ converges. We can write this series in the form $\frac{1}{k(k+1)}$ by shifting the index to $n = k + 1$. (Be sure that the index never results in division by zero; for example, if n were to start from 1 in the left series below, that would result in division by zero.)

$$\sum_{n=2}^{\infty} \frac{1}{n(n-1)} = \sum_{k=1}^{\infty} \frac{1}{k(k+1)} \quad \text{converges}$$

- Series with terms of the form $\frac{1}{n^2}$ converge. See Example 1.

$$\sum_{n=1}^{\infty} \frac{1}{n^2} \quad \text{converges}$$

- Series with terms of the form $\frac{1}{n^p}$ converge if $p > 1$ (see Chapter 6). Note that the series above corresponds to the special case $p = 2$.

$$\sum_{n=1}^{\infty} \frac{1}{n^p} \quad \text{converges if } p > 1$$

Following are examples of common infinite series that are known to **diverge**.

- Recall from Chapter 4 that a **geometric** series with terms of the form ar^{n-1} diverges if $|r| > 1$ (assuming $a \neq 0$). If we let $q = \frac{1}{r}$, we can write the series with terms of the form $\frac{a}{q^{n-1}}$; in this case, the series converges if $|q| < 1$. We can also shift the index to $k = n - 1$ to express the terms in the form $\frac{a}{q^k}$.

$$\sum_{n=1}^{\infty} ar^{n-1} = \sum_{n=1}^{\infty} \frac{a}{q^{n-1}} = \sum_{k=0}^{\infty} \frac{a}{q^k} \quad \text{diverges if } |r| > 1, |q| < 1, a \neq 0$$

- The **harmonic** series, which has terms of the form $\frac{1}{n}$, diverges (see Chapter 6).

$$\sum_{n=1}^{\infty} \frac{1}{n} \quad \text{diverges}$$

- Series with terms of the form $\frac{1}{n^p}$ diverge if $p \leq 1$ (see Chapter 6). Note that the harmonic series corresponds to the special case $p = 1$.

$$\sum_{n=1}^{\infty} \frac{1}{n^p} \quad \text{diverges if } p \leq 1$$

Now we will define a few terms relating to sequences, which will be relevant for making comparison tests.

- For a **nondecreasing** sequence, $a_{n+1} \geq a_n$ for every value of n. For example, $\frac{1}{2}, \frac{2}{3}, \frac{3}{4}, \frac{4}{5}, \frac{5}{6}, \ldots, \frac{n}{n+1}, \ldots$ is a nondecreasing sequence since no fraction is smaller than the previous fraction. Note that it's okay for consecutive terms to be equal in a nondecreasing sequence, as in the example 1, 1, 2, 3, 5, 8, …, which begins with two 1's. (As a counterexample, 1, 2, 1, 2, 1, 2, 1, 2, … isn't nondecreasing.)
- For a **nonincreasing** sequence, $a_{n+1} \leq a_n$ for every value of n. For example, $1, \frac{1}{2}, \frac{1}{4}, \frac{1}{8}, \frac{1}{16}, \frac{1}{32}, \frac{1}{64}, \ldots$ is a nonincreasing sequence.
- A sequence is said to be **bounded** from **above** if there exists some maximum value that it never exceeds. A sequence is said to be **bounded** from **below** if there exists some minimum value that it never falls below.
- If a sequence never exceeds the value N_U, then N_U is an **upper bound**, meaning that $a_n \leq N_U$ for any term a_n.

- If L_U is the **least upper bound**, then it isn't possible to find a number smaller than L_U that is an upper bound. For example, consider the sequence $\frac{1}{3}, \frac{2}{5}, \frac{3}{7}, \frac{4}{9}, \frac{5}{11}, \frac{6}{13}, \frac{7}{15}, \frac{8}{17}, \frac{9}{19}, \ldots, \frac{n}{2n+1}, \ldots$ This sequence approaches the limiting value $L = \frac{1}{2}$ (recall Chapter 2). Observe that 1, 2, $\sqrt{3}$, π, and 5 are also upper bounds, whereas the **least** upper bound is $L_U = \frac{1}{2}$ because no value smaller than $\frac{1}{2}$ is an upper bound.
- Similarly, if a sequence never gets lower than the value N_L, then N_L is a **lower bound**. If L_L is the **greatest lower bound**, then it isn't possible to find a number greater than L_L that is a lower bound.

Here is why the above definitions are helpful:

- If a series **doesn't have any negative terms**, it has a **nondecreasing** sequence of partial sums. That is, if $a_n \geq 0$ for every term in the series $\sum a_n$, then the partial sums $s_n = a_1 + a_2 + a_3 + \cdots + a_n$ form a nondecreasing sequence because $s_{n+1} = s_n + a_n$ (that is, since $a_n \geq 0$, it follows that $s_{n+1} \geq s_n$).
- If a **nondecreasing** sequence is **bounded from above**, the sequence **converges**. Since this statement is regarding **sequences**, this should be clear from what we learned in Chapter 2; such a sequence clearly converges to a finite limiting value.
- Combining these two points together, if a series doesn't have negative terms, if the sequence of **partial sums** is **bounded from above**, the series **converges**. This time, we're using the definition from Chapter 4; the sequence of partial sums determines whether or not a **series** converges.

If we want to determine whether or not a series converges, if no terms are negative, we just need to determine whether or not the sequence of partial sums is bounded from above. (What if the series has negative terms? See Chapters 10-11.)

How can you tell whether or not the sequence of partial sums is bounded from above? One way is to **compare** the series to a series that is known to converge or diverge.

Consider the two series below. On the left is a geometric series with $a = 1$ and $r = \frac{1}{2}$. Since $|r| < 1$, we know that the left series converges.

$$\sum_{n=1}^{\infty} \frac{1}{2^n} = \sum_{n=1}^{\infty} \left(\frac{1}{2}\right)^n \quad , \quad \sum_{n=1}^{\infty} \frac{e^{-n}}{2^n}$$

The series on the right also features the factor of $\frac{1}{2^n}$. Neither series has any negative terms (since $\frac{1}{2^n}$ and e^{-n} are strictly positive for all n), so for each series the sequence of partial sums is nondecreasing. Thus, if we can show that the right sum is bounded from above, we will know that the right series converges. Let's try comparing the two series. How does each term, $\frac{e^{-n}}{2^n}$, of the right series compare with each term, $\frac{1}{2^n}$, of the left series? Since $e^{-n} \leq 1$ for all n (since n begins at 1), it follows that $\frac{e^{-n}}{2^n} \leq \frac{1}{2^n}$. That is, we can clearly see that every term of the right series is smaller than the corresponding term of the left series. (It doesn't matter that $\sum \frac{1}{2^n}$ isn't in standard form, since $\sum \frac{1}{2^{m-1}}$ and $\sum \frac{1}{2^n}$ both converge; shifting the index from m to $n = m - 1$ won't make it diverge.) Since the left series is known to converge, it follows that the right series also converges.

The **comparison test** is formulated as follows, where the series $\sum a_n$, $\sum c_n$, and $\sum d_n$ do not have any negative terms for all $n > N$.

- If a series $\sum c_n$ is known to converge and if $a_n \leq c_n$ for all $n > N$, then the series $\sum a_n$ **converges**.
- If a series $\sum d_n$ is known to diverge and if $a_n \geq d_n$ for all $n > N$, then the series $\sum a_n$ **diverges**.

The **limit comparison test** formulates the comparison in terms of a limit, where the series $\sum a_n$, $\sum c_n$, and $\sum d_n$ do not have any negative terms for all $n > N$.

- If a series $\sum c_n$ is known to converge and if $\lim\limits_{n\to\infty} \frac{a_n}{c_n}$ is finite, then $\sum a_n$ **converges**.
- If a series $\sum d_n$ is known to diverge and if $\lim\limits_{n\to\infty} \frac{a_n}{d_n} > 0$, then $\sum a_n$ **diverges**.

Observe that if $0 < \lim\limits_{n\to\infty} \frac{a_n}{b_n} < \infty$, then either $\sum a_n$ and $\sum b_n$ both converge or $\sum a_n$ and $\sum b_n$ both diverge. The basic idea behind the limit comparison test is that the **tail** of the series is what determines whether the series converges or diverges. (We assume that every term a_n is individually finite; for example, n must start at 4 or higher if $a_n = \frac{1}{n-3}$ in order to avoid a problem with division by zero.)

Example 1. Use a comparison test to determine if the series converges or diverges.

$$\sum_{n=2}^{\infty} \frac{1}{n^2}$$

Solution: Look at the list of bullet points with series that we know converge or diverge. (Yes, the series above is on that list, but we can't compare this series with itself.) One of the series known to converge is:

$$\sum_{n=2}^{\infty} \frac{1}{n(n-1)} = \sum_{n=2}^{\infty} \frac{1}{n^2 - n}$$

We distributed the n to write $n(n-1) = n^2 - n$, which makes it easier to compare the two series. Compare the terms of the given series, $a_n = \frac{1}{n^2}$, with the terms of the series that we know converges, $c_n = \frac{1}{n^2-n}$. For a given value of n, between a_n and c_n, which is bigger? Observe that $a_n < c_n$ for all n. (Since c_n has the smaller denominator, c_n is larger than a_n.) For example, for $n = 3$, $a_3 = \frac{1}{3^2} = \frac{1}{9} \approx 0.111$ is smaller than $c_3 = \frac{1}{3^2-3} = \frac{1}{9-3} = \frac{1}{6} \approx 0.167$. Since neither series has any negative terms, $\sum c_n$ is known to converge, and $a_n \leq c_n$ for all n, it follows that the series $\sum a_n$ **converges**.

Example 2. Use a comparison test to determine if the series converges or diverges.

$$\sum_{k=3}^{\infty} \frac{k}{k^4 - 9}$$

Solution: Since $\frac{k}{k^4} = \frac{1}{k^3}$, it makes sense to compare the given series to the series below, which is known to converge. (More generally, $\sum \frac{1}{k^p}$ converges if $p > 1$.)

$$\sum_{k=1}^{\infty} \frac{1}{k^3}$$

We can't use the same comparison test as in the previous example because $a_k = \frac{k}{k^4-9}$ isn't less than $c_k = \frac{1}{k^3}$. For example, for $k = 3$, $a_3 = \frac{3}{3^4-9} = \frac{3}{81-9} = \frac{3}{72} = \frac{1}{24} \approx 0.0417$ is bigger than $c_3 = \frac{1}{3^3} = \frac{1}{27} \approx 0.0370$. Since a_k isn't smaller than c_k, we will use the **limit** comparison test:

$$\lim_{k\to\infty}\frac{a_n}{c_n}=\lim_{k\to\infty}\frac{\frac{k}{k^4-9}}{\frac{1}{k^3}}=\lim_{k\to\infty}\left(\frac{k}{k^4-9}\div\frac{1}{k^3}\right)=\lim_{k\to\infty}\left(\frac{k}{k^4-9}\times\frac{k^3}{1}\right)=\lim_{k\to\infty}\frac{k^4}{k^4-9}$$

$$=\lim_{k\to\infty}\frac{1}{1-\frac{9}{k^4}}=\frac{1}{1-0}=\frac{1}{1}=1$$

Remember that the way to divide by a fraction is to multiply by its reciprocal. After carrying out the division, we divided the numerator and denominator each by k^4, using the method of Examples 9-10 from Chapter 2 (see the bullet points following Example 10). Since neither series has any negative terms, $\sum c_n$ is known to converge, and $\lim\limits_{n\to\infty}\frac{a_n}{c_n}=1$ (which is finite), it follows that the series $\sum a_n$ **converges**. (Note that the fact that the series have different start values doesn't affect the convergence.)

Example 3. Use a comparison test to determine if the series converges or diverges.

$$\sum_{i=2}^{\infty}\frac{1}{\sqrt{i+6}}$$

Solution: We will compare this series to the harmonic series, which we know diverges.[1]

$$\sum_{i=1}^{\infty}\frac{1}{i}$$

For a given value of i, between $a_i=\frac{1}{\sqrt{i+6}}$ and $d_i=\frac{1}{i}$, which is bigger? Observe that $a_i\geq d_i$ for all $i\geq 3$. For $i=3$, $a_3=\frac{1}{\sqrt{3+6}}=\frac{1}{\sqrt{9}}=\frac{1}{3}$ is equal $d_3=\frac{1}{3}$, while for $i>3$, a_i is greater than d_i. For example, for $i=4$, $a_4=\frac{1}{\sqrt{4+6}}=\frac{1}{\sqrt{10}}\approx 0.316$ is greater than $d_4=\frac{1}{4}=0.25$ and for $i=5$, $a_5=\frac{1}{\sqrt{5+6}}=\frac{1}{\sqrt{11}}\approx 0.302$ is greater than $d_5=\frac{1}{5}=0.2$. Since neither series has any negative terms, $\sum d_n$ is known to diverge, and $a_n\geq d_n$ for all $i>2$, it follows that the series $\sum a_n$ **diverges**.

---

[1] You could compare it with $\sum\frac{1}{i^{1/2}}=\sum\frac{1}{\sqrt{i}}$, which is known to converge since the power $p=\frac{1}{2}$ is less than or equal to one. In that case, you would use the limit comparison test since $\frac{1}{\sqrt{i+6}}$ isn't bigger than $\frac{1}{\sqrt{i}}$.

Example 4. Use a comparison test to determine if the series converges or diverges.

$$\sum_{\ell=1}^{\infty} \frac{\ell(2^\ell)}{\ell+1}$$

Solution: Compare this series to the geometric series below, where $a = 1$ and $r = 2$. Since $|r| \geq 1$, the geometric series below diverges. (As noted earlier in this chapter, it doesn't matter that it isn't in the standard form $\sum 2^{m-1}$. We don't need to shift the index to $\ell = m - 1$ to know that the series diverges.)

$$\sum_{\ell=1}^{\infty} 2^\ell$$

We can't use the same comparison test as in the previous example because $a_\ell = \frac{\ell(2^\ell)}{\ell+1}$ isn't bigger than $d_\ell = 2^\ell$. (Contrast this with Example 2. The first comparison test wants a_ℓ smaller than c_ℓ to test for convergence, but wants a_ℓ bigger than d_ℓ to test for divergence. Either way, you can still use the corresponding limit comparison test.) For example, for $\ell = 3$, $a_3 = \frac{3(2^3)}{3+1} = \frac{3(8)}{4} = \frac{24}{4} = 6$ isn't bigger than $c_3 = 2^3 = 8$. Since a_ℓ isn't bigger than d_ℓ, we will use the **limit** comparison test.

$$\lim_{\ell\to\infty} \frac{a_\ell}{d_\ell} = \lim_{\ell\to\infty} \left(\frac{\ell(2^\ell)}{\ell+1} \div 2^\ell\right) = \left(\lim_{\ell\to\infty} \frac{\ell(2^\ell)}{\ell+1} \times \frac{1}{2^\ell}\right) = \lim_{\ell\to\infty} \frac{\ell}{\ell+1} = \lim_{\ell\to\infty} \frac{1}{1+\frac{1}{\ell}} = \frac{1}{1+0} = \frac{1}{1} = 1$$

Remember that the way to divide by a fraction is to multiply by its reciprocal. After carrying out the division, we divided the numerator and denominator each by ℓ, using the method of Examples 9-10 from Chapter 2 (see the bullet points following Example 10). Since neither series has any negative terms, $\sum d_n$ is known to diverge, and $\lim\limits_{n\to\infty} \frac{a_n}{d_n} = 1$ (which is positive), it follows that the series $\sum a_n$ **diverges**. (Contrast this with Example 2. When using the limit comparison test for convergence, we want the limit to be **finite**, but when using the limit comparison test for divergence, we want the limit to be **positive**.)

Note: You could solve this problem with the simple divergence test, except that the instructions ask you to use a comparison test.

Chapter 5 Problems

Directions: Determine whether each series converges or diverges.

(1)

$$\sum_{n=1}^{\infty} \frac{1}{n^2 + 3}$$

(2)

$$\sum_{i=2}^{\infty} \frac{4}{i - \sqrt{i}}$$

(3)

$$\sum_{m=1}^{\infty} \frac{5m + 1}{m^3 - 2m}$$

(4)

$$\sum_{j=4}^{\infty} \frac{9}{6 + \sqrt{j}}$$

(5)

$$\sum_{\ell=1}^{\infty} \frac{3 - 4^{\ell+1}}{5^{\ell-1}}$$

(6)

$$\sum_{k=3}^{\infty} \frac{2}{k - \ln k}$$

(7)

$$\sum_{i=1}^{\infty} \frac{1}{i^2} \sin\left(\frac{1}{i}\right)$$

(8)

$$\sum_{m=1}^{\infty} \frac{\cos^2 m}{m^2}$$

(9)

$$\sum_{n=0}^{\infty} \frac{2^n - 5}{5(2^n - 3)}$$

(10)

$$\sum_{k=2}^{\infty} \frac{k^3 - 4k + 5}{3k^5 + 7k^3 - 1}$$

(11)

$$\sum_{j=1}^{\infty} \frac{\ln j}{j^2}$$

(12)

$$\sum_{\ell=0}^{\infty} \frac{\ell^5}{e^\ell}$$

(13)

$$\sum_{n=3}^{\infty} \sqrt{\frac{n^2 + 5n + 2}{n^4 - 4n^2 + 3}}$$

(14)

$$\sum_{i=1}^{\infty} \left(\frac{i - 1}{2i + 3}\right)\left(\frac{1}{4^i}\right)$$

(15)

$$\sum_{k=2}^{\infty} \frac{4}{(k - 1)\sqrt{k + 1}}$$

(16)

$$\sum_{m=1}^{\infty} \frac{3m-1}{6^m(m+2)}$$

(17)

$$\sum_{j=0}^{\infty} \frac{1}{j!}$$

(18)

$$\sum_{\ell=2}^{\infty} \frac{3\ell+8}{(\ell-1)^2}$$

(19)

$$\frac{1}{2\sqrt{2}}+\frac{2}{3\sqrt{3}}+\frac{3}{8}+\frac{4}{5\sqrt{5}}+\frac{5}{6\sqrt{6}}+\frac{6}{7\sqrt{7}}+\frac{7}{16\sqrt{2}}+\frac{8}{27}+\cdots$$

(20)

$$\frac{1}{3}+\frac{1}{5}+\frac{1}{9}+\frac{1}{17}+\frac{1}{33}+\frac{1}{65}+\cdots$$

(21)

$$13+\frac{15}{2}+\frac{17}{6}+\frac{19}{24}+\frac{7}{40}+\frac{23}{720}+\frac{5}{1008}+\frac{3}{4480}+\frac{29}{362{,}880}+\cdots$$

6 Integral Test

The integral test can help to test for the convergence or divergence of a **nonincreasing** infinite series that doesn't have negative terms. Suppose that the series below is a nonincreasing series, meaning that $a_{n+1} \leq a_n$, and that no a_n is negative for $n > N$. (Of course, any **decreasing** function is nonincreasing.) Here, n_0 is the starting value and N is some value for which and $a_{n+1} \leq a_n$ and $a_n \geq 0$ for all $n > N$.

$$\sum_{n=n_0}^{\infty} a_n$$

If we let $a_n = f(n)$ and $f(x)$ is a continuous, nonincreasing function of x for $x > N$, then according to the **integral test**, the sum and integral below either both converge or both diverge. (The sum above will also converge, assuming the terms from n_0 to N are each finite.)

$$\sum_{n=N}^{\infty} a_n \quad , \quad \int_{x=N}^{\infty} f(x)\,dx$$

One way to understand why the integral test works is to write out each expression above as follows and remember that a definite integral over a function represents the **area** below the curve.

$$\sum_{n=N}^{\infty} a_n = a_N + a_{N+1} + a_{N+2} + a_{N+3} + \cdots$$

$$\int_{x=N}^{\infty} f(x)\,dx = \int_{x=N}^{N+1} f(x)\,dx + \int_{x=N+1}^{N+2} f(x)\,dx + \int_{x=N+2}^{N+3} f(x)\,dx + \int_{x=N+3}^{N+4} f(x)\,dx + \cdots$$

Since $f(x)$ is a nonincreasing function, in the first integral, $f(x) \leq f(N)$; the first integral can't be larger than the area of a rectangle with a height of $f(N)$ and a width of $N + 1 - N = 1$, which has an area of $(1)f(N) = f(N) = a_N$. Similarly, the second integral can't be larger than the area of a rectangle with a height of $f(N + 1)$ and a width of $N + 2 - (N + 1) = N + 2 - N - 1 = 1$, which has an area of $(1)f(N + 1) = f(N + 1) = a_{N+1}$. No integral above can be greater than the corresponding term of the sum, simply because the function is nonincreasing. The sum of the integrals above is less than or equal to the sum of the a_n's. If the sum of the integrals diverges, then the series of a_n's also diverges. (Similarly, if the series of a_n's converges, then the sum of the integrals converges.)

We haven't yet shown that if the integral converges, then the series converges. We will show this by dropping the first term of the series and comparing again.

$$a_{N+1} + a_{N+2} + a_{N+3} + \cdots$$

$$\int_{x=N}^{\infty} f(x)\,dx = \int_{x=N}^{N+1} f(x)\,dx + \int_{x=N+1}^{N+2} f(x)\,dx + \int_{x=N+2}^{N+3} f(x)\,dx + \int_{x=N+3}^{N+4} f(x)\,dx + \cdots$$

In the comparison above, unlike on the previous page, we omitted a_N; the actual series includes a_N, but it is useful to ignore it for a moment. In this comparison, the first integral must be larger than the area of a rectangle with a height of $f(N+1)$ and a width of $N + 1 - N = 1$, which has an area of $(1)f(N+1) = f(N+1) = a_{N+1}$. The second integral must be larger than the area of a rectangle with a height of $f(N+2)$ and a width of $N + 2 - (N+1) = N + 2 - N - 1 = 1$, which has an area of $(1)f(N+2) = f(N+2) = a_{N+2}$. This sum of integrals is greater than the above series (which is missing the first term compared to the original series). If the sum of the integrals converges, then the series of a_n's also converges (adding the first term back on won't change this convergence). (Similarly, if the series of a_n's diverges, then the sum of the integrals diverges.)

Putting this all together, we see that either the series and the (full) integral both converge or both diverge, which allows us to use the (full) integral to test for both convergence and divergence. It's important that the function in the integrand be **nonincreasing**; a strictly **decreasing** function is fine (but you don't want a function that increases at any point after reaching the lower limit $x = N$).

Furthermore, comparing the sum of the integrals to the series with or without the first term provides useful bounds on the sum of a series in the case of convergence:

$$\sum_{n=N+1}^{\infty} a_n \leq \int_{x=N}^{\infty} f(x)\,dx \leq \sum_{n=N}^{\infty} a_n$$

The left series above begins with a_{N+1}, whereas the right series begins with a_N.

As an example, consider the **harmonic** series below.

$$\sum_{n=1}^{\infty} \frac{1}{n}$$

Let $f(x) = \frac{1}{x}$ and $a_n = f(n) = \frac{1}{n}$. Observe that $\frac{1}{x}$ is continuous and decreasing over the interval $x \geq 1$ (where the 1 corresponds to the initial value of the sum). In this case, there is no reason to choose N different from $n_0 = 1$. Apply the integral test.

$$\int_{x=1}^{\infty} f(x)\,dx = \int_{x=1}^{\infty} \frac{dx}{x} = [\ln x]_{x=1}^{\infty} = \ln \infty - \ln 1 = \ln \infty - 0 = \ln \infty = \infty$$

This integral and the harmonic series both **diverge**.

As a second example, consider the series below.

$$\sum_{n=1}^{\infty} \frac{1}{n^2}$$

Let $f(x) = \frac{1}{x^2}$ and $a_n = f(n) = \frac{1}{n^2}$. Again, $\frac{1}{x^2}$ is continuous and decreasing over the interval $x \geq 1$. Apply the integral test.

$$\int_{x=1}^{\infty} f(x)\,dx = \int_{x=1}^{\infty} \frac{dx}{x^2} = \left[-\frac{1}{x}\right]_{x=1}^{\infty} = -\frac{1}{\infty} - \left(-\frac{1}{1}\right) = 0 + 1 = 1$$

This integral and the corresponding series above both **converge**.

Note: We know that they both converge and we know that the integral converges to 1, but we don't know which value the sum converges to. The integral test tells us that the series converges, but doesn't give us a value for the sum. However, integrals can help to determine lower and upper bounds for series that converge (see Chapter 12). For example, if we use the inequality that we found earlier, we can say that

$$\sum_{n=2}^{\infty} \frac{1}{n^2} \leq 1 \leq \sum_{n=1}^{\infty} \frac{1}{n^2}$$

Since the first term is $\frac{1}{1^2} = \frac{1}{1} = 1$, we can rewrite this inequality as

$$\sum_{n=2}^{\infty} \frac{1}{n^2} \leq 1 \leq 1 + \sum_{n=2}^{\infty} \frac{1}{n^2}$$

Since the sum beginning from $n = 2$ converges to a value less than or equal to 1, the right-hand side converges to a value less than or equal to 2 (since $1 + 1 = 2$). Putting this all together, we have both lower and upper bounds for the full series:

$$1 \leq \sum_{n=1}^{\infty} \frac{1}{n^2} \leq 2$$

It turns out that this series converges to $\frac{\pi^2}{6} \approx 1.645$ (Chapter 15).

Example 1. Use the integral test to show that the series below converges if $p > 1$ and diverges if $p \leq 1$.

$$\sum_{n=1}^{\infty} \frac{1}{n^p}$$

Solution: Let $f(x) = \frac{1}{x^p}$ and $a_n = f(n) = \frac{1}{n^p}$. This function is nonincreasing provided that p isn't negative. (If p is negative, the series will diverge according to the simple divergence test from Chapter 4.) Apply the integral test. Recall that $x^{-p} = \frac{1}{x^p}$.

$$\int_{x=1}^{\infty} f(x)\,dx = \int_{x=1}^{\infty} \frac{dx}{x^p} = \int_{x=1}^{\infty} x^{-p}\,dx$$

We need to work out different cases because $\int \frac{dx}{x}$ is special as it gives a natural log.

- If $p \neq 1$, we get

  $$\int_{x=1}^{\infty} x^{-p}\,dx = \left[\frac{x^{1-p}}{1-p}\right]_{x=1}^{\infty} = \lim_{x\to\infty} \frac{x^{1-p}}{1-p} - \frac{1^{1-p}}{1-p} = \lim_{x\to\infty} \frac{x^{1-p}}{1-p} - \frac{1}{1-p}$$

  If $p < 1$, then $1 - p > 0$, such that the exponent of x^{1-p} is positive. In this case, $\lim\limits_{x\to\infty} \frac{x^{1-p}}{1-p} = \infty$, and the integral and series **diverge**.

  If $p > 1$, then $1 - p < 0$, such that the exponent of x^{1-p} is negative. In this case, $\lim\limits_{x\to\infty} \frac{x^{1-p}}{1-p} = 0$ and the integral is $\int_{x=1}^{\infty} x^{-p}\,dx = 0 - \frac{1}{1-p} = -\frac{1}{1-p} = \frac{1}{p-1}$, which is finite. In this case, the integral and series **converge**.

- If $p = 1$, we get the harmonic series, which **diverges**, as shown below.

$$\int_{x=1}^{\infty} \frac{dx}{x^1} = \int_{x=1}^{\infty} \frac{dx}{x} = [\ln x]_{x=1}^{\infty} = \ln \infty - \ln 1 = \ln \infty - 0 = \ln \infty = \infty$$

Combining these results together, the series **converges** if $p > 1$ and **diverges** if $p \leq 1$.

Example 2. Use the integral test to determine if the series converges or diverges.

$$\sum_{n=1}^{\infty} \frac{n}{e^n}$$

Solution: Let $f(x) = \frac{x}{e^x}$ and $a_n = f(n) = \frac{n}{e^n}$. By graphing the function $f(x)$, you can verify that it decreases for $x > 1$.[1] Apply the integral test. Note that $\frac{1}{e^x} = e^{-x}$.

$$\int_{x=1}^{\infty} f(x)\, dx = \int_{x=1}^{\infty} \frac{x}{e^x} dx = \int_{x=1}^{\infty} xe^{-x}\, dx$$

One way to perform this integral is to use integration by parts. Identifying $xe^{-x}\, dx$ as $u\, dv$, we can let $u = x$ and $dv = e^{-x}\, dx$. Implicitly differentiate both sides of $u = x$ to get $du = dx$ and integrate both sides of $dv = e^{-x}\, dx$ to get $v = -e^{-x}$. Use the formula for **integration by parts**. (You may want to review techniques of integration if you don't feel fluent with this. The next example will remind you of partial fractions.)

$$\int_{i}^{f} u\, dv = [uv]_i^f - \int_{i}^{f} v\, du$$

Replace u with x, v with $-e^{-x}$, du with dx, and dv with $e^{-x}\, dx$.

$$\int_{i}^{f} u\, dv = \int_{x=1}^{\infty} xe^{-x}\, dx = [-xe^{-x}]_{x=1}^{\infty} - \int_{x=1}^{\infty} (-e^{-x})\, dx = [-xe^{-x}]_{x=1}^{\infty} - [e^{-x}]_{x=1}^{\infty}$$

$$= -\lim_{x\to\infty} xe^{-x} - (-1e^{-1}) - \lim_{x\to\infty} e^{-x} + e^{-1}$$

$$= -\lim_{x\to\infty} xe^{-x} + e^{-1} - \lim_{x\to\infty} e^{-x} + e^{-1} = -\lim_{x\to\infty} xe^{-x} + 2e^{-1} - \lim_{x\to\infty} e^{-x}$$

[1] Or you can use calculus. Set the first derivative of $f(x) = xe^{-x}$ equal to zero to find the values of x corresponding to critical values, and use the second derivative test to determine whether critical points are relative maxima or minima. You will see that this function has a relative maximum at $x = 1$.

Recall that $e^{-x} = \frac{1}{e^x}$.

$$\int_{x=1}^{\infty} xe^{-x}\,dx = -\lim_{x\to\infty}\frac{x}{e^x} + \frac{2}{e} - \lim_{x\to\infty}\frac{1}{e^x}$$

The last limit equals zero. Apply l'Hôpital's (Chapter 2) rule to evaluate the first limit, since it is an indeterminate form (since x and e^x each grow to infinity).

$$\int_{x=1}^{\infty} xe^{-x}\,dx = -\frac{\left.\frac{d}{dx}x\right|_{x\to\infty}}{\left.\frac{d}{dx}e^x\right|_{x\to\infty}} + \frac{2}{e} + 0 = -\frac{1|_{x\to\infty}}{e^x|_{x\to\infty}} + \frac{2}{e} = -\frac{1}{\infty} + \frac{2}{e} = 0 + \frac{2}{e} = \frac{2}{e}$$

Since the integral is finite, the corresponding series **converges**. (Note: It's easy to make sign mistakes when integrating. One sign mistake that students often make in the calculation above is getting $\frac{1}{e} - \frac{1}{e}$ or $-\frac{1}{e} + \frac{1}{e}$ instead of the correct $\frac{1}{e} + \frac{1}{e} = \frac{2}{e}$. It's easy to see that the $\frac{1}{e}$'s can't cancel out because the integral can't be zero; since the function starts out positive and decreases towards zero, the area under the curve has to be positive. If you can understand concepts like this, it can help you catch and fix some math mistakes that are easy to make along the way.)

Example 3. Use the integral test to determine if the series converges or diverges.

$$\sum_{n=2}^{\infty}\frac{1}{(n+2)(n+3)}$$

Solution: Let $f(x) = \frac{1}{(x+2)(x+3)}$ and $a_n = f(n) = \frac{1}{(n+2)(n+3)}$. This function decreases. Apply the integral test.

$$\int_{x=2}^{\infty} f(x)\,dx = \int_{x=2}^{\infty}\frac{dx}{(x+2)(x+3)}$$

Use the method of **partial fractions**. (Recall Chapter 4.)

$$\frac{1}{(x+2)(x+3)} = \frac{p}{x+2} + \frac{q}{x+3}$$

Multiply both sides by $(x+2)(x+3)$.

$$1 = p(x+3) + q(x+2) = px + 3p + qx + 2q = (p+q)x + 3p + 2q$$

Both sides of this equation will be equal for any value of x if $p + q = 0$ and $3p + 2q = 1$. The first equation gives $q = -p$. Plug this into the second equation to find that $3p + 2(-p) = 3p - 2p = p = 1$. The solution is $p = 1$ and $q = -p = -1$. Plug these values for p and q into the equation of partial fractions above and use this in the integral.

$$\int_{x=2}^{\infty} \frac{dx}{(x+2)(x+3)} = \int_{x=2}^{\infty} \frac{dx}{x+2} - \int_{x=2}^{\infty} \frac{dx}{x+3} = [\ln(x+2)]_{x=2}^{\infty} - [\ln(x+3)]_{x=2}^{\infty}$$

Be careful here. If you evaluate each integral at the upper limit ($x \to \infty$), each integral will be infinite. But there is a minus sign. Do the infinities cancel out? One way to find out is to subtract the natural logarithms before evaluating over the limits. Recall the identity $\ln\left(\frac{t}{u}\right) = \ln t - \ln u$.

$$\int_{x=2}^{\infty} \frac{dx}{(x+2)(x+3)} = [\ln(x+2) - \ln(x+3)]_{x=2}^{\infty} = \left[\ln\left(\frac{x+2}{x+3}\right)\right]_{x=2}^{\infty}$$

When we evaluate at the upper limit, we will apply the trick from Chapter 2 where we first divide the numerator and denominator each by x.

$$\int_{x=2}^{\infty} \frac{dx}{(x+2)(x+3)} = \lim_{x\to\infty}\left[\ln\left(\frac{x+2}{x+3}\right)\right] - \ln\left(\frac{2+2}{2+3}\right)$$

$$= \lim_{x\to\infty}\left[\ln\left(\frac{1+\frac{2}{x}}{1+\frac{3}{x}}\right)\right] - \ln\left(\frac{4}{5}\right) = \ln\left(\frac{1+0}{1+0}\right) - \ln\left(\frac{4}{5}\right)$$

$$= \ln(1) - \ln\left(\frac{4}{5}\right) = 0 - \ln\left(\frac{4}{5}\right) = -\ln\left(\frac{4}{5}\right)$$

Recall that $\ln 1 = 0$. Since the integral is finite, the corresponding series **converges**.

Note: This integral is positive: $-\ln\left(\frac{4}{5}\right) = -\ln 0.8 \approx -(-0.223) \approx 0.223$.

Chapter 6 Problems

Directions: Use the integral test to determine if each series converges or diverges.

(1)

$$\sum_{n=1}^{\infty} \frac{1}{n^2 + 1}$$

(2)

$$\sum_{k=2}^{\infty} \frac{k}{k^2 + 3}$$

(3)

$$\sum_{m=0}^{\infty} \frac{1}{\sqrt{m+1}}$$

(4)

$$\sum_{j=4}^{\infty} \frac{6}{j\sqrt{j}}$$

(5)

$$\sum_{i=4}^{\infty} \frac{8}{i(i+4)}$$

(6)

$$\sum_{\ell=3}^{\infty} \frac{3\ell}{(\ell+2)(\ell+5)}$$

(7) Note: c is a constant. Work out different cases for different possible values of c.

$$\sum_{k=1}^{\infty} e^{ck}$$

(8)

$$\sum_{n=1}^{\infty} \frac{1}{\sqrt{e^n}}$$

(9)

$$\sum_{m=1}^{\infty} \frac{m}{e^{m^2}}$$

(10)

$$\sum_{j=2}^{\infty} \frac{1}{j \ln j}$$

7 Ratio Test

The ratio test examines the ratio between two consecutive terms in the limit that the index grows to infinity. To use the **ratio test**, first find

$$L = \lim_{n \to \infty} \left| \frac{a_{n+1}}{a_n} \right|$$

- If $L < 1$, the series $\sum a_n$ converges (absolutely; see Chapter 10).
- If $L > 1$, the series $\sum a_n$ diverges.
- If $L = 1$, the test is inconclusive; you need to use a different test.

The reason that the ratio test works relates to a geometric series, which has the form $\sum g_n = \sum ar^{n-1}$. For a geometric series, $\frac{g_{n+1}}{g_n} = r$. A geometric series converges if $|r| < 1$ and diverges if $|r| \geq 1$ (recall Chapter 4). Note that $g_2 = g_1 r$, $g_3 = g_2 r = g_1 r^2$, $g_4 = g_3 r = g_1 r^3$, etc., such that

$$\sum_{n=1}^{\infty} g_n = g_1 + g_1 r + g_1 r^2 + g_1 r^3 + \cdots + g_1 r^n + \cdots$$
$$= g_1(1 + r + r^2 + r^3 + \cdots + r^n + \cdots)$$

Compare this with

$$\sum_{n=1}^{\infty} a_n = a_1 + a_2 + a_3 + a_4 + \cdots + a_n + \cdots$$

If the series $\sum a_n$ has $L < 1$, there exists some value of $|r| < 1$ such that $a_n \leq g_1 r^{n-1}$ (meaning that $a_n \leq g_n$) for sufficiently large enough values of n. (As usual, this means to start the comparison at $n > N$.) The comparison limit test can be used to show that $\sum a_n$ converges in this case. If the series $\sum a_n$ has $L > 1$, there exists some value of $|r| > 1$ such that $a_n \geq g_1 r^{n-1}$ (meaning that $a_n \geq g_n$) for sufficiently large enough values of $n > N$. The comparison limit test can be used to show that $\sum a_n$ diverges in this case.

There exist both converging series and diverging series for which $L = 1$ (as shown in Example 2). This is why the ratio test is inconclusive when $L = 1$.

Example 1. What does the ratio test tell you about the series below?

$$\sum_{n=2}^{\infty} \frac{1}{n!}$$

Solution: Since $a_n = \frac{1}{n!}$, it follows that $a_{n+1} = \frac{1}{(n+1)!} = \frac{1}{(n+1)n!}$. For example, when a_n equals $\frac{1}{8!}$, the next term is $\frac{1}{9!} = \frac{1}{9(8!)}$ (where the denominator is 9 times 8!).

$$L = \lim_{n\to\infty} \left|\frac{a_{n+1}}{a_n}\right| = \lim_{n\to\infty} \left|\frac{1}{(n+1)n!} \div \frac{1}{n!}\right| = \lim_{n\to\infty} \left|\frac{1}{(n+1)n!} \times \frac{n!}{1}\right|$$

$$L = \lim_{n\to\infty} \frac{1}{n+1} = \lim_{n\to\infty} \frac{1/n}{1 + \frac{1}{n}} = \frac{0}{1+0} = \frac{0}{1} = 0$$

Since $L < 1$, the series <u>**converges**</u>.

Example 2. What does the ratio test tell you about the series below, where p is a constant?

$$\sum_{k=1}^{\infty} \frac{1}{k^p}$$

Solution: Since $a_k = \frac{1}{k^p}$, it follows that $a_{k+1} = \frac{1}{(k+1)^p}$. In the second line, we evaluate the limit by dividing the numerator and denominator each by k (recall Chapter 2).

$$L = \lim_{k\to\infty} \left|\frac{a_{k+1}}{a_k}\right| = \lim_{k\to\infty} \left|\frac{1}{(k+1)^p} \div \frac{1}{k^p}\right| = \lim_{k\to\infty} \left|\frac{1}{(k+1)^p} \times \frac{k^p}{1}\right|$$

$$L = \lim_{k\to\infty} \frac{k^p}{(k+1)^p} = \lim_{k\to\infty} \left(\frac{k}{k+1}\right)^p = \lim_{k\to\infty} \left(\frac{1}{1+\frac{1}{k}}\right)^p = \left(\frac{1}{1+0}\right)^p = \left(\frac{1}{1}\right)^p = 1^p = 1$$

Since $L = 1$, the ratio test is <u>**inconclusive**</u>. You should recognize this series from Chapter 5. When $p \leq 1$, the series diverges (for example, when $p = 1$, we get the harmonic series $\sum \frac{1}{k}$), and when $p > 1$, the series converges (for example, when $p = 2$, we get $\sum \frac{1}{k^2}$). In each case, $L = 1$. Since $L = 1$ for $\sum \frac{1}{k}$, which we know diverges, and $L = 1$ for $\sum \frac{1}{k^2}$, which we know converges, this demonstrates that $L = 1$ isn't enough information to determine whether a series converges or diverges.

Example 3. What does the ratio test tell you about the series below?

$$\sum_{n=2}^{\infty} \frac{2^n}{n^2}$$

Solution: Since $a_n = \frac{2^n}{n^2}$, it follows that $a_{n+1} = \frac{2^{n+1}}{(n+1)^2}$.

$$L = \lim_{n\to\infty} \left|\frac{a_{n+1}}{a_n}\right| = \lim_{n\to\infty} \left|\frac{2^{n+1}}{(n+1)^2} \div \frac{2^n}{n^2}\right| = \lim_{n\to\infty} \left|\frac{2^{n+1}}{(n+1)^2} \times \frac{n^2}{2^n}\right| = \lim_{n\to\infty} \left|\frac{2^{n+1}}{2^n} \times \left(\frac{n}{n+1}\right)^2\right|$$

Note that $\frac{2^{n+1}}{2^n} = 2$. For example, $\frac{2^7}{2^6} = 2^{7-6} = 2^1 = 2$.

$$L = \lim_{n\to\infty} \left|2 \times \left(\frac{1}{1+\frac{1}{n}}\right)^2\right| = 2\left(\frac{1}{1+0}\right)^2 = 2\left(\frac{1}{1}\right)^2 = 2(1)^2 = 2(1) = 2$$

Since $L > 1$, the series **diverges**.

Chapter 7 Problems

Directions: What does the ratio test tell you about each series below?

(1)

$$\sum_{j=1}^{\infty} \frac{j!}{(2j)!}$$

(2)

$$\sum_{n=2}^{\infty} \frac{(2n)!\,n!}{(3n)!}$$

(3)

$$\sum_{k=0}^{\infty} \frac{3^k}{k!}$$

(4) This one is a bit tricky. Hint: Review Example 15 and Exercise 40 in Chapter 2.

$$\sum_{m=1}^{\infty} \frac{m^m}{m!}$$

(5)

$$\sum_{\ell=1}^{\infty} \frac{e^{-\ell}}{\sqrt{\ell}}$$

(6)

$$\sum_{i=2}^{\infty} i!\, e^{-i}$$

(7)

$$\sum_{k=1}^{\infty} \frac{(2k)!}{k!\,k!\,(4^k)}$$

(8) The formula is given for the $(k+1)^{\text{th}}$ term.

$$1 + \frac{1}{3} + \frac{1}{4} + \frac{1}{4} + \frac{7}{24} + \frac{3}{8} + \frac{33}{64} + \frac{143}{192} + \frac{143}{128} + \cdots + \frac{2k-1}{k+2} a_k$$

(9) The formula is given for the $(n+1)^{\text{th}}$ term.

$$2 + \frac{2}{3} + \frac{1}{3} + \frac{1}{5} + \frac{2}{15} + \frac{2}{21} + \frac{1}{14} + \frac{1}{18} + \frac{2}{45} + \cdots + \frac{n}{n+2} a_n + \cdots$$

(10) The formula is given for the $(k+1)^{\text{th}}$ term.

$$6 + 6 + 4 + \frac{20}{9} + \frac{10}{9} + \frac{14}{27} + \frac{56}{243} + \frac{8}{81} + \frac{10}{243} + \cdots + \frac{k+2}{3k} a_k + \cdots$$

8 Root Test

For a series $\sum a_n$, the root test examines the nth root of a_n in the limit that the index grows to infinity. To use the **root test**, first find

$$L = \lim_{n\to\infty} |a_n|^{1/n} = \lim_{n\to\infty} \sqrt[n]{|a_n|}$$

- If $L < 1$, the series $\sum a_n$ converges (absolutely; see Chapter 10).
- If $L > 1$, the series $\sum a_n$ diverges.
- If $L = 1$, the test is inconclusive; you need to use a different test.

The reason that the root test works relates to a geometric series, which has the form $\sum g_n = \sum ar^{n-1}$. For a geometric series, $\frac{g_{n+1}}{g_n} = r$. A geometric series converges if $|r| < 1$ and diverges if $|r| \geq 1$ (recall Chapter 4). Note that $g_2 = g_1 r, g_3 = g_2 r = g_1 r^2, g_4 = g_3 r = g_1 r^3$, etc., such that

$$\sum_{n=1}^{\infty} g_n = g_1 + g_1 r + g_1 r^2 + g_1 r^3 + \cdots + g_1 r^n + \cdots$$

The nth term of a geometric series thus has the form $g_n = g_1 r^n$. If you take the nth root of both sides, you get $g_n^{1/n} = \sqrt[n]{g_n} = r(\sqrt[n]{g_1})$, where g_1 is a constant (so is r). The nth term of $\sum a_n$ is a_n, for which the nth root is $a_n^{1/n} = \sqrt[n]{a_n}$. The main idea is that if the series $\sum a_n$ has $L < 1$, there exists some value of $|r| < 1$ such that $a_n \leq g_1 r^n$, which equates to $\sqrt[n]{a_n} \leq r\sqrt[n]{g_1}$, for sufficiently large enough values of n. (As usual, this means to start the comparison at $n > N$.) A formal proof of this uses the comparison limit test to show that $\sum a_n$ converges in this case. On the other hand, if the series $\sum a_n$ has $L > 1$, there exists some value of $|r| > 1$ such that $a_n \geq g_1 r^n$, which equates to $\sqrt[n]{a_n} \geq r\sqrt[n]{g_1}$, for sufficiently large enough values of n. A formal proof of this uses the comparison limit test to show that $\sum a_n$ diverges in this case.

There exist both converging series and diverging series for which $L = 1$ (as shown in Example 2). This is why the root test is inconclusive when $L = 1$.

The root test is often useful when the index appears in an **exponent**; see the examples.

Example 1. What does the root test tell you about the series below?

$$\sum_{n=1}^{\infty} \frac{1}{n^n}$$

Solution: Apply the root test with $a_n = \frac{1}{n^n}$. Apply the rule $(x^n)^{1/n} = x^{n/n} = x^1 = x$.

$$L = \lim_{n\to\infty} |a_n|^{1/n} = \lim_{n\to\infty} \left|\frac{1}{n^n}\right|^{1/n} = \lim_{n\to\infty} \frac{1}{(n^n)^{\frac{1}{n}}} = \lim_{n\to\infty} \frac{1}{n^1} = \lim_{n\to\infty} \frac{1}{n} = 0$$

Since $L < 1$, the series **converges**.

Example 2. What does the root test tell you about the series below, where p is a constant?

$$\sum_{k=1}^{\infty} \frac{1}{k^p}$$

Solution: Apply the root test with $a_k = \frac{1}{k^p}$. The index is k, whereas p is a constant. (Unlike the previous example, this example doesn't have the index in an exponent.) We'll use the property $(x^a)^b = x^{ab} = (x^b)^a$.

$$L = \lim_{k\to\infty} |a_k|^{1/k} = \lim_{k\to\infty} \left|\frac{1}{k^p}\right|^{1/k} = \lim_{k\to\infty} \frac{1}{(k^p)^{1/k}} = \lim_{k\to\infty} \frac{1}{k^{p/k}} = \lim_{k\to\infty} \frac{1}{(k^{1/k})^p} = \frac{1}{1^p} = 1$$

Note that $\lim_{k\to\infty} k^{1/k} = 1$ (see Exercise 39 of Chapter 2). Since $L = 1$, the root test is **inconclusive**. Since $p = 1$ gives the harmonic series $\sum \frac{1}{k}$, which diverges, while $p = 2$ gives $\sum \frac{1}{k^2}$, which converges, and in each case $L = 1$, this demonstrates that $L = 1$ isn't enough information to determine whether a series converges or diverges.

Example 3. What does the root test tell you about the series below?

$$\sum_{n=2}^{\infty} \frac{2^n}{n^4}$$

Solution: Apply the root test with $a_n = \frac{2^n}{n^4}$. Apply the rule $(x^n)^{1/n} = x^{n/n} = x^1 = x$.

$$L = \lim_{n\to\infty} |a_n|^{1/n} = \lim_{n\to\infty} \left|\frac{2^n}{n^4}\right|^{1/n} = \lim_{n\to\infty} \frac{(2^n)^{1/n}}{(n^4)^{1/n}} = \lim_{n\to\infty} \frac{2^1}{n^{4/n}} = \lim_{n\to\infty} \frac{2}{(n^{1/n})^4} = \frac{2}{1} = 2$$

Note that $\lim_{n\to\infty} n^{1/n} = 1$ (see Exercise 39 of Chapter 2). Since $L > 1$, the series **diverges**.

Chapter 8 Problems

Directions: What does the root test tell you about each series below?

(1)

$$\sum_{n=2}^{\infty} \left(\frac{6}{n}\right)^n$$

(2)

$$\sum_{m=0}^{\infty} \frac{m^8}{2^m}$$

(3)

$$\sum_{k=1}^{\infty} k^4 e^{-k}$$

(4)

$$\sum_{j=1}^{\infty}\left(\frac{j}{j^2+3}\right)^j$$

(5)

$$\sum_{n=2}^{\infty}\left(1-\frac{1}{2n}\right)^{2n}$$

(6)

$$\sum_{\ell=0}^{\infty}\left(\frac{8\ell+19}{6\ell+11}\right)^{\ell+2}$$

(7)

$$\sum_{i=2}^{\infty} \left(\frac{2}{3}\right)^i$$

(8)

$$\sum_{k=0}^{\infty} k \left(\frac{4}{3}\right)^k$$

(9)

$$10 + 25 + \frac{1000}{27} + \frac{625}{16} + 32 + \frac{15{,}625}{729} + \frac{10{,}000{,}000}{823{,}543} + \frac{390{,}625}{65{,}536} + \cdots$$

9 Convergence Tests

Previous chapters focused on one convergence test at a time. This chapter will help build your fluency with the different convergence tests.

Here is a brief review of the convergence/divergence tests we've encountered for $\sum a_n$, where a_n isn't negative for $n > N$. (Chapters 10-11 will discuss negative terms.)

- First see if $\sum a_n$ is a **geometric series** like ar^n (which converges if $|r| < 1$) or a **series of powers** like $\frac{a}{n^p}$ (which converges if $p > 1$).
- Apply the **simple divergence test** if you expect $\lim_{n\to\infty} a_n$ to be nonzero. If $\lim_{n\to\infty} a_n \neq 0$, the $\sum a_n$ diverges. However, if $\lim_{n\to\infty} a_n = 0$, the test is inconclusive.
- Apply the **comparison test** if you can compare a_n with a geometric series of the form ar^n (which converges if $|r| < 1$ and diverges if $|r| \geq 1$), a series of powers like $\frac{a}{n^p}$ (which converges if $p > 1$ and diverges if $p \leq 1$), or another series that you already know converges or diverges. If $\sum c_n$ converges and if $\lim_{n\to\infty} \frac{a_n}{c_n}$ is finite, then $\sum a_n$ converges. If $\sum d_n$ diverges and if $\lim_{n\to\infty} \frac{a_n}{d_n}$ is positive, then $\sum a_n$ diverges.
- Apply the **integral test** to a positive nonincreasing series it seems feasible to integrate $f(x)$ where $f(n) = a_n$. The sum $\sum a_n$ and integral $\int f(x)\,dx$ both converge or both diverge.
- Apply the **ratio test** if you expect $L = \lim_{n\to\infty} \left|\frac{a_{n+1}}{a_n}\right|$ to be different from one. If $L < 1$, $\sum a_n$ converges. If $L > 1$, $\sum a_n$ diverges. If $L = 1$, the test is inconclusive. The ratio test is often helpful for factorials like $n!$ or powers like r^n.
- Apply the **root test** if you expect $L = \lim_{n\to\infty} |a_n|^{1/n}$ to be different from one. If $L < 1$, $\sum a_n$ converges. If $L > 1$, $\sum a_n$ diverges. If $L = 1$, the test is inconclusive. The root test is often helpful if the index appears in an exponent.
- If needed, you can always apply the **definition of convergence**. Examine the sequence of partial sums, $\{S_n\}$, where $S_n = a_1 + a_2 + \cdots + a_n$. If $\{S_n\}$ converges to a finite limiting value L, $\sum a_n$ converges to L. If $\{S_n\}$ diverges, $\sum a_n$ diverges.

Chapter 9 Problems

Examples: Where are the examples? We already did several examples of each test in the previous chapters. To see examples of these tests, review Chapters 4-8.

Directions: Determine whether each series converges or diverges.

Challenge: Try to use every test meaningfully in at least four different solutions.

(1)

$$\sum_{k=1}^{\infty} \frac{8}{\sqrt{2k}}$$

(2)

$$\sum_{n=2}^{\infty} \left(\frac{7}{9}\right)^{n+1}$$

(3)

$$\sum_{\ell=1}^{\infty} \frac{\ell}{\ell^3 - 3}$$

(4)

$$\sum_{i=3}^{\infty} \frac{9}{(2i)!}$$

(5)

$$\sum_{m=0}^{\infty} \frac{8m}{m^2 + 9}$$

(6)

$$\sum_{j=1}^{\infty} j^{-j}$$

(7)

$$\sum_{n=0}^{\infty} \frac{3n^2 - 5}{2n^2 + 3n + 4}$$

(8)

$$\sum_{k=2}^{\infty} k^5 e^{-k}$$

(9)

$$\sum_{i=1}^{\infty} \frac{\sqrt{i}}{i^2}$$

(10)

$$\sum_{\ell=0}^{\infty} \frac{\ell^8}{6^\ell}$$

(11)

$$\sum_{m=1}^{\infty} \left(1 - \frac{1}{m^2}\right)^3$$

(12)

$$\sum_{j=2}^{\infty}\left(\frac{8j-7}{7j+8}\right)^{j}$$

(13)

$$\sum_{n=1}^{\infty}\frac{7}{2^{n}-n}$$

(14)

$$\sum_{k=2}^{\infty}\frac{6}{k(k+3)}$$

(15)

$$\sum_{m=3}^{\infty}\left(\frac{5}{4}\right)^{m-2}$$

(16)

$$\sum_{i=1}^{\infty} \frac{e^i}{i!}$$

(17)

$$\sum_{\ell=0}^{\infty} \ell^6 \left(\frac{2}{3}\right)^{\ell}$$

(18)

$$\sum_{j=2}^{\infty} \frac{5j}{j^2 - \ln j}$$

(19)

$$\sum_{k=1}^{\infty} \frac{k!}{k^k}$$

(20)

$$\sum_{n=0}^{\infty} \frac{n^4 - 3n^2 + 8}{n^6 + 9n^3 - 6}$$

(21)

$$\sum_{j=1}^{\infty} \frac{j^j}{100^j}$$

(22)

$$\sum_{m=3}^{\infty} 4\sin^2\left(\frac{m\pi}{6}\right)$$

(23)

$$\sum_{i=0}^{\infty} i^3 e^{-i}$$

(24)

$$\sum_{\ell=1}^{\infty} \sqrt{e^{-\ell}}$$

(25)

$$\sum_{n=0}^{\infty} \frac{n}{n^2 - 2n + 6}$$

(26)

$$\sum_{k=9}^{\infty} \sqrt{\frac{k-9}{9k+1}}$$

(27)

$$\sum_{j=1}^{\infty} \frac{3^j + j}{4^j}$$

(28)

$$\sum_{m=1}^{\infty} me^{-m^2}$$

(29)

$$\sum_{\ell=0}^{\infty} \frac{6}{(\ell+2)\sqrt{\ell+4}}$$

(30)

$$\sum_{i=1}^{\infty} \left(\frac{i^3+2i}{i^4-3i^2+1}\right)^i$$

(31)

$$\sum_{n=1}^{\infty} \frac{(2n)!}{n!\,n!}$$

(32)

$$\sum_{j=4}^{\infty} \sqrt{j}\, e^{-j}$$

(33)

$$\sum_{k=6}^{\infty} \frac{2k+4}{7^k(k-5)}$$

(34)

$$\frac{1}{2} + \frac{3}{2} + \frac{5}{4} + \frac{7}{12} + \frac{3}{16} + \frac{11}{240} + \frac{13}{1440} + \frac{1}{672} + \frac{17}{80{,}640} + \cdots$$

(35)

$$\frac{2}{7} + \frac{20}{89} + \frac{5}{27} + \frac{20}{127} + \frac{10}{73} + \frac{4}{33} + \frac{5}{46} + \frac{20}{203} + \frac{10}{111} + \cdots$$

10 Series with Negative Terms

Thus far, we've focused on series that don't have any negative terms, or which don't have negative terms once the index exceeds some finite value (i.e. $n > N$). In the latter case, we're basically separating the series into two parts:

$$\sum_{n=n_0}^{\infty} a_n = \sum_{n=n_0}^{N} a_n + \sum_{n=N+1}^{\infty} a_n$$

The first sum is a finite sum and the second sum is an infinite sum. Whether the whole series converges or diverges just depends on whether the infinite sum converges or diverges; the first sum is finite (assuming all of its terms are individually finite; that is, it doesn't contain a term like $\frac{1}{n-3}$ if n starts at 3 or lower and doesn't contain a term like $\ln n$ if n starts at 0).

Of course, if **every** term is negative, or if every term for which $n > N$ is negative, we can simply factor out an overall minus sign and then the series will be no different from those that we've already considered.

But what if there are **both positive and negative terms** throughout the series? This is what we will consider in this and the next chapter. One common series that includes both positive and negative terms throughout is an alternating series, which is the subject of Chapter 11, but it's not the only kind of series which has both positive and negative terms. In this chapter, we'll focus on a general series that includes positive and negative terms throughout the series; it could be an alternating series, but doesn't need to be.

There are two reasons that we've waited until now to discuss series that include both positive and negative terms throughout.

- First, if $\sum |a_n|$ converges, then $\sum a_n$ also converges. (When $\sum |a_n|$ converges, we say that the series converges **absolutely**.)
- If $\sum a_n$ converges, but $\sum |a_n|$ diverges (in this case, we say that the series $\sum a_n$ converges **conditionally**), seemingly 'strange' things can happen, as we'll see.

If a series includes both positive and negative terms, if it converges, it's helpful to determine whether the series converges absolutely or conditionally because these two classes of series behave quite differently.

A series converges **absolutely** if $\sum|a_n|$ converges. To test for absolute convergence, ignore any minus signs. If $\sum|a_n|$ converges, then $\sum a_n$ also converges. To understand why, suppose that $\sum|a_n| = |a_1| + |a_2| + |a_3| + \cdots = L$. If we now remove the absolute values and some of the terms are now negative, the new sum will be less than or equal to L. In general, $-\sum|a_n| \leq \sum a_n \leq \sum|a_n|$.

Beware that if $\sum a_n$ converges, $\sum|a_n|$ might not converge. If $\sum a_n$ converges, but $\sum|a_n|$ diverges, we say that $\sum a_n$ converges **conditionally**. As an example, the **alternating harmonic series** $\sum \frac{(-1)^{n+1}}{n} = 1 - \frac{1}{2} + \frac{1}{3} - \frac{1}{4} + \frac{1}{5} - \frac{1}{6} + \cdots$ converges, even though the harmonic series $\sum \frac{1}{n} = 1 + \frac{1}{2} + \frac{1}{3} + \frac{1}{4} + \frac{1}{5} + \frac{1}{6} + \cdots$ diverges (Chapter 5). Therefore, the alternating harmonic series converges conditionally.

Compare the four cases below:

- If $\sum|a_n|$ converges, then $\sum a_n$ also converges. This series converges **absolutely**.
- If $\sum a_n$ converges, this isn't enough information to know if $\sum|a_n|$ converges.
- If $\sum a_n$ diverges, then $\sum|a_n|$ also diverges.
- If $\sum|a_n|$ diverges, this isn't enough information to know if $\sum a_n$ diverges. If $\sum a_n$ converges and $\sum|a_n|$ diverges, we say that $\sum a_n$ converges **conditionally**.

The distinction between absolute convergence and conditional convergence stands out when we consider **rearranging terms**. Suppose that $\sum a_n$ is an infinite series with both positive and negative terms throughout. Let $\{b_n\}$ be a sequence that is made by simply rearranging the terms of the sequence $\{a_n\}$; the terms of these two sequences are the same, but the order of the terms is different.

- If $\sum a_n$ converges **absolutely**, then $\sum a_n = \sum b_n$. If the convergence is absolute, rearranging the terms has no effect on what value the series converges to.
- If $\sum a_n$ converges **conditionally**, then $\sum a_n$ and $\sum b_n$ may converge to two different values. In fact, when the convergence is conditional, the terms can be rearranged in such a way that the series converges to **any** desired real number.

When adding or subtracting a finite number of terms, the order in which the terms are added or subtracted doesn't matter. Addition is both commutative – for example, $8 - 3 = -3 + 8$ – and associative – for example, $6 + (9 - 2) = (6 + 9) - 2$. However, when adding or subtracting an infinite number of terms, the order in which the terms are added or subtracted matters if the series converges **conditionally** (but doesn't matter if it converges absolutely). We'll illustrate this with an example shortly.

Why does conditional or absolute convergence affect whether or not **rearrangement** affects the sum of a series? To understand this, let's separate the terms of the series $\sum a_n$ into positive and negative terms. Let $\sum b_n$ be the sum of the positive terms and let $\sum c_n$ be the sum of the absolute values of the negative terms. Another way to say the same thing is: If the nth term, a_n, is positive, then $b_n = a_n$ and $c_n = 0$, but if a_n is negative, then $c_n = |a_n|$ and $b_n = 0$. The three series are thus related by:

$$\sum a_n = \sum b_n - \sum c_n$$

Since all of the b_n's and c_n's are nonnegative, it follows that

$$\sum |a_n| = \sum b_n + \sum c_n$$

If $\sum a_n$ converges **absolutely**, $\sum |a_n|$ converges. Since $\sum b_n$ and $\sum c_n$ are inherently nonnegative, this means that $\sum b_n$ and $\sum c_n$ each converge in this case. In this case, $\sum a_n$ is formed by subtracting two series that converge (absolutely, since all of the b_n's and c_n's are nonnegative). In contrast, if $\sum a_n$ converges **conditionally**, $\sum |a_n|$ diverges. In order for $\sum a_n$ to converge while $\sum |a_n|$ diverges, $\sum b_n$ and $\sum c_n$ must both diverge in this case. In this case, $\sum a_n$ is formed by subtracting two diverging series; $\sum a_n$ only converges because there is cancellation between the terms of two series that individually form infinite sums. We're basically cancelling infinities when we subtract $\sum b_n - \sum c_n$ in the case of conditional convergence in order for $\sum a_n$ to yield a finite sum. It's ultimately this subtraction of infinities to form a finite sum that makes it possible to rearrange the terms of a series that converges conditionally with the result that the sum then converges to a different value.

Now we will look at a specific example where we rearrange the terms of an infinite series that converges conditionally and see that the sum converges to a different value.

Consider the alternating harmonic series below, which includes both positive and negative terms throughout the series. We'll see in Chapter 15 that this series converges to $\ln 2 \approx 0.693$ when the terms are arranged in the order below. Since $\sum |a_n| = \sum \frac{1}{n}$ diverges (Chapter 5), the alternating harmonic series converges **conditionally**.

$$\sum_{n=1}^{\infty} a_n = \sum_{n=1}^{\infty} \frac{(-1)^{n+1}}{n} = 1 - \frac{1}{2} + \frac{1}{3} - \frac{1}{4} + \frac{1}{5} - \frac{1}{6} + \frac{1}{7} - \frac{1}{8} + \frac{1}{9} - \frac{1}{10} + \cdots = \ln 2$$

Let's reorder the terms as follows. We start with 1. Subtract fractions with even denominators until the sum falls below 0.3. After subtracting 1/2 and 1/4, the sum is at 0.25. Add 1/3 to put the sum back above 0.3; now it's at ≈ 0.583. Subtract 1/6 and 1/8 to put the sum back below 0.3; now it's at ≈ 0.292. Add 1/5 to put the sum back above 0.3; now it's at ≈ 0.492. We need to subtract 1/10, 1/12, and 1/14 to put the sum back below 0.3; now it's at ≈ 0.237. Add 1/7 to put the sum back above 0.3; now it's at ≈ 0.380. Subtract 1/16 and 1/18 to put the sum back below 0.3; now it's at ≈ 0.262. Add 1/9 to make the sum ≈ 0.373. Subtract 1/20 and 1/22 to make the sum ≈ 0.277. Add 1/11 to make the sum ≈ 0.368. With this rearrangement, the series will now converge to 0.3 (which equals 3/10), which is different from the original value of ln 2. Using this method, we can make the series converge to **any** real number. (It may help to review Chapter 4, specifically how a sequence of **partial sums** is used to establish the convergence of an infinite series.)

$$1 - \frac{1}{2} - \frac{1}{4} + \frac{1}{3} - \frac{1}{6} - \frac{1}{8} + \frac{1}{5} - \frac{1}{10} - \frac{1}{12} - \frac{1}{14} + \frac{1}{7} - \frac{1}{16} - \frac{1}{18} + \frac{1}{9} - \frac{1}{20} - \frac{1}{22} + \frac{1}{11} + \cdots = 0.3$$

This shows that the order in which the terms appear is important for a series that converges conditionally. If a series converges **conditionally**, **don't rearrange the terms.** (However, if a series converges **absolutely**, you may rearrange the terms.)

Example 1. Test the absolute convergence of the series below. What, if anything, can you conclude from this test regarding the convergence or divergence of $\sum a_n$?

$$\sum_{n=1}^{\infty} a_n = \sum_{n=1}^{\infty} \frac{1}{n \cos\left(\frac{n\pi}{3}\right)}$$

Solution: Put each term in absolute values. Only the cosine function can be negative.

$$\sum_{n=1}^{\infty}|a_n| = \sum_{n=1}^{\infty}\left|\frac{1}{n\cos\left(\frac{n\pi}{3}\right)}\right| = \sum_{n=1}^{\infty}\frac{1}{n\left|\cos\left(\frac{n\pi}{3}\right)\right|}$$

We will compare $\sum|a_n|$ with $\sum d_n = \frac{1}{n}$. Since $0 \leq \left|\cos\left(\frac{n\pi}{3}\right)\right| \leq 1$, it follows that $\frac{1}{\left|\cos\left(\frac{n\pi}{3}\right)\right|} \geq 1$ such that $|a_n| = \frac{1}{n\left|\cos\left(\frac{n\pi}{3}\right)\right|} \geq \frac{1}{n} = d_n$. Since $|a_n| \geq d_n$ for all n and $\sum d_n$ is known to diverge (recall Chapter 5), it follows that $\sum|a_n|$ **diverges**. When $\sum|a_n|$ diverges, **more information is needed** to determine whether or not $\sum a_n$ converges or diverges; **we can't conclude anything** about the convergence of $\sum a_n$ from this test of $\sum|a_n|$.

Example 2. Test the absolute convergence of the series below. What, if anything, can you conclude from this test regarding the convergence or divergence of $\sum b_k$?

$$\sum_{k=1}^{\infty} b_k = \frac{1}{2}+\frac{1}{4}+\frac{1}{8}-\frac{1}{16}+\frac{1}{32}+\frac{1}{64}+\frac{1}{128}-\frac{1}{256}+\frac{1}{512}+\frac{1}{1024}+\frac{1}{2048}-\frac{1}{4096}+\cdots$$

Solution: Put each term in absolute values.

$$\sum_{k=1}^{\infty} |b_k| = \frac{1}{2}+\frac{1}{4}+\frac{1}{8}+\frac{1}{16}+\frac{1}{32}+\frac{1}{64}+\frac{1}{128}+\frac{1}{256}+\frac{1}{512}+\frac{1}{1024}+\frac{1}{2048}+\frac{1}{4096}+\cdots$$

This is a geometric series with $r = \frac{a_{k+1}}{a_k} = \frac{1}{2}$. Since $|r| \leq 1$, this series **converges**. Since $\sum|b_k|$ converges, $\sum b_k$ also **converges**. This series converges **absolutely**.

Example 3. Test the convergence of the series below. What, if anything, can you conclude from this test regarding the convergence or divergence of $\sum|a_i|$?

$$\sum_{i=4}^{\infty} a_i = (-1)^{i+1}\frac{i-1}{i-3}$$

Solution: Apply the simple divergence test. Since $\lim_{i\to\infty}\frac{i-1}{i-3} = \lim_{i\to\infty}\frac{1-\frac{1}{i}}{1-\frac{3}{i}} = \frac{1-0}{1-0} = \frac{1}{1} = 1$, the sequence $\left\{(-1)^{i+1}\frac{i-1}{i-3}\right\}$ doesn't converge to zero. Therefore, $\sum a_i$ **diverges**. It follows that $\sum|a_i|$ also **diverges**. It is instructive to compare Examples 1 and 3.

Chapter 10 Problems

Directions: Either test the convergence or the absolute convergence of each series. What, if anything, can you conclude from the test?

(1)

$$\sum_{n=1}^{\infty} a_n = \sum_{n=1}^{\infty} \frac{\sin\left(\frac{n\pi}{6}\right)}{n^2}$$

(2)

$$\sum_{k=1}^{\infty} b_k = \frac{1}{5} + \frac{1}{15} - \frac{1}{45} - \frac{1}{135} + \frac{1}{405} + \frac{1}{1215} - \frac{1}{3645} - \frac{1}{10{,}935} + \frac{1}{32{,}805} + \cdots$$

(3)

$$\sum_{m=1}^{\infty} a_m = -1 + \frac{1}{4} + \frac{1}{9} - \frac{1}{16} + \frac{1}{25} + \frac{1}{36} - \frac{1}{49} + \frac{1}{64} + \frac{1}{81} - \frac{1}{100} + \frac{1}{121} + \cdots$$

(4)

$$\sum_{j=4}^{\infty} b_j = \sum_{j=4}^{\infty} \frac{(-1)^j}{\sqrt{j}}$$

(5)

$$\sum_{\ell=0}^{\infty} a_\ell = \sum_{\ell=0}^{\infty} \frac{(-1)^{\ell+1}}{5^\ell}$$

(6)

$$\sum_{i=2}^{\infty} b_i = \sum_{i=2}^{\infty} (-1)^i \frac{2i^2 - 6}{i(4i+3)}$$

(7)

$$\sum_{k=1}^{\infty} a_k = \sum_{k=1}^{\infty} \left(-\frac{1}{4}\right)^{k+1}$$

(8)

$$\sum_{n=0}^{\infty} b_n = \sum_{n=0}^{\infty} e^{-n} \sin n$$

(9)

$$\sum_{j=1}^{\infty} a_j = \sum_{j=1}^{\infty} \frac{j^4}{(-2)^j}$$

(10)

$$\sum_{m=1}^{\infty} b_m = \frac{1!}{2!} - \frac{2!}{4!} - \frac{3!}{6!} - \frac{4!}{8!} + \frac{5!}{10!} - \frac{6!}{12!} - \frac{7!}{14!} - \frac{8!}{16!} + \frac{9!}{18!} - \frac{10!}{20!} - \frac{11!}{22!} - \frac{12!}{24!} + \cdots$$

(11)

$$\sum_{m=0}^{\infty} a_m = \sum_{m=0}^{\infty} \frac{(-1)^m}{\sqrt{m^2 + 5m + 3}}$$

11 Alternating Series

In an **alternating series**, consecutive terms change sign. In summation notation, one way that an alternating series can arise is to include a factor of $(-1)^n$ or $(-1)^{n+1}$.

- $(-1)^n$ is negative when n is odd and positive when n is even; $(-1)^{n+2}$, $(-1)^{n-2}$, $(-1)^{n+4}$, $(-1)^{n-36}$, etc. have the same effect.

$$\sum_{n=1}^{\infty} \frac{(-1)^n}{n^2} = -1 + \frac{1}{4} - \frac{1}{9} + \frac{1}{16} - \frac{1}{25} + \frac{1}{36} - \frac{1}{49} + \frac{1}{64} - \frac{1}{81} + \cdots$$

- $(-1)^{n+1}$ is negative when n is even and positive when n is odd; $(-1)^{n-1}$, $(-1)^{n+3}$, $(-1)^{n+19}$, etc. have the same effect.

$$\sum_{n=1}^{\infty} \frac{(-1)^{n+1}}{n^2} = 1 - \frac{1}{4} + \frac{1}{9} - \frac{1}{16} + \frac{1}{25} - \frac{1}{36} + \frac{1}{49} - \frac{1}{64} + \frac{1}{81} + \cdots$$

If other expressions also have minus signs, you need to account for those, too. For example, $\frac{1-n}{n}$ is negative for all $n > 1$. There series below includes $(-1)^n$, but unlike the series at the top of the page, its terms are negative when n is even and positive when n is odd. For example, $a_2 = (-1)^2 \frac{(1-2)}{2} = (1)\frac{(-1)}{2} = -\frac{1}{2}$.

$$\sum_{n=1}^{\infty} (-1)^n \frac{(1-n)}{n} = 0 - \frac{1}{2} + \frac{2}{3} - \frac{3}{4} + \frac{4}{5} - \frac{5}{6} + \frac{6}{7} - \frac{7}{8} + \cdots$$

As to whether or not the 'first' term of the series is positive, this also depends on the starting value. Compare the two series below.

$$\sum_{n=1}^{\infty} \frac{(-1)^n}{\sqrt{n}} = -1 + \frac{1}{\sqrt{2}} - \frac{1}{\sqrt{3}} + \frac{1}{2} - \frac{1}{\sqrt{5}} + \frac{1}{\sqrt{6}} - \frac{1}{\sqrt{7}} + \cdots$$

$$\sum_{n=2}^{\infty} \frac{(-1)^n}{\sqrt{n}} = \frac{1}{\sqrt{2}} - \frac{1}{\sqrt{3}} + \frac{1}{2} - \frac{1}{\sqrt{5}} + \frac{1}{\sqrt{6}} - \frac{1}{\sqrt{7}} + \frac{1}{\sqrt{8}} + \cdots$$

There are other ways to make alternating series besides using $(-1)^n$ and $(-1)^{n+1}$. For example, consider adding π radians (or 180°) to the angle of the sine function:

$$\sum_{n=1}^{\infty} \frac{1}{n} \sin\left(\frac{\pi}{3} + n\pi\right) = -\frac{\sqrt{3}}{2} + \frac{\sqrt{3}}{4} - \frac{\sqrt{3}}{6} + \frac{\sqrt{3}}{8} - \frac{\sqrt{3}}{10} + \frac{\sqrt{3}}{12} - \frac{\sqrt{3}}{14} + \cdots$$

According to **Leibniz's theorem**, an **alternating series** with either of the two forms below **converges** if it satisfies both of the conditions outlined below. The theorem still applies when an alternating series is made some other way that is equivalent to $(-1)^n$ or $(-1)^{n+1}$, such as $\sin\left(\frac{\pi}{3}+n\pi\right)$ which was mentioned in the previous paragraph. Note that the second series below is simply the negative of the first series.[1]

$$|a_1| - |a_2| + |a_3| - |a_4| + |a_5| - |a_6| + |a_7| - |a_8| + \cdots = \sum_{n=1}^{\infty} |a_n|(-1)^{n+1}$$

$$-|a_1| + |a_2| - |a_3| + |a_4| - |a_5| + |a_6| - |a_7| + |a_8| + \cdots = \sum_{n=1}^{\infty} |a_n|(-1)^{n}$$

- The sequence of absolute values is **nonincreasing**: $|a_n| \geq |a_{n+1}|$. For example, in the series $\frac{1}{2}-\frac{1}{3}+\frac{1}{4}-\frac{1}{5}+\frac{1}{6}-\frac{1}{7}+\cdots$, we have $\frac{1}{2}>\frac{1}{3}$ and $\frac{1}{3}>\frac{1}{4}$. Consecutive terms may be equal in absolute values, but $|a_{n+1}|$ can't be larger than $|a_n|$.
- As always, the sequence $\{|a_n|\}$ must **converge to zero**: $\lim_{n\to\infty} |a_n| = 0$. Otherwise, the series diverges according to the simple divergence test (Chapter 4).

Note that Leibniz's theorem specifically tests for **convergence** and that **both** points must be satisfied. The first point can't be used to test for divergence; Leibniz's theorem is **inconclusive** if the first condition isn't satisfied but the second condition is. Since the second condition is just the simple divergence test, the second condition can be used to test for **divergence**.

The main idea behind Leibniz's theorem is that the sequence of partial sums oscillates around the limit $\lim_{n\to\infty} s_n = L$, while the amplitude of oscillation decreases as $n \to \infty$ (if both conditions are satisfied). Let's explore this for the case where $a_1 > 0$. (If $a_1 < 0$, you could factor out an overall minus sign from the series.)

[1] It's common for textbooks to use the notation $a_1 - a_2 + a_3 - a_4 + a_5 - a_6 + \cdots$ for Leibniz's theorem, where suddenly the textbook is using a_n to be a nonnegative value, which contradicts the usage of a_n from other chapters. (For example, if the textbook uses absolute values in the ratio test $\left|\frac{a_{n+1}}{a_n}\right|$, or if a textbook specifies that a_n is nonnegative in a particular test, then that textbook is ordinarily allowing for a_n to be positive or negative in general; otherwise, they wouldn't need to use absolute values or say that a_n is nonnegative.) It seems to make more sense to use $|a_n|$ for an absolute value, which is what we're doing.

- The first partial sum is $s_1 = |a_1|$. (Again, we're working out the case where the first term happens to be positive.)
- The second partial sum, $s_2 = |a_1| - |a_2|$; s_2 is less positive than s_1.
- The third partial sum is $s_3 = |a_1| - |a_2| + |a_3|$; s_3 is more positive than s_2, but not as positive as s_1. (This follows when the first condition is satisfied.)
- The fourth partial sum is $s_4 = |a_1| - |a_2| + |a_3| - |a_4|$; s_4 is less positive than s_3, but more positive than s_2.
- Continuing in this fashion, the series will converge to a limit $\lim_{n\to\infty} s_n = L$, where $s_1, s_3, s_5, \ldots$ are greater than L and where $s_2, s_4, s_6, \ldots$ are less than L. (This is true when the second condition is satisfied.)

Example 1. For each series below, determine if it is an alternating series.

$$\sum_{n=1}^{\infty} a_n = \sum_{n=1}^{\infty} \frac{(-1)^n}{\sqrt{n}} \quad , \quad \sum_{k=1}^{\infty} b_k = \sum_{k=1}^{\infty} \frac{\frac{1}{2} + \sin^2 k}{(-1)^{k+1}} \quad , \quad \sum_{i=1}^{\infty} c_i = \sum_{i=1}^{\infty} \frac{1}{i} \cos\left(\frac{i\pi}{6}\right)$$

Solution: $\sum a_n$ is an alternating series; when n is odd, a_n is negative, and when n is even, a_n is positive. $\sum b_k$ is an alternating series; $\frac{1}{2} + \sin^2 k \geq \frac{1}{2}$ is strictly positive; when k is odd, b_k is positive, and when k is even, b_k is negative. $\sum c_i$ isn't an alternating series. For example, $c_1 = \frac{1}{1}\cos\left(\frac{\pi}{6}\right) = 1\left(\frac{\sqrt{3}}{2}\right) = \frac{\sqrt{3}}{2}$ and $c_2 = \frac{1}{2}\cos\left(\frac{2\pi}{6}\right) = \frac{1}{2}\cos\left(\frac{\pi}{3}\right) = \frac{1}{2}\left(\frac{1}{2}\right) = \frac{1}{4}$ are both positive. (The angle is in **radians**.) In order to be an alternating series, consecutive terms must alternate sign throughout the series.

Example 2. Is the series below absolutely convergent, conditionally convergent, or divergent?

$$\sum_{n=1}^{\infty} a_n = \sum_{n=1}^{\infty} \frac{(-1)^n}{n} = -1 + \frac{1}{2} - \frac{1}{3} + \frac{1}{4} - \frac{1}{5} + \frac{1}{6} + \cdots$$

Solution: This is an **alternating harmonic series**. Observe that $\{|a_n|\}$ is a nonincreasing sequence: $1 > \frac{1}{2} > \frac{1}{3} > \frac{1}{4} > \cdots$ with $\lim_{n\to\infty} |a_n| = \lim_{n\to\infty} \frac{1}{n} = 0$. Since $\sum a_n$ is an alternating series that satisfies both conditions for Leibniz's theorem, $\sum a_n$ converges. However, the series with absolute values, $\sum |a_n| = 1 + \frac{1}{2} + \frac{1}{3} + \frac{1}{4} + \frac{1}{5} + \frac{1}{6} + \cdots$, diverges because the (non-alternating) harmonic series diverges (Chapter 6). Since $\sum a_n$ converges but $\sum|a_n|$ diverges, $\sum a_n$ **converges conditionally** (Chapter 10). In Chapter 15, we'll see which value the alternating harmonic series converges to.

Example 3. Is the series below absolutely convergent, conditionally convergent, or divergent?

$$\sum_{n=3}^{\infty} a_k = \sum_{k=3}^{\infty} \frac{(-1)^{k+1}}{k^2} = \frac{1}{9} - \frac{1}{16} + \frac{1}{25} - \frac{1}{36} + \frac{1}{49} - \frac{1}{64} + \cdots$$

Solution: Since $\sum|a_k| = \sum\frac{1}{k^2}$ converges (Chapter 6), $\sum a_k$ is **absolutely convergent** (Chapter 10). Thus, $\sum a_k$ also converges. (If $\sum|a_k|$ converges, the series is absolutely convergent and you don't need to bother with Leibniz's theorem. Recall Chapter 10.)

Example 4. Is the series below absolutely convergent, conditionally convergent, or divergent?

$$\sum_{n=6}^{\infty} a_j = \sum_{j=6}^{\infty} (-1)^j \frac{j-5}{j} = \frac{1}{6} - \frac{2}{7} + \frac{3}{8} - \frac{4}{9} + \frac{5}{10} - \frac{6}{11} + \frac{7}{12} - \frac{8}{13} + \cdots$$

Solution: This series doesn't satisfy either condition for Leibniz's theorem. Observe that $\frac{3}{8} = 0.375$ is less than $\frac{4}{9} \approx 0.444$, for example. There doesn't exist a value of N for which $a_j \geq a_{j+1}$ for $n > N$. Also, $\{|a_j|\}$ doesn't converge to zero.

$$\lim_{j\to\infty} |a_j| = \lim_{j\to\infty} \left|\frac{j-5}{j}\right| = \lim_{j\to\infty} \left|\frac{1-\frac{5}{j}}{1}\right| = \frac{1-0}{1} = \frac{1}{1} = 1$$

An alternating series must satisfy both conditions for Leibniz's theorem in order to converge, but this series doesn't satisfy either condition. This series **diverges** according to the simple divergence test (Chapter 4): $\lim_{j\to\infty} |a_j| = 1$ doesn't converge to zero. (The second condition can be used to test for divergence because it's the simple divergence test, but the first condition by itself doesn't tell you whether or not an alternating series diverges. Beware that if an alternating series doesn't satisfy the first condition but does satisfy the second condition, Leibniz's theorem would be inconclusive.)

Chapter 11 Problems

Directions: Determine whether or not each series is an alternating series.

(1)

$$\sum_{m=1}^{\infty} \frac{2}{(-1)^{m+5}}$$

(2)

$$\sum_{j=2}^{\infty} (-4)^j$$

(3)

$$\sum_{k=0}^{\infty} \frac{(-1)^{2k+1}}{k+3}$$

(4)

$$\sum_{n=2}^{\infty} \frac{(-4)^{n+8}}{(-7)^{n-2}}$$

(5) Note: The angle is in radians.

$$\sum_{i=1}^{\infty} \sin\left(\frac{i\pi}{4}\right)$$

(6) Note: The angle is in radians.

$$\sum_{\ell=0}^{\infty} \cos\left(\frac{\pi + 3\ell\pi}{3}\right)$$

(7)

$$\sum_{j=1}^{\infty} \frac{(-1)^{j+1} + 2}{2}$$

(8)

$$\sum_{k=2}^{\infty} \left[(-1)^k + \frac{1}{3}\right]$$

Directions: Determine if each series is absolutely convergent, conditionally convergent, or divergent.

(9)

$$\sum_{m=0}^{\infty} \frac{1}{(-3)^m}$$

(10)

$$\sum_{i=1}^{\infty} \frac{(-1)^{i+1}}{i\sqrt{i}}$$

(11)

$$\sum_{n=2}^{\infty} \frac{(-1)^n}{\sqrt{n}}$$

(12)

$$\sum_{\ell=1}^{\infty} \frac{(-2)^{\ell}}{\ell^2}$$

(13)

$$\sum_{k=2}^{\infty} \frac{(k^2-3)(-1)^{k+4}}{2k^2+6k+5}$$

(14)

$$\sum_{j=2}^{\infty} \frac{\ln j}{j(-1)^{j-1}}$$

(15)

$$\sum_{i=1}^{\infty} \frac{i!}{(-3)^i}$$

(16)

$$\sum_{n=2}^{\infty} (-1)^n \frac{2+\sqrt{n}}{\sqrt{2n+1}}$$

(17)

$$2 - \frac{3}{4} + \frac{4}{9} - \frac{5}{16} + \frac{6}{25} - \frac{7}{36} + \frac{8}{49} - \frac{9}{64} + \frac{10}{81} - \frac{11}{100} + \cdots$$

(18)

$$1 - \frac{1}{3} + \frac{1}{2} - \frac{1}{6} + \frac{1}{3} - \frac{1}{9} + \frac{1}{4} - \frac{1}{12} + \frac{1}{5} - \frac{1}{15} + \frac{1}{6} - \frac{1}{18} + \cdots$$

12 Power Series

Until now, we've only discussed series where every term is a constant. In Chapters 12-14, we will consider series where each term is a **function** of a **variable**, like the example below where $u_0(x), u_1(x), u_2(x), u_3(x)$, ... are functions of x.

$$\sum_{n=0}^{\infty} u_n(x) = u_0(x) + u_1(x) + u_2(x) + u_3(x) + \cdots$$

Specifically, we will consider a simple type of series of functions called power series, which turn out to be particularly useful in calculus and its applications. In a **power series**, the functions involve simple powers of the variable, namely $u_n(x) = a_n x^n$. The coefficients, a_n, form a constant sequence (like those of Chapters 1-2).

$$\sum_{n=0}^{\infty} a_n x^n = a_0 + a_1 x + a_2 x^2 + a_3 x^3 + a_4 x^4 + \cdots$$

A power series may be more generally expressed as follows, where the argument of the function is $x - a$ rather than simply x. The above series is the special case where a happens to be zero. We call the general case below a power series **about** $x = a$. When x equals a, all of the terms are zero except for the first; for this particular value of x, the power series converges to a_0.[1]

$$\sum_{n=0}^{\infty} a_n (x-a)^n = a_0 + a_1(x-a) + a_2(x-a)^2 + a_3(x-a)^3 + a_4(x-a)^4 + \cdots$$

A power series is a **function**, $f(x)$, which behaves like a **polynomial** with an infinite number of terms. Its **domain** is the interval over which the power series converges.

$$f(x) = \sum_{n=0}^{\infty} a_n (x-a)^n$$

[1] When $x = a$, we get the peculiarity of $a_0(x-a)^0 = a_0(0)^0$. When $x \neq a$, the first term is $a_0(1) = a_0$, since any nonzero number raised to the zero power equals one, but when $x \neq 0$, we get 0^0, which is indeterminate. Why? Consider $\frac{y^m}{y^m} = y^{m-m} = y^0 = 1$. This equals one for nonzero y since $\frac{y^m}{y^m} = 1$, but if $y = 0$, you run into a problem with division by zero. Nonetheless, with power series, it's conventional to write the first term as a_0, even in the special case $x = a$, where we get $a_0(x-a)^0 = a_0(0)^0$.

All power series converge for some value(s) of x. Any power series is guaranteed to converge for the value of $x = a$, since every term beyond a_0 is zero, as every term involves $(x-a)^n$. The **interval of convergence** provides all of the values of x for which the power series converges; this interval includes $x = a$. A common alternative to the interval of convergence is the **radius of convergence**. Given the radius of convergence R, the power series converges if $|x-a| < R$ and diverges if $|x-a| > R$.

As an example, consider the relatively simple power series for which $a_n = 1$ for all n and where $a = 0$. (Be careful not to confuse the constant a that appears in $x - a$ with the coefficients a_n; these are different a's.)

$$f(x) = \sum_{n=0}^{\infty} 1(x-0)^n = \sum_{n=0}^{\infty} x^n = 1 + x + x^2 + x^3 + x^4 + \cdots$$

This should look familiar. It's an infinite geometric series with $r = x$. From Chapter 4, we know that this series converges if $|r| = |x| < 1$. The interval of convergence is $-1 < x < 1$. The corresponding radius of convergence is $R = 1$. (It may help to think of $x = a$ as the 'center'; in this example, $a = 0$.) Recall from Chapter 4 that[2]

$$\sum_{k=1}^{\infty} r^{k-1} = \frac{1}{1-r}$$

Since the power series begins at $n = 0$ instead of $k = 1$, we will shift the sum for the geometric series to $n = k - 1$ (recall Chapter 3).

$$\sum_{n=0}^{\infty} r^n = \frac{1}{1-r}$$

Since $r = x$ for the power series in this example, it follows that

$$\frac{1}{1-x} = 1 + x + x^2 + x^3 + x^4 + \cdots \quad (-1 < x < 1)$$

If we replace x with $-x$ in the above formula, we find the similar relation

$$\frac{1}{1+x} = 1 - x + x^2 - x^3 + x^4 + \cdots \quad (-1 < x < 1)$$

[2] A general geometric series may also have a coefficient, which we called a in Chapter 4. Here, that coefficient equals one, but let's not call it a in the context of power series, since we're using a with $x - a$, which is different from the a from geometric series; a is a common and convenient symbol.

Note that these series converge rather quickly. For example, let $x = \frac{1}{4}$. Then $\frac{1}{1-1/4} = \frac{1}{3/4} = \frac{4}{3} \approx 1.333$ and $\frac{1}{1+1/4} = \frac{1}{5/4} = \frac{4}{5} = 0.8$. If we only keep the first 5 terms of each series, look how closely the sums approximate the actual values:

$$\frac{1}{1-1/4} \approx 1 + \frac{1}{4} + \frac{1}{16} + \frac{1}{64} + \frac{1}{256} = 1.332$$

$$\frac{1}{1-1/4} \approx 1 - \frac{1}{4} + \frac{1}{16} - \frac{1}{64} + \frac{1}{256} = 0.801$$

These numerical examples suggest that power series might be useful for approximating functions. This is indeed the case, as we'll see in Chapters 13-14.

Consider the general power series below. There is an important theorem regarding its convergence, which we state below.

$$\sum_{n=0}^{\infty} a_n (x-a)^n = a_0 + a_1(x-a) + a_2(x-a)^2 + a_3(x-a)^3 + \cdots$$

According to the <u>**convergence theorem for power series**</u>:

- If the power series converges for $x = x_c$ (where $x_c \neq a$), then the power series also converges absolutely for all $|x-a| < x_c$.
- If the power series diverges for $x = x_d$, then the power series diverges for all $|x-a| > x_d$.

Why? It basically follows from the comparison test (Chapter 4). Let $w = x - a$. Then the power series takes the form

$$\sum_{n=0}^{\infty} a_n w^n = a_0 + a_1 w + a_2 w^2 + a_3 w^3 + \cdots$$

Note that $|x-a| < x_c$ is equivalent to $|w| < x_c$ and $|x-a| > x_d$ is equivalent to $|w| > x_d$. Every term in the power series has a power of w, like w^3 or w^8. If the power series converges for x_c, if $|w| < x_c$, then $|w|^n < x_c^n$ (for $n > 0$) and the absolute value of every term is smaller than $|x_c|^n$, so by the comparison test, $\sum |a_n w^n|$ converges. On the other hand, if the power series diverges for x_d, if $|w| > x_d$, then $|w|^n > x_d^n$ (for $n > 0$) and the absolute value of every term is larger than $|x_d|^n$. Note that the <u>**radius of convergence**</u>, R, is special in that the power series converges if $|w| < R$ and diverges if $|w| > R$; any other value of x would only serve as x_c or x_d, not both.

As with other series, multiplying an entire power series by a finite constant doesn't change whether the power series converges. If $\sum a_n(x-a)^n$ converges over $0 \le x < R$, then the series $\sum ca_n(x-a)^n$ (where c is a constant) also converges over $0 \le x < R$.

Suppose that the two power series below converge to L_a and L_b, respectively, over $0 \le x < R$. If the two power series are added together, the combined series converges over $0 \le x < R$ and the combined sum equals $L_a + L_b$.

$$\sum_{n=0}^{\infty} a_n(x-a)^n = L_a \quad , \quad \sum_{n=0}^{\infty} b_n(x-b)^n = L_b$$

$$\sum_{n=0}^{\infty} [a_n(x-a)^n + b_n(x-b)^n] = L_a + L_b$$

The **series multiplication theorem**, which applies to series that converge absolutely, also applies to power series. Suppose that the two power series below converge. (Note that $a = b = 0$ below, unlike the similar series above.)

$$\sum_{n=0}^{\infty} a_n x^n = L_a \quad , \quad \sum_{n=0}^{\infty} b_n x^n = L_b$$

The coefficients of these three series are related as follows:

- $c_0 = a_0 b_0$
- $c_1 = a_0 b_1 + a_1 b_0$
- $c_2 = a_0 b_2 + a_1 b_1 + a_2 b_0$
- $c_3 = a_0 b_3 + a_1 b_2 + a_2 b_1 + a_3 b_0$
- $c_4 = a_0 b_4 + a_1 b_3 + a_2 b_2 + a_3 b_1 + a_4 b_0$
- $c_n = a_0 b_n + a_1 b_{n-1} + a_2 b_{n-2} + \cdots + a_{n-2} b_2 + a_{n-1} b_1 + a_n b_0$

According to the **series multiplication theorem**, if $\sum a_n x^n$ and $\sum b_n x^n$ both converge over $0 \le x < R$, then $\sum c_n x^n$ also converges and it converges to

$$\sum_{n=0}^{\infty} c_n x^n = \left(\sum_{n=0}^{\infty} a_n x^n\right)\left(\sum_{k=0}^{\infty} b_k x^k\right) = L_a L_b$$

If a power series converges over $0 \le x < R$, its **derivative** and **integral** exist, have the same radius of convergence, and may be found term by term as follows:

$$f(x) = \sum_{n=0}^{\infty} a_n (x-a)^n$$

$$\frac{df}{dx} = \sum_{n=0}^{\infty} n a_n (x-a)^{n-1}$$

$$\int f(x)\,dx = \sum_{n=0}^{\infty} \frac{a_n (x-a)^{n+1}}{n+1} + c$$

As usual, an indefinite integral may have a constant of integration, c.

Example 1. Find the radius and interval of convergence for the power series below.

$$\sum_{n=1}^{\infty} \frac{(x-2)^n}{n^2}$$

Solution: Apply the ratio test (Chapter 7).

$$L = \lim_{n\to\infty} \left| \frac{(x-2)^{n+1}}{(n+1)^2} \div \frac{(x-2)^n}{n^2} \right| = \lim_{n\to\infty} \left| \frac{(x-2)^{n+1}}{n^2+2n+1} \times \frac{n^2}{(x-2)^n} \right| = \lim_{n\to\infty} \left| \frac{(x-2)n^2}{n^2+2n+1} \right|$$

$$= \lim_{n\to\infty} \left| \frac{(x-2)}{1+\frac{2}{n}+\frac{1}{n^2}} \right| = \left| \frac{x-2}{1+0+0} \right| = \left| \frac{x-2}{1} \right| = |x-2|$$

The power series converges if $L = |x-2| < 1$, which equates to $-1 < x-2 < 1$ or $1 < x < 3$. The radius of convergence is one-half of this interval: $R = 1$. We also need to check the case $|x-2| = 1$, which equates to $x = 1$ or 3, since the ratio test is inconclusive for $L = 1$. When $x = 1$, the power series is $\sum \frac{(-1)^n}{n^2}$, and when $x = 3$, the power series is $\sum \frac{1}{n^2}$. Both of these cases converge absolutely since $\sum \frac{1}{n^2}$ converges (recall Chapters 10-11). The complete interval is $1 \le x \le 3$. (We changed the $<$ signs to $\le$ signs once we learned that the series converges at the endpoints, $x = 1$ or 3.)

Tips: To find the radius of convergence for a power series, it's often helpful to apply either the **ratio test** (Chapter 7) or the **root test** (Chapter 8). Remember to **check the endpoints** (as in Examples 1-2) after finding the radius of convergence because the ratio test and root test are inconclusive when $L = 1$.

Example 2. Find the radius and interval of convergence for the power series below.

$$\sum_{n=0}^{\infty} \frac{x^n}{2^n}$$

Solution: Apply the root test (Chapter 8). Recall that $(x^n)^{1/n} = x^1 = x$.

$$L = \lim_{n \to \infty} \left|\frac{x^n}{2^n}\right|^{1/n} = \lim_{n \to \infty} \left|\frac{x}{2}\right| = \left|\frac{x}{2}\right|$$

The power series converges if $L = \left|\frac{x}{2}\right| < 1$, which equates to $-1 < \frac{x}{2} < 1$ or $-2 < x < 2$. The radius of convergence is one-half of this interval: $R = 2$. We also need to check the case $\left|\frac{x}{2}\right| = 1$, which equates to $x = -2$ or 2, since the root test is inconclusive for $L = 1$. When $x = -2$, the power series is $\sum \frac{(-2)^n}{2^n} = \sum \left(-\frac{2}{2}\right)^n = \sum(-1)^n$, and when $x = 2$, the power series is $\sum \frac{2^n}{2^n} = \sum 1$. Both of these cases diverge according to the simple divergence test (Chapter 4) since $\lim_{n \to \infty} a_n$ isn't zero. Unlike the previous example, here the complete interval is $-2 < x < 2$ (without using any $\leq$ signs).

Example 3. For the power series below, find (A) a derivative with respect to x and (B) the indefinite integral over x.

$$\sum_{n=0}^{\infty} \frac{(x-3)^n}{n+1}$$

Solution: (A) If you're having trouble seeing the derivative with the expression above, try writing the series out longhand.

$$\sum_{n=0}^{\infty} \frac{(x-3)^n}{n+1} = 1 + \frac{x-3}{2} + \frac{(x-3)^2}{3} + \frac{(x-3)^3}{4} + \frac{(x-3)^4}{5} + \cdots$$

Consider the term $\frac{(x-3)^4}{5}$. The long way to take the derivative is to multiply $(x-3)$ by itself repeatedly. The easy way is to let $u = x - 3$ and $f(u) = \frac{u^4}{5} = \frac{(x-3)^4}{5}$. Recall from first-semester calculus that the way to find $\frac{df}{dx}$ is to apply the **chain rule** as follows:

$$\frac{df}{dx} = \frac{df}{du}\frac{du}{dx} = \frac{d}{du}\frac{u^4}{5}\frac{d}{dx}(x-3) = \frac{4u^3}{5}(1) = \frac{4u^3}{5} = \frac{4(x-3)^3}{5}$$

If we take a derivative of the nth term, which is $\frac{(x-3)^n}{n+1}$, with respect to x, the n will come down to multiply the coefficient and the exponent will reduce to $n-1$.

$$\frac{d}{dx}\sum_{n=0}^{\infty}\frac{(x-3)^n}{n+1} = \sum_{n=0}^{\infty}\frac{n(x-3)^{n-1}}{n+1}$$

Observe that this agrees with the rule below, which applies to power series.

$$\frac{df}{dx} = \sum_{n=0}^{\infty}\frac{d}{dx}a_n(x-a)^n = \sum_{n=0}^{\infty}na_n(x-a)^{n-1}$$

In this example, $a = 3$ and $a_n = \frac{1}{n+1}$.

(B) This is like Part A, except for integrating rather than differentiating. Apply the rule below, which applies to power series. (The $+c$ is separate from the sum, not part of it.)

$$\int f(x)\,dx = \sum_{n=0}^{\infty}\int a_n(x-a)^n\,dx = \sum_{n=0}^{\infty}\frac{a_n(x-a)^{n+1}}{n+1} + c$$

Set $a = 3$ and $a_n = \frac{1}{n+1}$ to get

$$\sum_{n=0}^{\infty}\int\frac{(x-3)^n}{n+1}dx = \sum_{n=0}^{\infty}\frac{(x-3)^{n+1}}{(n+1)^2} + c$$

Example 4. Multiply the two series below, only keeping terms up to x^4 in your answer.

$$\left(1 + x + \frac{x^2}{4} + \frac{x^3}{9} + \frac{x^4}{16}\cdots\right)\left(1 - x + \frac{x^2}{2} - \frac{x^3}{3} + \frac{x^4}{4} + \cdots\right)$$

Solution: Use the **series multiplication theorem** with $a_0 = 1$, $a_1 = 1$, $a_2 = \frac{1}{4}$, $a_3 = \frac{1}{9}$, $a_4 = \frac{1}{16}$, $b_0 = 1$, $b_1 = -1$, $b_2 = \frac{1}{2}$, $b_3 = -\frac{1}{3}$, and $b_4 = \frac{1}{4}$ (the **coefficients** of each series).

$$c_0 = a_0b_0 = 1(1) = 1$$

$$c_1 = a_0b_1 + a_1b_0 = 1(-1) + 1(1) = -1 + 1 = 0$$

$$c_2 = a_0b_2 + a_1b_1 + a_2b_0 = 1\left(\frac{1}{2}\right) + 1(-1) + \frac{1}{4}(1) = \frac{1}{2} - 1 + \frac{1}{4} = \frac{3}{4} - 1 = -\frac{1}{4}$$

$$c_3 = a_0b_3 + a_1b_2 + a_2b_1 + a_3b_0 = 1\left(-\frac{1}{3}\right) + 1\left(\frac{1}{2}\right) + \frac{1}{4}(-1) + \frac{1}{9}(1)$$

$$c_3 = -\frac{1}{3} + \frac{1}{2} - \frac{1}{4} + \frac{1}{9} = \frac{-12 + 18 - 9 + 4}{36} = \frac{1}{36}$$

$$c_4 = a_0 b_4 + a_1 b_3 + a_2 b_2 + a_3 b_1 + a_4 b_0 = 1\left(\frac{1}{4}\right) + 1\left(-\frac{1}{3}\right) + \frac{1}{4}\left(\frac{1}{2}\right) + \frac{1}{9}(-1) + \frac{1}{16}(1)$$

$$c_4 = \frac{1}{4} - \frac{1}{3} + \frac{1}{8} - \frac{1}{9} + \frac{1}{16} = \frac{36 - 48 + 18 - 16 + 9}{144} = -\frac{1}{144}$$

$$\sum_{n=0}^{\infty} c_n x^n = \left(\sum_{n=0}^{\infty} a_n x^n\right)\left(\sum_{k=0}^{\infty} b_k x^k\right) = c_0 + c_1 x + c_2 x^2 + c_3 x^3 + c_4 x^4 + \cdots$$

$$= 1 + 0x - \frac{1}{4}x^2 + \frac{1}{9}x^3 - \frac{1}{144}x^4 + \cdots = 1 - \frac{x^2}{4} + \frac{x^3}{9} - \frac{x^4}{144} + \cdots$$

You don't need to use the formulas for the coefficients. You can just multiply the two series together as you would normally multiply polynomials in algebra. Since the instructions say to keep terms up to x^4, we ignore anything with x^5 or higher (those terms aren't really ignored; they're included in the $+\cdots$).

$$\left(1 + x + \frac{x^2}{4} + \frac{x^3}{9} + \frac{x^4}{16} \cdots\right)\left(1 - x + \frac{x^2}{2} - \frac{x^3}{3} + \frac{x^4}{4} + \cdots\right)$$

$$= 1 - x + \frac{x^2}{2} - \frac{x^3}{3} + \frac{x^4}{4} + x - x^2 + \frac{x^3}{2} - \frac{x^4}{3} + \frac{x^2}{4} - \frac{x^3}{4} + \frac{x^4}{8} + \frac{x^3}{9} - \frac{x^4}{9} + \frac{x^4}{16} + \cdots$$

$$= 1 + \left(\frac{1}{2} - 1 + \frac{1}{4}\right)x^2 + \left(-\frac{1}{3} + \frac{1}{2} - \frac{1}{4} + \frac{1}{9}\right)x^3 + \left(\frac{1}{4} - \frac{1}{3} + \frac{1}{8} - \frac{1}{9} + \frac{1}{16}\right)x^4 + \cdots$$

$$= 1 + \frac{2 - 4 + 1}{4}x^2 + \frac{-12 + 18 - 9 + 4}{36}x^3 + \frac{36 - 48 + 18 - 16 + 9}{144}x^4 + \cdots$$

$$= 1 - \frac{x^2}{4} + \frac{x^3}{9} - \frac{x^4}{144} + \cdots$$

The answer holds for values of x within the interval of convergence that is common to **both** series. (This example didn't ask us to work out what this interval is, but we could using the technique from Examples 1-2 for each series separately and then see what the overlapping interval is.)

Chapter 12 Problems

Directions: Find the radius and interval of convergence for each power series below.

(1)

$$\sum_{n=1}^{\infty} \frac{x^n}{n}$$

(2)

$$\sum_{k=2}^{\infty} \frac{x^k}{k^2}$$

(3)

$$\sum_{m=0}^{\infty} \frac{x^m}{m!}$$

(4)

$$\sum_{j=2}^{\infty} \frac{x^j}{\sqrt{j}}$$

(5)

$$\sum_{i=0}^{\infty} \frac{x^i}{5^i}$$

(6)

$$\sum_{\ell=1}^{\infty} \ell^2 x^\ell$$

(7)

$$\sum_{n=0}^{\infty} x^n e^{-n}$$

(8)

$$\sum_{k=2}^{\infty} \frac{(x-4)^k}{k+3}$$

(9)

$$\sum_{m=0}^{\infty} \frac{(x+6)^m}{4^m}$$

(10)

$$\sum_{j=2}^{\infty} \frac{(2x-3)^j}{x^4}$$

Directions: Find the derivative or indefinite integral of each series, as indicated.

(11)

$$\frac{d}{dx}\sum_{i=1}^{\infty}\frac{x^i}{i^2}$$

(12)

$$\int\sum_{\ell=0}^{\infty}\frac{(x-4)^\ell}{\ell+1}dx$$

(13)

$$\frac{d}{dx}\sum_{k=2}^{\infty}\frac{(x+3)^k}{5k}$$

(14)

$$\int\sum_{n=1}^{\infty}\frac{(3x+2)^n}{\sqrt{n}}dx$$

Directions: Simplify each expression, keeping terms up to x^5.

(15)

$$\left(x+\frac{x^2}{2}+\frac{x^3}{3}+\frac{x^4}{4}+\frac{x^5}{5}+\cdots\right)\left(1-\frac{x}{4}+\frac{x^2}{9}-\frac{x^3}{16}+\frac{x^4}{25}-\frac{x^5}{36}+\cdots\right)$$

(16)

$$\left(1-x+\frac{x^2}{2}-\frac{x^3}{6}+\frac{x^4}{24}-\frac{x^5}{120}+\cdots\right)\left(x-\frac{x^2}{2}+\frac{x^3}{3}-\frac{x^4}{4}+\frac{x^5}{5}+\cdots\right)$$

(17)

$$\frac{x-\frac{x^3}{6}+\frac{x^5}{120}+\cdots}{1-\frac{x^2}{2}+\frac{x^4}{24}+\cdots}$$

13 Taylor Series

In this chapter, we will consider functions $f(x)$ with the following features, where the region around the point $x = a$ is a **region of interest**. We may wish to approximate $f(x)$ at values of x that are near (but not exactly equal to) a, for example.

- The function $f(x)$ and all orders of its **derivatives**, $\frac{df}{dx}, \frac{d^2f}{dx^2}, \frac{d^3f}{dx^3}, \ldots$, exist and are continuous and differentiable over an open interval containing $x = a$.
- If c is a constant between a and x, then

$$\lim_{k\to\infty}\left[\frac{(x-a)^k}{k!}\left.\frac{d^kf(x)}{dx^k}\right|_{x=c}\right] = 0$$

where $\left.\frac{d^kf(x)}{dx^k}\right|_{x=c}$ means to take the kth derivative and evaluate it at $x = c$.

As an example, consider $f(x) = \ln x$. In this case, we can't let a be zero because $\frac{d}{dx}\ln x = \frac{1}{x}$ results in division by zero if x equals zero, but we could let a be a positive constant.

If a function $f(x)$ satisfies both of the conditions above, then the **Taylor series** (which is said to be an expansion of the function about the point $x = a$) below converges to $f(x)$ for all values of x within the radius of convergence (that is, $|x - a| < R$).

$$f(x) = \sum_{k=0}^{\infty}\frac{(x-a)^k}{k!}\left.\frac{d^kf(x)}{dx^k}\right|_{x=a}$$

The kth derivative of $f(x)$ evaluated at the point $x = a$ is often expressed with more concise notation as

$$f^{(k)}(a) = \left.\frac{d^kf(x)}{dx^k}\right|_{x=a}$$

Using this notation for the derivative, the Taylor series has the form

$$f(x) = \sum_{k=0}^{\infty}\frac{(x-a)^k}{k!}f^{(k)}(a) = \sum_{k=0}^{\infty}\frac{f^{(k)}(a)}{k!}(x-a)^k$$

Also, note that the first few derivatives are often expressed using primes rather than $\frac{df}{dx}, \frac{d^2f}{dx^2}, \frac{d^3f}{dx^3}$, etc., as indicated below.

$$f'(x) = \frac{df}{dx} \quad , \quad f''(x) = \frac{d^2f}{dx^2} \quad , \quad f'''(x) = \frac{d^3f}{dx^3} \quad \dots \quad f^{(k)}(x) = \frac{d^kf}{dx^k}$$

Taylor series are particularly useful in calculus and its applications (especially, physics) because they can help to **approximate functions with polynomials**. For example, in physics, calculation of the period of a simple pendulum is greatly simplified by using the Taylor series (or more specifically, the Maclaurin series) expansion for $\sin\theta$. As another example, consider the indefinite integral $\int e^{-x^2}\,dx$. No algebraic anti-derivative exists for e^{-x^2}; you can't find an algebraic function $g(x)$ for which you would be able to take the derivative $\frac{dg}{dx}$ and obtain e^{-x^2} as a result.[1] The Taylor series expansion for e^{-x^2} allows us to integrate the Taylor series instead of integrating the function itself. (We'll see the Maclaurin series expansions for $\sin\theta$ and e^{-x^2} in Chapter 14.)

The reason that a Taylor series expansion is useful for approximating functions is the second bullet point from the beginning of this chapter. We'll rewrite that limit here using the notation $f^{(k)}(c) = \left.\frac{d^kf(x)}{dx^k}\right|_{x=c}$.

$$\lim_{k\to\infty}\left[\frac{f^{(k)}(c)}{k!}(x-a)^k\right] = 0$$

Because of this limit, higher-order terms (those with larger values of k) become very small as k grows large. In many cases, the Taylor series converges quite rapidly, such that the approximation is very good by keeping only a few terms. We'll explore this numerically later in this chapter and much more in Chapter 15.

Let's take a moment to try to understand how Taylor series came about. Suppose that we have a function $f(x)$. It could be a trig function, an exponential, a logarithm, it could be something like $\frac{1}{\sqrt{1+x^2}}$, it could be a combination of standard functions like $e^{-x}\sin(3x^2-4)$, or something entirely different. Suppose that we would like to make

---

[1] The problem is the x^2. For example, $\frac{d}{dx}e^{-x^2} = -2xe^{-x^2}$, so in contrast to e^{-x} (for which it is very easy to find the anti-derivative), e^{-x^2} doesn't have an algebraic anti-derivative.

a polynomial function $p(x)$ that closely approximates $f(x)$ for values of x that are near a. Which properties should $p(x)$ have?

- The function $p(x)$ should match $f(x)$ when it is evaluated at $x = a$. That is, it would be desirable for $p(a) = f(a)$. (If these two functions aren't equal at this value of x, how could we expect $p(x)$ to approximate $f(x)$ at values of x near a?)
- The derivative of $p(x)$ with respect to x should also match the derivative of $f(x)$ when it is evaluated at $x = a$. If we plot $p(x)$ and $f(x)$, although the two curves may differ when $x \neq a$ (since one is just an approximation to the other), they should have the same slope when $x = a$. That is, it would be desirable for $p'(a) = f'(a)$, where the prime (′) indicates the first derivative with respect to x. (Again, if the slopes don't match at $x = a$, it wouldn't seem likely for $p(x)$ to approximate $f(x)$ at values of x near a.)
- Going a step further, the concavity of $p(x)$ and $f(x)$ should match at $x = a$. This means that $p''(a) = f''(a)$, where the double prime indicates a second derivative.
- Continuing the pattern, all orders of derivatives should match at $x = a$. We should have $p^{(n)}(a) = f^{(n)}(a)$ for the nth derivative.

This is exactly what a Taylor series does. Consider the following polynomial:

$$p(x) = a_0 + a_1(x-a) + a_2(x-a)^2 + a_3(x-a)^3 + a_4(x-a)^4 \ldots + a_n(x-a)^n$$

- Observe that when x equals a, at that point $p(a) = a_0$. If we let $a_0 = f(a)$, then $p(a) = f(a)$.
- The first derivative of $p(x)$ with respect to x is:

$$p'(x) = \frac{dp}{dx} = a_1 + 2a_2(x-a) + 3a_3(x-a)^2 + 4a_4(x-a)^3 + \cdots + na_n(x-a)^{n-1}$$

  At the point $x = a$, we get $p'(a) = a_1$. If we let $a_1 = f'(a)$, then $p'(a) = f'(a)$.
- The second derivative is:

$$p''(x) = \frac{d^2p}{dx^2} = 2a_2 + 6a_3(x-a) + 12a_4(x-a)^2 + \cdots + n(n-1)a_n(x-a)^{n-2}$$

  At $x = a$, we get $p''(a) = 2a_2$. If we let $a_2 = \frac{f''(a)}{2}$, then $p''(a) = f''(a)$.

- The third derivative is:

$$p'''(x) = \frac{d^3p}{dx^3} = 6a_3 + 24a_4(x-a) + \cdots + n(n-1)(n-2)a_n(x-a)^{n-3}$$

At $x = a$, we get $p'''(a) = 6a_3$. If we let $a_3 = \frac{f'''(a)}{6}$, then $p'''(a) = f'''(a)$.

- The fourth derivative is:

$$p^{(4)}(x) = \frac{d^3p}{dx^3} = 24a_4 + \cdots + n(n-1)(n-2)(n-3)a_n(x-a)^{n-3}$$

At $x = a$, we get $p^{(4)}(a) = 24a_4$. If we let $a_4 = \frac{f^{(4)}(a)}{24}$, then $p^{(4)}(a) = f^{(4)}(a)$.

- The nth derivative is:

$$p^{(n)}(x) = \frac{d^np}{dx^n} = n!\, a_n$$

At $x = a$, we get $p^{(n)}(a) = n!\, a_n$. If we let $a_n = \frac{f^{(n)}(a)}{n!}$, then $p^{(n)}(a) = f^{(n)}(a)$.

When we put all of these bullet points together, we get a **Taylor polynomial** of degree n:

$$p(x) = a_0 + a_1(x-a) + a_2(x-a)^2 + a_3(x-a)^3 + a_4(x-a)^4 + \cdots + a_n(x-a)^n$$

$$p(x) = f(a) + f'(a)(x-a) + \frac{f''(a)}{2}(x-a)^2 + \frac{f'''(a)}{6}(x-a)^3 + \frac{f^{(4)}(a)}{24}(x-a)^4 + \cdots + \frac{f^{(n)}(a)}{n!}(x-a)^n$$

$$p(x) = \sum_{k=0}^{n} \frac{f^{(k)}(a)}{k!}(x-a)^k$$

In the limit that $n \to \infty$, the Taylor polynomial becomes the infinite Taylor series. For a finite value of n, the Taylor polynomial serves as a good approximation to $f(x)$ for values of x near a. As $n \to \infty$, the Taylor series converges to $f(x)$.

Let's look at one example of a Taylor series and see numerically how well it converges. Consider the function $f(x) = \cos x$. Suppose that we are interested in values of x near $\frac{\pi}{2}$ radians (corresponding to 90°). In order to find the Taylor series for $f(x) = \cos x$, we will need to find its derivatives, so we will find these first.

$$f'(x) = -\sin x \quad , \quad f''(x) = -\cos x \quad , \quad f'''(x) = \sin x \quad , \quad f^{(4)}(x) = \cos x$$

These derivatives repeat forever, since every fourth derivative is the same. As an example, $f'(x) = f^{(5)}(x) = f^{(9)}(x) = f^{(13)}(x) = \cdots = -\sin x$. For the Taylor series, we need to evaluate each derivative at $x = \frac{\pi}{2}$ radians, so next we will plug $x = \frac{\pi}{2}$ into the derivatives that we found.

$$f'\left(\frac{\pi}{2}\right) = -\sin\frac{\pi}{2} = -1\,, f''\left(\frac{\pi}{2}\right) = -\cos\frac{\pi}{2} = 0\,, f'''\left(\frac{\pi}{2}\right) = \sin\frac{\pi}{2} = 1\,, f^{(4)}\left(\frac{\pi}{2}\right) = \cos\frac{\pi}{2} = 0$$

Plug these derivatives into the formula for a Taylor series expansion of $f(x)$ about a.

$$f(x) = \sum_{k=0}^{\infty} \frac{f^{(k)}\left(\frac{\pi}{2}\right)}{k!}\left(x - \frac{\pi}{2}\right)^k$$

$$= f\left(\frac{\pi}{2}\right) + f'\left(\frac{\pi}{2}\right)\left(x - \frac{\pi}{2}\right) + \frac{f''\left(\frac{\pi}{2}\right)}{2}\left(x - \frac{\pi}{2}\right)^2 + \frac{f'''\left(\frac{\pi}{2}\right)}{6}\left(x - \frac{\pi}{2}\right)^3 + \frac{f^{(4)}\left(\frac{\pi}{2}\right)}{24}\left(x - \frac{\pi}{2}\right)^4 + \cdots$$

$$= \cos\frac{\pi}{2} - \left(x - \frac{\pi}{2}\right) - \frac{0}{2}\left(x - \frac{\pi}{2}\right)^2 + \frac{1}{6}\left(x - \frac{\pi}{2}\right)^3 + \frac{0}{24}\left(x - \frac{\pi}{2}\right)^4 - \frac{1}{120}\left(x - \frac{\pi}{2}\right)^5 + \cdots$$

$$= 0 - \left(x - \frac{\pi}{2}\right) - 0 + \frac{1}{6}\left(x - \frac{\pi}{2}\right)^3 + 0 - \frac{1}{120}\left(x - \frac{\pi}{2}\right)^5 + \cdots$$

$$= -\left(x - \frac{\pi}{2}\right) + \frac{1}{6}\left(x - \frac{\pi}{2}\right)^3 - \frac{1}{120}\left(x - \frac{\pi}{2}\right)^5 + \cdots$$

The next term will be $\frac{1}{7!}\left(x - \frac{\pi}{2}\right)^7$, then $-\frac{1}{9!}\left(x - \frac{\pi}{2}\right)^9$, etc. Let's see how good this series is at approximating $\cos 75° = \cos\left(\frac{5\pi}{12}\right)$. If your trig skills are really good, you can work out that the exact[2] answer is $\frac{\sqrt{6}-\sqrt{2}}{4}$. At least, you can check on a calculator that the answer is approximately 0.258819045. Note that the Taylor series above is only valid if the angle is in **radians** (not degrees). Let's plug $x = \frac{5\pi}{12}$ into the first few Taylor polynomials and see how close these are to the correct answer of 0.2588.

- Keeping only one term, the first Taylor polynomial is $p_1 = -\left(x - \frac{\pi}{2}\right)$. The first-order approximation is $p_1\left(\frac{5\pi}{12}\right) = -\left(\frac{5\pi}{12} - \frac{\pi}{2}\right) = -\left(\frac{5\pi}{12} - \frac{6\pi}{12}\right) = \frac{\pi}{12} \approx 0.2618$. It's amazing that the very first term is already so close to 0.2588. As we add more terms, the Taylor series will converge rather rapidly in this example.

[2] Most students get an answer in some other form, like $\sqrt{\frac{1}{2} - \frac{\sqrt{3}}{4}}$, which is equivalent to $\frac{\sqrt{6}-\sqrt{2}}{4}$ (as you can verify numerically with a calculator, since each is about 0.2588), which has one square root inside of another. It takes some nifty algebra (like completing the square) to obtain the form $\frac{\sqrt{6}-\sqrt{2}}{4}$.

- Keeping two terms, we have $p_3 = -\left(x - \frac{\pi}{2}\right) + \frac{1}{6}\left(x - \frac{\pi}{2}\right)^3$. (Why are we calling the second one p_3 instead of p_2? Technically, $p_2 = 0$ since the next term was zero. All the even terms are zero, and only the odd terms are nonzero in this Taylor series.) We get $p_3\left(\frac{5\pi}{12}\right) = -\left(\frac{5\pi}{12} - \frac{\pi}{2}\right) + \frac{1}{6}\left(\frac{5\pi}{12} - \frac{\pi}{2}\right)^3 = \frac{\pi}{12} + \frac{1}{6}\left(-\frac{\pi}{12}\right)^3 = \frac{\pi}{12} - \frac{\pi^3}{10{,}368} \approx 0.258808813$. This matches the first 4 digits of the exact answer.
- With three terms, $p_5 = -\left(x - \frac{\pi}{2}\right) + \frac{1}{6}\left(x - \frac{\pi}{2}\right)^3 - \frac{1}{120}\left(x - \frac{\pi}{2}\right)^5$. We get $p_5\left(\frac{5\pi}{12}\right)$ $= -\left(\frac{5\pi}{12} - \frac{\pi}{2}\right) + \frac{1}{6}\left(\frac{5\pi}{12} - \frac{\pi}{2}\right)^3 - \frac{1}{120}\left(x - \frac{\pi}{2}\right)^5 = \frac{\pi}{12} + \frac{1}{6}\left(-\frac{\pi}{12}\right)^3 - \frac{1}{120}\left(-\frac{\pi}{12}\right)^5 = \frac{\pi}{12} - \frac{\pi^3}{10{,}368} + \frac{\pi^5}{29{,}859{,}840} \approx 0.258819062$. Compare this with 0.258819045.

This shows that Taylor series can converge quite rapidly. In our example, we got the first 6 digits of the answer with just three (nonzero) terms in the Taylor series.

Taylor's theorem helps to determine the remainder (or error) when the polynomial $p_n(x)$, which has $n + 1$ terms because n begins from zero, is used to approximate the function $f(x)$. According to **Taylor's theorem**, if $f(x)$ and all orders of its **derivatives** exist and are continuous over $[a, b]$ and if $f^{(n)}(x)$ is differentiable over (a, b), then it is possible to find a value c between a and b such that

$$f(b) = f(a) + f'(a)(b-a) + \frac{f''(a)}{2}(b-a)^2 + \frac{f'''(a)}{6}(b-a)^3 + \frac{f^{(4)}(a)}{24}(b-a)^4 + \dots + \frac{f^{(n)}(a)}{n!}(b-a)^n + \frac{f^{(n+1)}(c)}{(n+1)!}(b-a)^n$$

Observe that $f(b) = p_n(b) + \frac{f^{(n+1)}(c)}{(n+1)!}(b-a)^n$. Here, $p_n(b)$ is Taylor's polynomial with a finite number of terms evaluated at $x = b$, which is used to approximate the function $f(x)$ at the value $x = b$ (which is in the neighborhood of a). The expression $\frac{f^{(n+1)}(c)}{(n+1)!}(b-a)^n$ represents the **remainder** (or error); it is the difference between the exact value, $f(b)$, and the approximating polynomial, $p_n(b)$. If we replace b with x in Taylor's theorem, we get:

$$f(x) = f(a) + f'(a)(x-a) + \frac{f''(a)}{2}(x-a)^2 + \frac{f'''(a)}{6}(x-a)^3 + \frac{f^{(4)}(a)}{24}(x-a)^4 +$$
$$\ldots + \frac{f^{(n)}(a)}{n!}(x-a)^n + \frac{f^{(n+1)}(c)}{(n+1)!}(x-a)^n$$

which can be written as

$$f(x) = p(x) + R_n(x)$$

where the $R_n(x)$ is the **remainder** (not to be confused with R, which is the radius of convergence; these are two different R's):

$$R_n(x) = \frac{f^{(n+1)}(c)}{(n+1)!}(x-a)^n$$

When $\lim_{k\to\infty}\left[\frac{(x-a)^k}{k!}\frac{d^k f(x)}{dx^k}\bigg|_{x=c}\right] = 0$ (recall the second condition from the beginning of this chapter), $R_n(x) \to 0$ as $n \to \infty$. That is, the infinite Taylor series converges to $f(x)$. This occurs in an interval of convergence (recall Chapter 12) around the point $x = a$. The remainder can be used to estimate the error when using a Taylor polynomial to approximate a function, as we'll see in Chapter 15.

In many applications of Taylor's theorem, the constant a equals zero. This is so common that this special case has its own name (Maclaurin series). We'll save the case where a is zero for the next chapter, and focus on $a \neq 0$ in the examples and exercises of this chapter.

Example 1. Find the Taylor series for $f(x) = e^{-x}$ expanded about $x = 1$. Also, find the radius and interval of convergence.

Solution: First work out the derivatives:

$$f'(x) = -e^{-x} \quad , \quad f''(x) = e^{-x} \quad , \quad f'''(x) = -e^{-x} \quad , \quad f^{(4)}(x) = e^{-x}$$

Evaluate the function and its derivatives at $x = a = 1$.

$$f(1) = e^{-1} = \frac{1}{e}, f'(1) = -e^{-1} = -\frac{1}{e}, f''(1) = e^{-1} = \frac{1}{e}$$

$$f'''(1) = -e^{-1} = -\frac{1}{e}, f^{(4)}(1) = e^{-1} = \frac{1}{e}$$

Plug these into the formula for a Taylor series expansion of $f(x)$ about $a = 1$.

$$f(x) = \sum_{k=0}^{\infty} \frac{f^{(k)}(1)}{k!}(x-1)^k$$

$$= f(1) + f'(1)(x-1) + \frac{f''(1)}{2}(x-1)^2 + \frac{f'''(1)}{6}(x-1)^3 + \frac{f^{(4)}(1)}{24}(x-1)^4 + \cdots$$

$$= \frac{1}{e} - \frac{1}{e}(x-1) + \frac{1}{2e}(x-1)^2 - \frac{1}{6e}(x-1)^3 + \frac{1}{24e}(x-1)^4 + \cdots$$

The nth term (starting with $n = 0$) is $(-1)^n \frac{(x-1)^n}{n!e}$. To find the radius of convergence, use the ratio test (Chapter 7).

$$L = \lim_{n\to\infty}\left|\frac{(x-1)^{n+1}}{(n+1)!\,e} \div \frac{(x-1)^n}{n!\,e}\right| = \lim_{n\to\infty}\left|\frac{(x-1)^{n+1}}{(n+1)n!\,e} \times \frac{n!\,e}{(x-1)^n}\right| = \lim_{n\to\infty}\left|\frac{x-1}{n+1}\right| = 0$$

Since this $L < 1$ for any value of x (since the denominator grows infinite), the radius of convergence is $R = \infty$ and the interval of convergence is $-\infty < x < \infty$.

Example 2. Find the Taylor series for $f(x) = \frac{1}{x}$ expanded about $x = 4$. Also, find the radius and interval of convergence.

Solution: First work out the derivatives. Note that $f(x) = \frac{1}{x} = x^{-1}$.

$$f'(x) = -x^{-2} = -\frac{1}{x^2} \quad , \quad f''(x) = 2x^{-3} = \frac{2}{x^3}$$

$$f'''(x) = -6x^{-4} = -\frac{6}{x^4} \quad , \quad f^{(4)}(x) = 24x^{-5} = \frac{24}{x^5}$$

Evaluate the function and its derivatives at $x = a = 4$.

$$f(4) = \frac{1}{4}, f'(4) = -\frac{1}{4^2} = -\frac{1}{16}, f''(4) = \frac{2}{4^3} = \frac{2}{64} = \frac{1}{32}$$

$$f'''(4) = -\frac{6}{4^4} = -\frac{6}{256} = -\frac{3}{128}, f^{(4)}(4) = \frac{24}{4^5} = \frac{24}{1024} = \frac{3}{128}$$

Plug these into the formula for a Taylor series expansion of $f(x)$ about $a = 4$.

$$f(x) = \sum_{k=0}^{\infty} \frac{f^{(k)}(4)}{k!}(x-4)^k$$

$$= f(4) + f'(4)(x-4) + \frac{f''(4)}{2}(x-4)^2 + \frac{f'''(4)}{6}(x-4)^3 + \frac{f^{(4)}(4)}{24}(x-4)^4 + \cdots$$

$$= \frac{1}{4} - \frac{1}{16}(x-4) + \frac{1}{32}\frac{1}{2}(x-4)^2 - \frac{3}{128}\frac{1}{6}(x-4)^3 + \frac{3}{128}\frac{1}{24}(x-4)^4 + \cdots$$

$$= \frac{1}{4} - \frac{1}{16}(x-4) + \frac{1}{64}(x-4)^2 - \frac{1}{256}(x-4)^3 + \frac{1}{1024}(x-4)^4 + \cdots$$

The nth term (starting with $n = 0$) is $(-1)^n \frac{(x-4)^n}{4^{n+1}}$. To find the radius of convergence, use the ratio test (Chapter 7).

$$L = \lim_{n\to\infty}\left|\frac{(x-4)^{n+1}}{4^{n+2}} \div \frac{(x-4)^n}{4^{n+1}}\right| = \lim_{n\to\infty}\left|\frac{(x-4)^{n+1}}{4^{n+2}} \times \frac{4^{n+1}}{(x-4)^n}\right| = \lim_{n\to\infty}\left|\frac{x-4}{4}\right| = \left|\frac{x-4}{4}\right|$$

This Taylor series converges if $\left|\frac{x-4}{4}\right| < 1$, which equates to $|x-4| < 4$, which can be written as $-4 < x - 4 < 4$, which simplifies to $0 < x < 8$. The radius of convergence is one-half of this interval: $R = 4$. We also need to check the endpoints (recall Chapter 12). When $x = 0$, the series is $\sum(-1)^n \frac{(-4)^n}{4^{n+1}} = \sum \frac{4^n}{4^{n+1}} = \sum \frac{1}{4}$, which diverges (since it has an infinite number of $\frac{1}{4}$'s added together). We used the rule $p^n q^n = (pq)^n$ to get $(-1)^n(-4)^n = (4)^n = 4^n$. When $x = 8$, the series is $\sum(-1)^n \frac{4^n}{4^{n+1}} = \sum \frac{(-1)^n}{4}$, which diverges according to the simple divergence test (Chapter 4); it alternates between $\frac{1}{4}$ and $-\frac{1}{4}$ without every converging to any one value (recall Chapters 2 and 4). (Though if you want to argue that this sum is finite and either equals $\frac{1}{4}$ or $-\frac{1}{4}$, it would seem hard to argue this point, but it doesn't '**converge**' to either value.) Since the series diverges at both endpoints, the complete interval of convergence is $0 < x < 8$. (If it had converged at an endpoint, we would change a $<$ sign into a $\leq$ sign, like we did in Chapter 12.)

Chapter 13 Problems

Directions: Find the Taylor series for each function expanded about the indicated point. Write out the terms up to $n = 5$. Also, find the radius and interval of convergence.

(1) $f(x) = \sin x$ about $x = \frac{\pi}{2}$

(2) $f(x) = e^x$ about $x = 1$

(3) $f(x) = \ln x$ about $x = 1$

(4) $f(x) = \sqrt{x}$ about $x = 9$

Note: You don't need to find the interval of convergence for this problem.

Challenge: Find a formula for the nth term of the series and use it to find the radius of convergence.

14 Maclaurin Series

In applications of series calculus, it is so common for Taylor series to be expanded about $x = 0$ that this special case has its own name. A **Maclaurin series** is a Taylor series expanded about $x = 0$. To obtain the general formula for a Maclaurin series, set the constant $a = 0$.

$$f(x) = \sum_{k=0}^{\infty} \frac{f^{(k)}(0)}{k!} x^k$$

Of course, $f^{(k)}(0)$ means to first take the kth derivative of $f(x)$ with respect to x and then plug in $x = 0$. (It would be a big mistake to first evaluate the function at $x = 0$ and then take the derivative.)

One very commonly used Maclaurin series is the **binomial series**. The **binomial series** is the Maclaurin series for the function $f(x) = (1 + x)^p$, where the exponent p is nonzero. (The exponent p could be a fraction or a whole number, and it could be positive or negative.) To find the Maclaurin series for $f(x) = (1 + x)^p$, we need to work out the pattern for the derivatives.

- $f'(x) = p(1 + x)^{p-1}$
- $f''(x) = p(p - 1)(1 + x)^{p-2}$
- $f'''(x) = p(p - 1)(p - 2)(1 + x)^{p-3}$
- $f^{(4)}(x) = p(p - 1)(p - 2)(p - 3)(1 + x)^{p-4}$

Now evaluate the function and each derivative at $x = 0$. Note that the $(1 + x)$ part will simply become 1.

- $f'(0) = p(1 + 0)^{p-1} = p$
- $f''(0) = p(p - 1)(1 + 0)^{p-2} = p(p - 1)$
- $f'''(0) = p(p - 1)(p - 2)(1 + 0)^{p-3} = p(p - 1)(p - 2)$
- $f^{(4)}(0) = p(p - 1)(p - 2)(p - 3)(1 + 0)^{p-4} = p(p - 1)(p - 2)(p - 3)$

Use the formula for a general Maclaurin series from the top of the page.

$$f(x) = f(0) + f'(0)x + \frac{f''(0)}{2}x^2 + \frac{f'''(0)}{6}x^3 + \frac{f^{(4)}(0)}{24}x^4 + \cdots$$

$$f(x) = 1 + px + \frac{p(p-1)}{2}x^2 + \frac{p(p-1)(p-2)}{6}x^3 + \frac{p(p-1)(p-2)(p-3)}{24}x^4 + \cdots$$

The series for $f(x)$ is the **binomial series**. In the special case that p happens to be a nonnegative integer, the series is finite. For example, for $p = 3$, we get

$$f(x) = 1 + 3x + \frac{3(2)}{2}x^2 + \frac{3(2)(1)}{6}x^3 = 1 + 3x + 3x^2 + x^3$$

because every higher-order term includes $(p-3)$, which equals zero in this case. This should make sense. The original function is a simple polynomial when $p = 3$; in this case, we get $(1+x)^3 = x^3 + 3x^2 + 3x + 1$, which is identical to the expression above. In the more general case, where p may be a fraction or may even be negative, the binomial series is **infinite**; this is the case with most applications of the binomial series. A few common special cases of the binomial series are listed below.

$$\frac{1}{1+x} = (1+x)^{-1} = 1 - x + x^2 - x^3 + x^4 + \cdots \quad (-1 < x < 1)$$

$$\frac{1}{(1+x)^2} = (1+x)^{-2} = 1 - 2x + 3x^2 - 4x^3 + 5x^4 + \cdots \quad (-1 < x < 1)$$

$$\sqrt{1+x} = (1+x)^{1/2}$$

$$= 1 + \frac{x}{2} - \frac{x^2}{(2)(4)} + \frac{(1)(3)}{(2)(4)(6)}x^3 - \frac{(1)(3)(5)}{(2)(4)(6)(8)}x^4 + \cdots \quad (-1 < x \leq 1)$$

$$\frac{1}{\sqrt{1+x}} = (1+x)^{-1/2}$$

$$= 1 - \frac{x}{2} + \frac{(1)(3)}{(2)(4)}x^2 - \frac{(1)(3)(5)}{(2)(4)(6)}x^3 + \frac{(1)(3)(5)(7)}{(2)(4)(6)(8)}x^4 + \cdots \quad (-1 < x \leq 1)$$

To find a binomial series with $1 - x$ instead of $1 + x$, simply reverse the signs of the odd terms. For example, compare the series for $\frac{1}{1-x}$ below with the series for $\frac{1}{1+x}$ above.

$$\frac{1}{1-x} = (1-x)^{-1} = 1 + x + x^2 + x^3 + x^4 + \cdots \quad (-1 < x < 1)$$

One method for finding the Maclaurin series for a given function $f(x)$ is to first take successive derivatives of $f(x)$, evaluate the function and derivatives at $x = 0$, and then use the Maclaurin series formula, like we did for the binomial series. However, sometimes there are more efficient ways to find the Maclaurin series for a function. Following are a handful of examples:

- If $f(x)$ is the product of two functions, it may be simpler to **multiply the series** for each function. For example, if $f(x) = e^{-x} \sin x$, instead of using the product rule repeatedly to take successive derivatives of $e^{-x} \sin x$, an alternative is to multiply the Maclaurin series for e^{-x} by the Maclaurin series for $\sin x$. (This applies the series multiplication theorem from Chapter 12.) See Example 3.
- If $f(x)$ is the ratio of two functions, it may be simpler to **divide** the series for each function. It's basically the same idea as the previous bullet point, since division is equivalent to multiplying by a reciprocal. For example, if $f(x) = \cot x$, you could divide the Maclaurin series for $\cos x$ by the Maclaurin series for $\sin x$ because $\cot x = \frac{\cos x}{\sin x}$. See Example 4.
- Sometimes, you can find the Maclaurin series for a function simply by making a **substitution** in the Maclaurin series for a related function. For example, if $f(x) = e^{-x^2}$, instead of finding successive derivatives of e^{-x^2} using the chain rule, you could just let $u = x^2$ and replace u with x^2 in the Maclaurin series for $f(u) = e^{-u}$, as shown in Example 5.
- Occasionally, it may be easier to find the Maclaurin series for the **derivative** or the **integral** of $f(x)$ than it would be to find the Maclaurin series for $f(x)$ itself. In that case, find what is easiest and then use calculus to obtain the desired series. (This applies the term-by-term strategies from Chapter 12 for finding derivatives or integrals of power series.) For example, consider $f(x) = \tan^{-1} x$. If you choose to differentiate $\tan^{-1} x$, the first derivative is $f'(x) = \frac{1}{1+x^2}$; for the second derivative, you will need to differentiate $\frac{1}{1+x^2}$, and so on. Alternatively, since $\int \frac{dx}{1+x^2} = \tan^{-1} x + c$, you could integrate each term of the Maclaurin series for $\frac{1}{1+x^2}$, as shown in Example 6.
- A **trig identity** may simplify the process of finding the Maclaurin series for a trig function. For example, for the function $f(x) = \cos^2 x$, you could apply the trig identity $\cos^2 x = \frac{1+\cos(2x)}{2}$. See Example 7.
- Note that some Maclaurin series are related by **complex numbers**. For example, by comparing the Maclaurin series for e^{ix}, $\cos x$, and $\sin x$, we get the relation $e^{ix} = \cos x + i \sin x$, where i is the imaginary number, as we'll explore shortly.

Following is a list of some common **Maclaurin** series:

$$e^x = \sum_{n=0}^{\infty} \frac{x^n}{n!} = 1 + x + \frac{x^2}{2} + \frac{x^3}{6} + \frac{x^4}{24} + \cdots \quad (-\infty < x < \infty)$$

$$e^{-x} = \sum_{n=0}^{\infty} \frac{(-1)^n x^n}{n!} = 1 - x + \frac{x^2}{2} - \frac{x^3}{6} + \frac{x^4}{24} + \cdots \quad (-\infty < x < \infty)$$

$$\sin x = \sum_{n=0}^{\infty} \frac{(-1)^n x^{2n+1}}{(2n+1)!} = x - \frac{x^3}{6} + \frac{x^5}{120} - \frac{x^7}{5040} + \cdots \quad (-\infty < x < \infty)$$

$$\cos x = \sum_{n=0}^{\infty} \frac{(-1)^n x^{2n}}{(2n)!} = 1 - \frac{x^2}{2} + \frac{x^4}{24} - \frac{x^6}{720} + \cdots \quad (-\infty < x < \infty)$$

$$\tan x = x + \frac{x^3}{3} + \frac{2x^5}{15} + \frac{17x^7}{315} + \cdots \quad \left(-\frac{\pi}{2} < x < \frac{\pi}{2}\right)$$

$$\sinh x = \frac{e^x - e^{-x}}{2} = \sum_{n=0}^{\infty} \frac{x^{2n+1}}{(2n+1)!} = x + \frac{x^3}{6} + \frac{x^5}{120} + \frac{x^7}{5040} + \cdots \quad (-\infty < x < \infty)$$

$$\cosh x = \frac{e^x + e^{-x}}{2} = \sum_{n=0}^{\infty} \frac{x^{2n}}{(2n)!} = 1 + \frac{x^2}{2} + \frac{x^4}{24} + \frac{x^6}{720} + \cdots \quad (-\infty < x < \infty)$$

$$\tanh x = \frac{e^x - e^{-x}}{e^x + e^{-x}} = x - \frac{x^3}{3} + \frac{2x^5}{15} - \frac{17x^7}{315} + \cdots \quad \left(-\frac{\pi}{2} < x < \frac{\pi}{2}\right)$$

$$\ln(1 + x) = \sum_{n=0}^{\infty} \frac{(-1)^n x^{n+1}}{n+1} = x - \frac{x^2}{2} + \frac{x^3}{3} - \frac{x^4}{4} + \cdots \quad (-1 < x \leq 1)^1$$

$$\ln(1 - x) = -\sum_{n=0}^{\infty} \frac{x^{n+1}}{n+1} = -x - \frac{x^2}{2} - \frac{x^3}{3} - \frac{x^4}{4} + \cdots \quad (-1 \leq x < 1)^1$$

$$\sin^{-1} x = x + \frac{1}{2}\frac{x^3}{3} + \frac{(1)(3)}{(2)(4)}\frac{x^5}{5} + \frac{(1)(3)(5)}{(2)(4)(6)}\frac{x^7}{7} + \cdots \quad (-1 < x < 1)$$

$$\cos^{-1} x = \frac{\pi}{2} - x - \frac{1}{2}\frac{x^3}{3} - \frac{(1)(3)}{(2)(4)}\frac{x^5}{5} - \frac{(1)(3)(5)}{(2)(4)(6)}\frac{x^7}{7} + \cdots \quad (-1 < x < 1)$$

[1] If you compare with other textbooks, some texts will have n instead of $n + 1$, will have $n - 1$ instead of n, or will have $2n - 1$ instead of $2n + 1$, for example. How can different texts have different formulas? Simple. Some texts start the index at $n = 1$, which means they have shifted the index as we learned how to do in Chapter 4. We chose to start all of the series on this page at $n = 0$ for consistency. However, in the examples and solutions, we may start n at other values to simplify the root or ratio test.

$$\tan^{-1} x = \sum_{n=0}^{\infty} \frac{(-1)^n x^{2n+1}}{2n+1} = x - \frac{x^3}{3} + \frac{x^5}{5} - \frac{x^7}{7} + \cdots \qquad (-1 \le x \le 1)^1$$

$$\tan^{-1} x = \frac{\pi}{2} - \frac{1}{x} + \frac{1}{3x^3} - \frac{1}{5x^5} + \frac{1}{7x^7} + \cdots \qquad (x > 1)$$

$$\tan^{-1} x = -\frac{\pi}{2} - \frac{1}{x} + \frac{1}{3x^3} - \frac{1}{5x^5} + \frac{1}{7x^7} + \cdots \qquad (x < -1)$$

Also, recall that we listed some common **binomial** series earlier in this chapter.

Note: Wondering why we listed series for $\ln(1 \pm x)$, but not $\ln x$? It's because $\ln x$ and its derivatives are undefined at $x = 0$. Although $\ln x$ doesn't have a Maclaurin series, you can get a Taylor series for it about a nonzero value, such as $a = 1$.

Now we'll see how **complex numbers** relate to Maclaurin series. Examine the Maclaurin series for e^x, $\cos x$, and $\sin x$. Observe that the series for e^x includes all of the terms from $\cos x$ and $\sin x$, except that the signs aren't right. Well, it turns out that there is a way to combine the series for $\cos x$ and $\sin x$ to make the series for e^x where the signs work out correctly. It has to do with the imaginary number i, where $i^2 = -1$. Observe that $i^3 = i^2 i = -i$, $i^4 = i^2 i^2 = 1$, $i^5 = i^4 i = i$, $i^6 = i^4 i^2 = -1$, etc. They form a pattern with every fourth power repeating:[2]

$$i^5 = i^9 = i^{13} = \cdots = i \quad , \quad i^2 = i^6 = i^{10} = \cdots = -1$$
$$i^3 = i^7 = i^{11} = \cdots = -i \quad , \quad i^4 = i^8 = i^{12} = \cdots = 1$$

If we replace x with ix in the Maclaurin series for e^x, we get a series for e^{ix}:

$$e^{ix} = 1 + ix + \frac{(ix)^2}{2} + \frac{(ix)^3}{6} + \frac{(ix)^4}{24} + \frac{(ix)^5}{120} + \frac{(ix)^6}{720} + \frac{(ix)^7}{5040} + \cdots$$

$$e^{ix} = 1 + ix - \frac{x^2}{2} - i\frac{x^3}{6} + \frac{x^4}{24} + i\frac{x^5}{120} - \frac{x^6}{720} + i\frac{x^7}{5040} + \cdots$$

Comparison for the above series with the Maclaurin series for $\cos x$ and $\sin x$ leads to Euler's formula below:

$$e^{ix} = \cos x + i \sin x$$

[2] Every fourth derivative of sine (or cosine) also repeats. Observe that $\frac{d}{dx}\sin x = \cos x$, $\frac{d^2}{dx^2}\sin x = -\sin x$, $\frac{d^3}{dx^3}\sin x = -\cos x$, $\frac{d^4}{dx^4}\sin x = \sin x$, $\frac{d^5}{dx^5}\sin x = \cos x$, $\frac{d^6}{dx^6}\sin x = -\sin x$, $\frac{d^7}{dx^7}\sin x = -\cos x$, etc.

If you compare the Maclaurin series for **hyperbolic** cosine and sine with the series for ordinary cosine and sine, you will observe that the hyperbolic functions and ordinary trig functions are related by the imaginary number as follows.[3]

$$\cos(ix) = \cosh x \quad , \quad \cosh(ix) = \cos x \quad , \quad \sin(ix) = i \sinh x \quad , \quad \sinh(ix) = i \sin x$$

Example 1. Find the Maclaurin series for $f(x) = e^{-x}$. Also, find the radius and interval of convergence.

Solution: First work out the derivatives:

$$f'(x) = -e^{-x} \quad , \quad f''(x) = e^{-x} \quad , \quad f'''(x) = -e^{-x} \quad , \quad f^{(4)}(x) = e^{-x}$$

Evaluate the function and its derivatives at $x = 0$.

$$f(0) = e^{-0} = e^0 = 1 \, , f'(0) = -e^{-0} = -e^0 = -1 \, , f''(0) = e^{-0} = e^0 = 1$$

$$f'''(0) = -e^{-0} = -e^0 = -1 \, , f^{(4)}(0) = e^{-0} = e^0 = 1$$

Plug these into the formula for a Maclaurin series expansion of $f(x)$.

$$f(x) = \sum_{k=0}^{\infty} \frac{f^{(k)}(0)}{k!} x^k$$

$$= f(0) + f'(0)x + \frac{f''(0)}{2} x^2 + \frac{f'''(0)}{6} x^3 + \frac{f^{(4)}(0)}{24} x^4 + \cdots$$

$$= 1 - x + \frac{x^2}{2} - \frac{x^3}{6} + \frac{x^4}{24} + \cdots$$

The nth term (starting with $n = 0$) is $(-1)^n \frac{x^n}{n!}$. To find the radius of convergence, use the ratio test (Chapter 7).

$$L = \lim_{n \to \infty} \left| \frac{x^{n+1}}{(n+1)!} \div \frac{x^n}{n!} \right| = \lim_{n \to \infty} \left| \frac{x^{n+1}}{(n+1)!} \times \frac{n!}{x^n} \right| = \lim_{n \to \infty} \left| \frac{x}{n+1} \right| = 0$$

Since this $L < 1$ for any value of x (since the denominator, $n + 1$, grows infinite), the radius of convergence is $R = \infty$ and the interval of convergence is $-\infty < x < \infty$.

[3] If you're curious about complex trigonometry, see the author's other book, Complex Numbers Essentials Math Workbook with Answers, Chapters 10-17 (especially, Chapter 12).

Example 2. Find the Maclaurin series for $f(x) = \sin x$. Also, find the radius and interval of convergence.

Solution: First work out the derivatives:

$f'(x) = \cos x$, $f''(x) = -\sin x$, $f'''(x) = -\cos x$, $f^{(4)}(x) = \sin x$, $f^{(5)}(x) = \cos x$

Evaluate the function and its derivatives at $x = 0$.

$$f(0) = \sin 0 = 0\,, f'(0) = \cos 0 = 1\,, f''(0) = -\sin 0 = 0$$

$$f'''(0) = -\cos 0 = -1\,, f^{(4)}(0) = \sin 0 = 0\,, f^{(5)}(0) = \cos 0 = 1$$

Plug these into the formula for a Maclaurin series expansion of $f(x)$.

$$f(x) = f(0) + f'(0)x + \frac{f''(0)}{2}x^2 + \frac{f'''(0)}{6}x^3 + \frac{f^{(4)}(0)}{24}x^4 + \frac{f^{(5)}(0)}{120}x^5 + \cdots$$

$$= 0 + x + 0 - \frac{x^3}{6} + 0 + \frac{x^5}{120} + \cdots = x - \frac{x^3}{6} + \frac{x^5}{120} + \cdots$$

The nth term (starting with $n = 0$)[4] is $\frac{(-1)^n x^{2n+1}}{(2n+1)!}$. To find the radius of convergence, use the ratio test (Chapter 7).

$$L = \lim_{n\to\infty}\left|\frac{x^{2(n+1)+1}}{[2(n+1)+1]!} \div \frac{x^{2n+1}}{(2n+1)!}\right| = \lim_{n\to\infty}\left|\frac{x^{2n+3}}{(2n+3)!} \times \frac{(2n+1)!}{x^{2n+1}}\right|$$

$$L = \lim_{n\to\infty}\left|\frac{x^{3-1}(2n+1)!}{(2n+3)(2n+2)(2n+1)!}\right| = \lim_{n\to\infty}\left|\frac{x^2}{(2n+3)(2n+2)}\right| = 0$$

Since this $L < 1$ for any value of x (since the denominator grows infinite), the radius of convergence is $R = \infty$ and the interval of convergence is $-\infty < x < \infty$.

Example 3. Find the Maclaurin series for $f(x) = e^{-x}\sin x$. Also, find the radius and interval of convergence.

Solution: Instead of using the product rule repeatedly to work out the successive derivatives, we'll apply the series multiplication theorem (Chapter 12), which allows us to simply multiply the Maclaurin series for e^{-x} and $\sin x$. We'll show terms up to x^5 in our answer; terms x^6 or higher will be lumped into the $+\cdots$

$$f(x) = e^{-x}\sin x = \left(1 - x + \frac{x^2}{2} - \frac{x^3}{6} + \frac{x^4}{24} + \cdots\right)\left(x - \frac{x^3}{6} + \frac{x^5}{120} + \cdots\right)$$

[4] This formula in terms of n, which starts at 0, doesn't include the zeroes. We're looking at the final form of the series, $x - \frac{x^3}{6} + \frac{x^5}{120} + \cdots$, which doesn't explicitly include terms that are equal to zero.

$$f(x) = x - \frac{x^3}{6} + \frac{x^5}{120} - x^2 + \frac{x^4}{6} + \frac{x^3}{2} - \frac{x^5}{12} - \frac{x^4}{6} + \frac{x^5}{24} + \cdots$$

$$f(x) = x - x^2 + \left(-\frac{1}{6} + \frac{1}{2}\right)x^3 + \left(\frac{1}{6} - \frac{1}{6}\right)x^4 + \left(\frac{1}{120} - \frac{1}{12} + \frac{1}{24}\right)x^5 + \cdots$$

$$f(x) = x - x^2 + \frac{x^3}{3} - \frac{x^5}{30} + \cdots$$

Since e^{-x} and $\sin x$ each have infinite radii of convergence, the radius of convergence for $e^{-x}\sin x$ is also $R = \infty$ and the interval of convergence is $-\infty < x < \infty$.

Example 4. Find the Maclaurin series for $f(x) = \cot x$.

Solution: We will divide two series. We're applying the series multiplication theorem from Chapter 12 with the idea that division equates to multiplying by a reciprocal. (It may help to review the solution to Exercise 17 in Chapter 12.) Since $\cot x = \frac{\cos x}{\sin x}$, we will divide the Maclaurin series for $\cos x$ by the Maclaurin series for $\sin x$.

$$\cot x = \frac{\cos x}{\sin x} = \frac{1 - \frac{x^2}{2} + \frac{x^4}{24} - \frac{x^6}{720} + \cdots}{x - \frac{x^3}{6} + \frac{x^5}{120} - \frac{x^7}{5040} + \cdots}$$

You could do long division with polynomials, if you know how. We'll work through an alternative method here. First, note that when the two series are divided, this will result in a series that begins with $\frac{1}{x}$. (Why? The least power of the numerator is 1 and the least power of the denominator is x.) Write the general form of the series that results from the division as shown below. Then multiply both sides by the denominator.

$$\frac{1 - \frac{x^2}{2} + \frac{x^4}{24} - \frac{x^6}{720} + \cdots}{x - \frac{x^3}{6} + \frac{x^5}{120} - \frac{x^7}{5040} + \cdots} = \frac{1}{x} + c_0 + c_1 x + c_2 x^2 + c_3 x^3 + c_4 x^4 + c_5 x^5 + \cdots$$

Multiply both sides by the denominator, which is $x - \frac{x^3}{6} + \frac{x^5}{120} - \frac{x^7}{5040} + \cdots$.

$$1 - \frac{x^2}{2} + \frac{x^4}{24} - \frac{x^6}{720} + \cdots = 1 - \frac{x^2}{6} + \frac{x^4}{120} - \frac{x^6}{5040} + c_0 x - \frac{c_0}{6}x^3 + \frac{c_0}{120}x^5$$
$$+ c_1 x^2 - \frac{c_1}{6}x^4 + \frac{c_1}{120}x^6 + c_2 x^3 - \frac{c_2}{6}x^5 + c_3 x^4 - \frac{c_3}{6}x^6 + c_4 x^5 + c_5 x^6 + \cdots$$

Factor out the coefficients on the right-hand side.

$$1-\frac{x^2}{2}+\frac{x^4}{24}-\frac{x^6}{720}+\cdots=1+c_0x+\left(-\frac{1}{6}+c_1\right)x^2+\left(-\frac{c_0}{6}+c_2\right)x^3$$
$$+\left(\frac{1}{120}-\frac{c_1}{6}+c_3\right)x^4+\left(\frac{c_0}{120}-\frac{c_2}{6}+c_4\right)x^5+\left(-\frac{1}{5040}+\frac{c_1}{120}-\frac{c_3}{6}+c_5\right)x^6+\cdots$$

Finally, equate coefficients on both sides.

$$c_0=0 \quad , \quad -\frac{1}{6}+c_1=-\frac{1}{2} \quad , \quad -\frac{c_0}{6}+c_2=0$$
$$\frac{1}{120}-\frac{c_1}{6}+c_3=\frac{1}{24} \quad , \quad \frac{c_0}{120}-\frac{c_2}{6}+c_4=0 \quad , \quad -\frac{1}{5040}+\frac{c_1}{120}-\frac{c_3}{6}+c_5=-\frac{1}{720}$$

All of the even coefficients must be zero: $c_0=c_2=c_4=\cdots=0$.

$$c_1=-\frac{1}{2}+\frac{1}{6}=-\frac{3}{6}+\frac{1}{6}=-\frac{2}{6}=-\frac{1}{3}$$
$$c_3=\frac{1}{24}-\frac{1}{120}+\frac{c_1}{6}=\frac{5}{120}-\frac{1}{120}-\frac{1}{3}\frac{1}{6}=\frac{4}{120}-\frac{1}{18}=\frac{1}{30}-\frac{1}{18}=\frac{3}{90}-\frac{5}{90}=-\frac{2}{90}=-\frac{1}{45}$$
$$c_5=-\frac{1}{720}+\frac{1}{5040}-\frac{c_1}{120}+\frac{c_3}{6}=-\frac{7}{5040}+\frac{1}{5040}-\left(-\frac{1}{3}\right)\frac{1}{120}-\frac{1}{45}\frac{1}{6}$$
$$c_5=-\frac{6}{5040}+\frac{1}{360}-\frac{1}{270}=-\frac{1}{840}+\frac{1}{360}-\frac{1}{270}=-\frac{9}{7560}+\frac{21}{7560}-\frac{28}{7560}$$
$$c_5=-\frac{16}{7560}=-\frac{2}{945}$$

Plug these coefficients into the series from the beginning.

$$\cot x=\frac{1}{x}-\frac{x}{3}-\frac{4}{45}x^3-\frac{2}{945}x^5+\cdots$$

(The radius and interval of convergence are more complicated with long division. If you visualize the periodic asymptotes of the cotangent function, you might not be surprised to learn that the interval of convergence is $0<|x|<\pi$; the cotangent function is undefined at $x=0$ or π.)

Example 5. Find the Maclaurin series for $f(x) = e^{-x^2}$. Also, find the radius and interval of convergence.

Solution: Instead of applying the chain rule repeatedly to find successive derivatives of e^{-x^2}, we will use the Maclaurin series for e^{-u} and then let $u = x^2$. From the list of common Maclaurin series, the series for e^{-u} is

$$e^{-u} = 1 - u + \frac{u^2}{2} - \frac{u^3}{6} + \frac{u^4}{24} + \cdots$$

Now replace u with x^2.

$$e^{-x^2} = 1 - x^2 + \frac{x^4}{2} - \frac{x^6}{6} + \frac{x^8}{24} + \cdots$$

The nth term (starting with $n = 0$) is $\frac{(-1)^n x^{2n}}{n!}$. To find the radius of convergence, use the ratio test (Chapter 7).

$$L = \lim_{n\to\infty} \left| \frac{x^{2(n+1)}}{(n+1)!} \div \frac{x^{2n}}{n!} \right| = \lim_{n\to\infty} \left| \frac{x^{2n+2}}{(n+1)!} \times \frac{n!}{x^{2n}} \right| = \lim_{n\to\infty} \left| \frac{x^2}{n+1} \right| = 0$$

Since this $L < 1$ for any value of x (since the denominator, $n + 1$, grows infinite), the radius of convergence is $R = \infty$ and the interval of convergence is $-\infty < x < \infty$.

Example 6. Find the Maclaurin series for $f(x) = \tan^{-1} x$. Also, find the radius and interval of convergence.

Solution: Instead of applying the chain rule repeatedly to find successive derivatives of $\tan^{-1} x$, we will use the fact that $\tan^{-1} x$ (plus a constant) is the anti-derivative of $\frac{1}{1+x^2}$. In particular, the following definite integral is equal to $\tan^{-1} x$ (this time, without a constant, since it's a definite integral).[5]

$$\tan^{-1} x = \int_{w=0}^{x} \frac{dw}{1 + w^2}$$

Note that the integrand is similar to the binomial series. (If needed, flip back to the beginning of the chapter to review the binomial series.)

[5] To perform this integral, let $w = \tan u$ such that $dw = \sec^2 u \, du$. The denominator will become $1 + w^2 = 1 + \tan^2 u = \sec^2 u$. (This trig identity is often helpful in calculus.) With this substitution, the integral becomes $\int du = u = \tan^{-1} w$ (the $\sec^2 u$'s cancel out). After evaluating it over the limits, the definite integral equals $\tan^{-1} x$.

$$\frac{1}{1+t} = (1+t)^{-1} = 1 - t + t^2 - t^3 + t^4 + \cdots$$

If we let $t = w^2$, we get

$$\frac{1}{1+w^2} = (1+w^2)^{-1} = 1 - w^2 + w^4 - w^6 + w^8 + \cdots$$

Now we just need to integrate this series to obtain the inverse tangent.

$$\tan^{-1} x = \int_{w=0}^{x} (1 - w^2 + w^4 - w^6 + w^8 + \cdots)\, dw$$

(Although this is a definite integral, it turns out that every term gives zero when evaluated at the lower limit.)

$$\tan^{-1} x = x - \frac{x^3}{3} + \frac{x^5}{5} - \frac{x^7}{7} + \frac{x^9}{9} + \cdots$$

The nth term (starting with $n = 0$) is $\frac{(-1)^n x^{2n+1}}{2n+1}$. To find the radius of convergence, use the ratio test (Chapter 7).

$$L = \lim_{n\to\infty} \left| \frac{x^{2(n+1)+1}}{2(n+1)+1} \div \frac{x^{2n+1}}{2n+1} \right| = \lim_{n\to\infty} \left| \frac{x^{2n+3}}{2n+3} \times \frac{2n+1}{x^{2n+1}} \right| = \lim_{n\to\infty} \left| x^2 \frac{2n+1}{2n+3} \right|$$

$$L = \lim_{n\to\infty} \left| x^2 \frac{2+\frac{1}{n}}{2+\frac{3}{n}} \right| = \left| x^2 \frac{2+0}{2+0} \right| = x^2 \left| \frac{2}{2} \right| = x^2|1| = x^2$$

Here, $L < 1$ if $x^2 < 1$, which means that $-1 < x < 1$. The radius of convergence is one-half of this interval: $R = 1$. Before stating the full interval of convergence, we need to test the endpoints (like we did in Chapters 12-13). When $x = -1$, the series is $\sum \frac{(-1)^n(-1)^{2n+1}}{2n+1} = -\sum \frac{(-1)^n}{2n+1}$ because $(-1)^{2n+1} = -1$ since $2n+1$ is always an odd number. This series converges conditionally because it is basically a version of the alternating harmonic series (Chapter 11). When $x = 1$, the series is similarly $\sum \frac{(-1)^n}{2n+1}$, which again converges conditionally. Therefore, the **complete interval of convergence** is $-1 \leq x \leq 1$ (with $\leq$ signs instead of $<$ signs).

Example 7. Find the Maclaurin series for $f(x) = \cos^2 x$.

Solution: There are a few different ways to solve this problem. One way is to take successive derivatives of $\cos^2 x$. Another way is to square the Maclaurin series for $\cos x$ using the series multiplication theorem (like Example 3). A third method is to use the trig identity $\cos^2 x = \frac{1+\cos(2x)}{2}$ (which is useful in calculus). Since we need $\cos(2x)$ in this formula, we'll begin by writing the Maclaurin series for $\cos y$, where $y = 2x$.

$$\cos y = 1 - \frac{y^2}{2} + \frac{y^4}{24} - \frac{y^6}{720} + \cdots$$

Now replace y with $2x$.

$$\cos(2x) = 1 - \frac{(2x)^2}{2} + \frac{(2x)^4}{24} - \frac{(2x)^6}{720} + \cdots = 1 - \frac{4x^2}{2} + \frac{16x^4}{24} - \frac{64x^6}{720} + \cdots$$

$$= 1 - 2x^2 + \frac{2x^4}{3} - \frac{4x^6}{45} + \cdots$$

Use the trig identity mentioned previously.

$$\cos^2 x = \frac{1 + \cos(2x)}{2} = \frac{1}{2} + \frac{1}{2}\cos(2x) = \frac{1}{2} + \frac{1}{2} - x^2 + \frac{x^4}{3} - \frac{2x^6}{45} + \cdots$$

$$\cos^2 x = 1 - x^2 + \frac{x^4}{3} - \frac{2x^6}{45} + \cdots$$

Check the answer: Plug in $x = 0.1$ rad and compare with $\cos^2(0.1) \approx 0.99003329$. Either put your calculator in radians mode or convert 0.1 rad to $\frac{18}{\pi} \approx 5.729577951°$. Then take the cosine and then square the result. (However, in the Maclaurin series, you must use radians, not degrees.)

$$\cos^2(0.1) = 1 - 0.1^2 + \frac{0.1^4}{3} - \frac{2(0.1)^6}{45} + \cdots$$

$$\approx 1 - 0.01 + 0.00003333 - 0.00000004 = 0.99003329$$

Chapter 14 Problems

Directions: Take successive derivatives to find the Maclaurin series for each function. Write out the terms up to x^5. Also, find the radius and interval of convergence.

(1) $f(x) = e^x$

(2) $f(x) = \cos x$

(3) $f(x) = \sinh x$

(4) $f(x) = \sqrt{x+4}$ Note: Don't worry about checking the endpoints for this series.

Directions: Use any method to find the Maclaurin series for each function. Write out the terms up to x^5 (but go above x^5 if there would otherwise be fewer than 3 terms).

(5) $f(x) = x^2 e^x$

(6) $f(x) = 2 \sin x^2$

(7) $f(x) = \ln(1 + x)$

(8) $f(x) = \ln(1 - x)$

(9) $f(x) = \sin^2\left(\frac{x}{2}\right)$

(10) $f(x) = e^x \cos x$

(11) $f(x) = \frac{\ln(1+x)}{\sqrt{1+x^2}}$

(12) $f(x) = \tan x$

(13) $f(x) = \frac{e^x}{\cos x}$

(14) $f(x) = \frac{1}{8x^3+1}$

(15) $f(x) = \sqrt{12x^2 + 3}$

(16) $f(x) = \sin^{-1} x$

(17)

$$f(x) = \int_{w=0}^{x} e^{-w^2} \, dw$$

(18)

$$f(x) = \frac{d}{dx} \frac{1}{\sqrt{1+x}}$$

15 Calculating Sums

If an infinite sum converges, how do we determine what the value of the sum equals? First, we'll recall a couple of infinite series where we can calculate an exact value for the sum, and then we'll discuss how to approximate a sum. Since there isn't a formula for finding the exact sum in general, we often need to use an approximation.

Recall from Chapter 4 that we already know how to find the sum for two kinds of infinite series:

- For an infinite **geometric series** with $r < 1$ starting at $n = 1$, the sum is

$$\sum_{n=1}^{\infty} ar^{n-1} = \frac{a}{1-r}$$

- For some infinite series, the method of **partial fractions** can be used to make pairs of terms that cancel in order to find the sum. In the example below, the series on the right is $\frac{1}{1} - \frac{1}{2} + \frac{1}{2} - \frac{1}{3} + \frac{1}{3} - \frac{1}{4} + \frac{1}{4} - \frac{1}{5} + \cdots$ Every term cancels out except for the first and last terms (but the 'last' term approaches zero).

$$\sum_{n=1}^{\infty} \frac{1}{n(n+1)} = \sum_{n=1}^{\infty} \left(\frac{1}{n} - \frac{1}{n+1}\right) = 1$$

Another way to find the sum of a series is if you recognize that the series is some form of a **Maclaurin series** (Chapter 14) or a more general **Taylor series** (Chapter 13). For example, consider the **alternating harmonic series** below.

$$\sum_{n=1}^{\infty} \frac{(-1)^{n+1}}{n} = 1 - \frac{1}{2} + \frac{1}{3} - \frac{1}{4} + \frac{1}{5} - \frac{1}{6} + \cdots$$

Looking at the common Maclaurin series in Chapter 14, which one could the above series be formed from? The answer is $\ln(1+x)$, which converges if $-1 < x \leq 1$.

$$\ln(1+x) = x - \frac{x^2}{2} + \frac{x^3}{3} - \frac{x^4}{4} + \frac{x^5}{5} - \frac{x^6}{6} + \cdots$$

For which value of x are the two series above the same? The answer is $x = 1$, which shows that the alternating harmonic series converges to $\ln(1+1) = \ln 2 \approx 0.693$. (Recall from Chapter 11 that the alternating harmonic series converges **conditionally**.)

The previous example illustrates what to do if you recognize a Maclaurin or Taylor series for a function that you know. What if a Maclaurin or Taylor series corresponds to a function that you don't know, or which doesn't have any known algebraic form. For example, consider the Maclaurin series below.

$$1 - \frac{1}{3} + \frac{1}{10} - \frac{1}{42} + \frac{1}{216} - \frac{1}{1320} + \cdots$$

The first step is to observe that the denominators are multiples of consecutive odd numbers: 10 is a multiple of 5, 42 is a multiple of 7, 216 is a multiple of 9, and 1320 is a multiple of 11. We can write the series as

$$\frac{1}{1(1)} - \frac{1}{3(1)} + \frac{1}{5(2)} - \frac{1}{7(6)} + \frac{1}{9(24)} - \frac{1}{11(120)} + \cdots$$

You should recognize the numbers in parentheses. These are factorials.

$$\frac{1}{1(0!)} - \frac{1}{3(1!)} + \frac{1}{5(2!)} - \frac{1}{7(3!)} + \frac{1}{9(4!)} - \frac{1}{11(5!)} + \cdots$$

We can think of this as a power series where x has been evaluated at one.

$$\frac{w}{1(0!)} - \frac{w^3}{3(1!)} + \frac{w^5}{5(2!)} - \frac{w^7}{7(3!)} + \frac{w^9}{9(4!)} - \frac{w^{11}}{11(5!)} + \cdots$$

These terms are anti-derivatives. That is, we can form the series above by integrating the series below.

$$\int \left(1 - w^2 + \frac{w^4}{2} - \frac{w^6}{6} + \frac{w^8}{24} - \frac{w^{10}}{120} + \cdots\right) dw = \frac{w}{1(0!)} - \frac{w^3}{3(1!)} + \frac{w^5}{5(2!)} - \frac{w^7}{7(3!)} + \frac{w^9}{9(4!)} - \frac{w^{11}}{11(5!)} + \cdots$$

If you let $u = w^2$, we can write the left series as

$$1 - u + \frac{u^2}{2} - \frac{u^3}{6} + \frac{u^4}{24} - \frac{u^5}{120} + \cdots$$

You should recognize this as the Maclaurin series for e^{-u}. Since $u = w^2$, this is the Maclaurin series for e^{-w^2}. We have to integrate the series for e^{-w^2}, which means that the original series was (recall the solution to Exercise 17 in Chapter 14)

$$f(1) = \int_{w=0}^{1} e^{-w^2}\, dw = 1 - \frac{1}{3} + \frac{1}{10} - \frac{1}{42} + \frac{1}{216} - \frac{1}{1320} + \cdots$$

Since e^{-w^2} doesn't have an algebraic anti-derivative, we can't simply evaluate the left-hand side in order to find the exact value that the series converges to. We can either approximate the integral numerically, or we can approximate the series numerically.

The way to approximate a Maclaurin series or a more general Taylor series is to calculate the sum of a finite number of terms and then estimate the remainder. If we approximate $f(x)$ with a finite number of terms, then

$$f(x) \approx f(a) + f'(a)(x-a) + \frac{f''(a)}{2!}(x-a)^2 + \frac{f'''(a)}{3!}(x-a)^3 + \cdots + \frac{f^{(n)}(a)}{n!}(x-a)^n$$

If $f(x)$ satisfies the conditions for **Taylor's theorem** (Chapter 13), then the difference between the actual sum and this approximation, which we call the **remainder**, is

$$R_n(x) = \frac{f^{n+1}(c)}{(n+1)!}(x-a)^{n+1}$$

where c lies between a and x. According to the **remainder estimation theorem**,

$$|R_n(x)| \leq q\frac{r^{n+1}|x-a|^{n+1}}{(n+1)!}$$

where q and r are positive constants such that $\left|f^{(n+1)}(w)\right| \leq qr^{n+1}$ for $a \leq w \leq x$. As an example, suppose that we wish to use the Maclaurin series for e^x to calculate the constant e to a large number of decimal places (using a computer program). We might want to know how many terms are needed in the series in order to get the desired number of decimal places. We begin with the Maclaurin series for e^x.

$$e^x = 1 + x + \frac{x^2}{2} + \frac{x^3}{6} + \frac{x^4}{24} + \frac{x^5}{120} + \frac{x^6}{720} + \cdots$$

In order for the series to converge to e, we need to evaluate it at $x = 1$.

$$e^1 = e = 1 + 1 + \frac{1}{2} + \frac{1}{6} + \frac{1}{24} + \frac{1}{120} + \frac{1}{720} + \cdots = 2 + \frac{1}{2} + \frac{1}{6} + \frac{1}{24} + \frac{1}{120} + \frac{1}{720} + \cdots$$

The derivative needed to find the remainder is $f^{(n+1)}(w) = \frac{d^{n+1}}{dw^{n+1}}e^w = e^w$ (since all derivatives of e^w equal e^w), where $0 \leq w \leq x$ (because $a = 0$ for a Maclaurin series). In this example, we're evaluating x at $x = 1$, so $0 \leq w \leq 1$. Since w can't exceed 1, the derivative $f^{(n+1)}(w) = e^w$ can't exceed e. If we choose $q = 3$ and $r = 1$, for example, we can be sure that $\left|f^{(n+1)}(w)\right| \leq qr^{n+1}$ since $qr^{n+1} = 3(1^{n+1}) = 3$ is greater than e. According to the remainder estimation theorem (where we will set x equal to 1),

$$|R_n(x)| \leq 3\frac{1^{n+1}|x-0|^{n+1}}{(n+1)!} = 3\frac{|x|^{n+1}}{(n+1)!} = \frac{3}{(n+1)!}$$

With a calculator, we can then find a value of n large enough to make $R_n(x)$ as small as desired. **The complete sum approximately equals the finite sum plus $R_n(x)$.**

What if the series doesn't originate from a Taylor series? There are a couple of other common ways to estimate remainders.

For a **nonincreasing series**, the **integral test** helps to establish a remainder. Recall from Chapter 6 that the basis for the integral test is the comparison between a series, $\sum a_n$, and an integral, $\int f(x)\,dx$, for a nonincreasing, positive function, $f(x)$, where $a_n = f(n)$. That same reasoning leads to the inequality below for the remainder, $R_n = a_{n+1} + a_{n+2} + a_{n+3} + \cdots$, which is the difference between the exact sum and the sum of a finite number of terms (up to a_n). See Example 3. **Add the remainder to the finite sum to approximate the complete sum**: $\sum_{k=1}^{\infty} a_k \approx R_n + \sum_{k=1}^{n} a_k$.

$$\int_{x=n+1}^{\infty} f(x)\,dx \le R_n \le \int_{x=n}^{\infty} f(x)\,dx$$

What if a series is nonincreasing for $n \ge N$, but not prior to that? You can calculate the finite sum up to and including N, use the integral to approximate the sum from $N + 1$ to ∞, and then add the finite sum and integral together.

Note that if a series isn't nonincreasing, the integral may still be a good approximation to the sum, but you won't be able to use the inequality above to find the remainder. Another way to estimate a sum is to apply the ratio test as follows.

If every term of the series has the same sign, the **ratio test** (Chapter 7) can help to determine the remainder. The ratio test states that if the ratio $\frac{a_{n+1}}{a_n}$ approaches a limit less than one, then the series converges. In this case, you may be able to find constants r_1 and r_2 such that $r_1 \le \frac{a_{n+1}}{a_n} \le r_2 < 1$ for $n \ge N$. This means that $r_1 a_n \le a_{n+1} \le r_2 a_n$. It similarly follows that $r_1 a_{n+1} \le a_{n+2} \le r_2 a_{n+1}$, that $r_1 a_{n+2} \le a_{n+3} \le r_2 a_{n+2}$, etc. The inequality $r_1 a_n \le a_{n+1} \le r_2 a_n$ leads to $r_1^2 a_n \le r_1 a_{n+1}$ and $r_2 a_{n+1} \le r_2^2 a_n$; plug these into $r_1 a_{n+1} \le a_{n+2} \le r_2 a_{n+1}$ to see that $r_1^2 a_n \le a_{n+2} \le r_2^2 a_n$. You can similarly show that $r_1^3 a_n \le a_{n+3} \le r_2^3 a_n$, that $r_1^4 a_n \le a_{n+4} \le r_2^4 a_n$, and so on. If we add these inequalities together, we get

$$r_1 a_n + r_1^2 a_n + r_1^3 a_n + \cdots \le a_{n+1} + a_{n+2} + a_{n+3} + \cdots \le r_2 a_n + r_2^2 a_n + r_2^3 a_n + \cdots$$
$$(r_1 + r_1^2 + r_1^3 + \cdots)a_n \le a_{n+1} + a_{n+2} + a_{n+3} + \cdots \le (r_2 + r_2^2 + r_2^3 + \cdots)a_n$$

The first and last series are geometric series with sums of $\frac{r_1}{1-r_1}$ and $\frac{r_2}{1-r_2}$ (Chapter 4).

$$\frac{r_1 a_n}{1-r_1} \leq \sum_{k=n+1}^{\infty} a_k \leq \frac{r_2 a_n}{1-r_2}$$

If the finite sum up to and including N is used to approximate the series, then the middle sum above represents the remainder:

$$R_n = \sum_{k=n+1}^{\infty} a_k$$

Again, **add the remainder to the finite sum to approximate the complete sum**: $\sum_{k=1}^{\infty} a_k \approx R_n + \sum_{k=1}^{n} a_k$. See Example 4.

The sum of an **alternating series** (Chapter 11) may be estimated using the following theorem. According to the **alternating series estimation theorem**, if an alternating series satisfies all of the conditions of Leibniz's theorem (Chapter 11), then if the finite sum below on the right is used to approximate the infinite sum below on the left, the error is less than the next term, $|a_{N+1}|$, and the remainder has the same sign as $(-1)^{n+1}|a_{N+1}|$. Note: This theorem can easily be adapted if a series has $(-1)^n$ instead of $(-1)^{n+1}$.

$$\sum_{n=1}^{\infty}(-1)^{n+1}|a_n| \quad , \quad \sum_{n=1}^{N}(-1)^{n+1}|a_n|$$

Finally, note that for most infinite series that converge, it is generally challenging to find the exact value for the sum. As an example, consider the sum below.

$$\sum_{n=1}^{\infty}\frac{1}{n^2}$$

We know that this sum converges, but it isn't easy to determine that it converges to the **exact value** of $\frac{\pi^2}{6} \approx 1.645$. If you add up the first 10 terms on your calculator, you will see that this finite sum equals 1.55, which is still 6% shy of the exact value. In the 1700's, Euler showed that the sum converges to $\frac{\pi^2}{6}$ using the following method. We begin with the Maclaurin series (Chapter 14) expansion for the sine function.

$$\sin x = x - \frac{x^3}{6} + \frac{x^5}{120} - \frac{x^7}{5040} + \cdots$$

Divide by x on both sides.

$$\frac{\sin x}{x} = 1 - \frac{x^2}{6} + \frac{x^4}{120} - \frac{x^6}{5040} + \cdots$$

For which values of x would the left-hand side of the equation be zero? The ratio $\frac{\sin x}{x}$ equals zero whenever $\sin x = 0$ and $x \neq 0$, which occurs when $x = \pm\pi, \pm 2\pi, \pm 3\pi$, etc. (Note that when x is zero, the ratio is 1 according to l'Hôpital's rule, so the ratio isn't zero when x equals zero.) Just as you can multiply the roots of a polynomial together to form a finite polynomial, like how $(x-1)(x+2)(x-4) = 0$ is equivalent to the equation $x^3 - 3x^2 - 6x + 8 = 0$, if we multiply the roots of $\frac{\sin x}{x}$ together, this should equal the right-hand side of the above equation.

$$\left(1 - \frac{x}{\pi}\right)\left(1 + \frac{x}{\pi}\right)\left(1 - \frac{x}{2\pi}\right)\left(1 + \frac{x}{2\pi}\right)\left(1 - \frac{x}{3\pi}\right)\left(1 + \frac{x}{3\pi}\right)\cdots = 1 - \frac{x^2}{6} + \frac{x^4}{120} + \cdots$$

When you multiply the infinite product on the left-hand side out, the first term is one, which matches the one on the right-hand side. Let's look at just the terms that are proportional to x^2 on the left-hand side, which must add up to the $-\frac{x^2}{6}$ from the right-hand side. In order to get a term proportional to x^2 from the infinite product, we need to have an infinite number of 1's multiplying two terms with $\frac{x}{n\pi}$'s. If the two terms that aren't ones have the same value of n, we get $-\frac{x^2}{n^2\pi^2}$. When the two terms that aren't ones don't have the same value of n, they cancel out. For example, look at the case involving $\frac{x}{2\pi}$ and $\frac{x}{3\pi}$. We'll get $-\frac{x}{2\pi}\left(-\frac{x}{3\pi}\right) - \frac{x}{2\pi}\left(\frac{x}{3\pi}\right) + \frac{x}{2\pi}\left(-\frac{x}{2\pi}\right) + \frac{x}{2\pi}\frac{x}{3\pi} = 0$. The non-zero terms proportional to x^2 on the left-hand side are:

$$-\frac{x^2}{\pi^2} - \frac{x^2}{4\pi^2} - \frac{x^2}{9\pi^2} - \frac{x^2}{16\pi^2} - \frac{x^2}{25\pi^2} + \cdots = -\frac{x^2}{6}$$

(As previously mentioned, these terms add up to the $-\frac{x^2}{6}$ from the right-hand side.) Multiply by π^2 and divide by $-x^2$ on both sides:

$$\frac{1}{1} + \frac{1}{4} + \frac{1}{9} + \frac{1}{16} + \frac{1}{25} + \cdots = \frac{\pi^2}{6}$$

Example 1. Find the exact sum for the series below.

$$1 - \frac{1}{2} + \frac{1}{24} - \frac{1}{720} + \frac{1}{40{,}320} - \frac{1}{3{,}628{,}800} + \cdots$$

Solution: Note that the denominators are factorials of even numbers: $2! = 2$, $4! = 24$, $6! = 720$, $8! = 40{,}320$, and $10! = 3{,}628{,}800$. Where have we seen a series where the denominators are factorials of even numbers? Recall the Maclaurin series (Chapter 14) for the cosine function:

$$\cos x = 1 - \frac{x^2}{2!} + \frac{x^4}{4!} - \frac{x^6}{6!} + \frac{x^8}{8!} - \frac{x^{10}}{10!} + \cdots$$

If you set $x = 1$ in the Maclaurin series for $\cos x$, you get the given series. Therefore, the given series converges to $\cos 1$, where the angle is in radians. Note that 1 radian is equal to $\frac{180}{\pi}$ degrees. Either put your calculator in radians mode to see that $\cos 1$ is approximately equal to 0.54, or take the cosine of $\frac{180}{\pi}$ degrees in degrees mode. If you add the six terms shown above, you'll see that this finite sum agrees with this value. The **exact** answer is the cosine of 1 radian, which is approximately 0.540302306.

Example 2. (A) Use the binomial series for $\sqrt{1+x}$ to approximate $\sqrt{2}$ by finding the sum of the first five terms with $x = 1$. (B) Estimate the remainder. (C) Estimate the complete (infinite) sum. Compare with your calculator's value for $\sqrt{2}$.

Solution: (A) Begin with the binomial series for $(1+x)^{1/2}$ from Chapter 14.

$$\sqrt{1+x} = (1+x)^{1/2} = 1 + \frac{x}{2} - \frac{x^2}{2(4)} + \frac{1(3)}{2(4)(6)}x^3 - \frac{1(3)(5)}{2(4)(6)(8)}x^4 + \cdots$$

Evaluate the series at $x = 1$. (Recall that the interval of convergence is $-1 < x \leq 1$.)

$$\sqrt{1+1} = \sqrt{2} = 1 + \frac{1}{2} - \frac{1}{8} + \frac{3}{48} - \frac{15}{384} + \cdots = 1 + \frac{1}{2} - \frac{1}{8} + \frac{1}{16} - \frac{5}{128} + \cdots$$

The sum of the first 5 terms is

$$\sqrt{2} \approx 1 + 0.5 - 0.125 + 0.0625 - 0.0390625 \approx 1.3984$$

(B) For a Maclaurin series (or a more general Taylor series), use the remainder estimation theorem. For this theorem, let $n = 0$ correspond to the term $f(a)$, which doesn't involve a derivative. Starting from 0, the last term in our example has $n = 4$. There are 5 terms, but $n = 4$ because we begin from zero. The remainder is based on the next term, $n + 1 = 4 + 1 = 5$, which involves the fifth derivative in this example.

$$f^{(4+1)}(w) = \frac{d^{4+1}}{dw^{4+1}}(1+w)^{1/2} = \frac{d^5}{dw^5}(1+w)^{1/2}$$

We need to take 5 successive derivatives. For the remainder theorem, $a \leq w \leq x$. In this case, $a = 0$ (for a Maclaurin series) and we're evaluating the function at $x = 1$, such that $0 \leq w \leq 1$. Note that $f(w) = \sqrt{1+w} = (1+w)^{1/2}$.

$$\frac{df}{dw} = \frac{1}{2}(1+w)^{-1/2} \quad , \quad \frac{d^2f}{dw^2} = -\frac{1}{4}(1+w)^{-3/2} \quad , \quad \frac{d^3f}{dw^3} = \frac{3}{8}(1+w)^{-5/2}$$

$$\frac{d^4f}{dw^4} = -\frac{15}{16}(1+x)^{-7/2} \quad , \quad \frac{d^5f}{dw^5} = \frac{105}{32}(1+w)^{-9/2}$$

The derivative $f^{(5)}(w) = \frac{105}{32}(1+w)^{-9/2} = \frac{105}{32(1+w)^{9/2}}$ has a maximum value when w is zero, which is $\frac{105}{32} \approx 3.28$. If we choose $q = 3.3$ and $r = 1$, for example, we can be sure that $\left|f^{(n+1)}(w)\right| \leq qr^{n+1}$ since $qr^{n+1} = 3.3(1^{n+1}) = 3.3(1) = 3.3$ is greater than $\frac{105}{32}$ ≈ 3.28. Use the remainder estimation theorem, setting x equal to 1 and $n = 4$ (where, again, $n = 4$ even though there are 5 terms because we start from zero).

$$|R_4(x)| \leq 3.3\frac{1^{4+1}|x-0|^{4+1}}{(4+1)!} = 3.3\frac{|x|^5}{5!} = \frac{3.3}{120} \approx 0.0275$$

(C) To estimate the complete (infinite) sum, add the remainder from Part B to the finite sum of Part A: $1.3984 + 0.0275 = 1.4259$. Using a calculator, $\sqrt{2} \approx 1.4142$. We see that adding the remainder gave an improved approximation. (The approximation would be better by keeping additional terms, as this series converges slowly.)

Example 3. (A) Find the sum of the first five terms of the series below. (B) What does the integral test say about the remainder? (C) Estimate the complete (infinite) sum.

$$\sum_{n=1}^{\infty} \frac{1}{n^2\sqrt{n}}$$

Solution: (A) Write out the first five terms and add them.

$$\frac{1}{1\sqrt{1}} + \frac{1}{4\sqrt{2}} + \frac{1}{9\sqrt{3}} + \frac{1}{16\sqrt{4}} + \frac{1}{25\sqrt{5}} \approx 1 + 0.1768 + 0.0642 + 0.0313 + 0.0179 = 1.2902$$

It should make sense that the sum is a bit less than $\frac{\pi^2}{6} \approx 1.64$, which is the sum of the series $\sum \frac{1}{n^2}$.

(B) Integrate the function $f(x) = \frac{1}{x^2\sqrt{x}}$. One integral will begin from 5 and the other will begin from $5 + 1 = 6$ because the finite sum went up to $n = 5$. Note that $x^2\sqrt{x} = x^2x^{1/2} = x^{5/2}$.

$$\int_{x=5}^{\infty} \frac{dx}{x^2\sqrt{x}} = \int_{x=5}^{\infty} \frac{dx}{x^{5/2}} = \int_{x=5}^{\infty} x^{-5/2}\,dx = \left[\frac{x^{-3/2}}{-3/2}\right]_{x=5}^{\infty} = \left[-\frac{2}{3x^{3/2}}\right]_{x=5}^{\infty} = \left[-\frac{2}{3x\sqrt{x}}\right]_{x=5}^{\infty}$$

$$= -\lim_{x\to\infty}\left(\frac{2}{3x\sqrt{x}}\right) + \frac{2}{3(5)\sqrt{5}} = -0 + \frac{2}{15\sqrt{5}} \approx 0.0596$$

For the integral starting at 6, simply change 5 to 6 in the last line.

$$\int_{x=6}^{\infty} \frac{dx}{x^2\sqrt{x}} = -0 + \frac{2}{3(6)\sqrt{6}} = \frac{2}{18\sqrt{6}} \approx 0.0454$$

Before using the remainder inequality associated with the integral test, first check that the function is nonincreasing over the interval. Observe that $\frac{1}{x^2\sqrt{x}}$ decreases over the interval $5 \leq x < \infty$. The remainder of the series (that is, the terms starting from $n = 6$) lies between 0.045 and 0.060. The average of these values gives an estimate of the remainder of 0.053.

(C) Add the sum of the first five terms from Part A to our estimate of the remainder from Part B to estimate the complete (infinite) sum: $1.29 + 0.05 = 1.34$.

Example 4. A series begins with $a_1 = 1$ and its terms satisfy $a_{n+1} = \frac{2n}{3n+4}a_n$. (A) Find the sum of the first five terms of the series. (B) What does the ratio test say about the remainder? (C) Estimate the complete (infinite) sum.

Solution: (A) Use the given formula to find the second thru fifth terms of the series.

$$a_2 = a_{1+1} = \frac{2(1)}{3(1)+4}a_1 = \frac{2}{7}(1) = \frac{2}{7}$$

$$a_3 = a_{2+1} = \frac{2(2)}{3(2)+4}a_2 = \frac{4}{10}\left(\frac{2}{7}\right) = \frac{4}{35}$$

$$a_4 = a_{3+1} = \frac{2(3)}{3(3)+4}a_3 = \frac{6}{13}\left(\frac{4}{35}\right) = \frac{24}{455}$$

$$a_5 = a_{4+1} = \frac{2(4)}{3(4)+4} = \frac{8}{16}\left(\frac{24}{455}\right) = \frac{12}{455}$$

Add the first five terms together.

$$1+\frac{2}{7}+\frac{4}{35}+\frac{24}{455}+\frac{12}{455} \approx 1+0.2857+0.1143+0.0527+0.0264=1.4791$$

Note: This series doesn't converge as rapidly as some of the others have. The fifth term, 0.0264, is pretty sizable. The finite sum would be a better approximation if we kept more terms. We're using five terms to illustrate how the process works.

(B) The ratio of any two consecutive terms is $\frac{a_{n+1}}{a_n} = \frac{2n}{3n+4}$. We want to find two constants r_1 and r_2 such that $r_1 \leq \frac{a_{n+1}}{a_n} \leq r_2 < 1$ for $n \geq 5$ (since in this problem, there are $N = 5$ terms in the finite series approximation), which means $r_1 \leq \frac{2n}{3n+4} \leq r_2 < 1$.

- Observe that $\left\{\frac{2n}{3n+4}\right\}$ is an increasing sequence: $\frac{2}{7}, \frac{4}{10}, \frac{6}{13}, \frac{8}{16}, \ldots$ For $n \geq 5$, we can be sure that $\frac{2n}{3n+4}$ is no smaller than $\frac{2(5)}{3(5)+19} = \frac{10}{19}$. We may thus choose $r_1 = \frac{10}{19}$.
- Observe that $\left\{\frac{2n}{3n+4}\right\}$ approaches the limiting value of $\frac{2}{3}$ (recall Chapter 2) as n grows infinite. Thus, $\frac{2n}{3n+4}$ is no greater than $\frac{2}{3}$. We may thus choose $r_2 = \frac{2}{3}$.

Every term of this series is positive; it's necessary to check this before using the remainder inequality associated with the ratio test.

$$\frac{r_1 a_n}{1-r_1} \leq R_n \leq \frac{r_2 a_n}{1-r_2}$$

Use $n = 5$, $a_5 = \frac{12}{455}$, $r_1 = \frac{10}{19}$, and $r_2 = \frac{2}{3}$.

$$\frac{\frac{10}{19}\frac{12}{455}}{1-\frac{10}{19}} \leq R_5 \leq \frac{\frac{2}{3}\frac{12}{455}}{1-\frac{2}{3}}$$

$$\frac{24}{1729} \div \frac{9}{19} \leq R_5 \leq \frac{8}{455} \div \frac{1}{3}$$

$$\frac{24}{1729} \times \frac{19}{9} \leq R_5 \leq \frac{8}{455} \times \frac{3}{1}$$

$$\frac{8}{273} \leq R_5 \leq \frac{24}{455}$$

$$0.029 \leq R_5 \leq 0.053$$

The remainder of the series (that is, the terms starting from $n = 6$) lies between 0.029 and 0.053. The average of these values gives an estimate of the remainder of 0.041.

(C) Add the sum of the first five terms from Part A to our estimate of the remainder from Part B to estimate the complete (infinite) sum: $1.48 + 0.04 = 1.52$.

Example 5. (A) Find the sum of the first six terms of the series below. (B) Estimate the error in approximating the complete sum with just the first six terms.

$$\sum_{n=1}^{\infty} (-1)^n \frac{\ln n}{n!}$$

Solution: (A) Write out the first six terms and find their sum. Recall that $0! = 1$.

$$-\frac{\ln 1}{1!} + \frac{\ln 2}{2!} - \frac{\ln 3}{3!} + \frac{\ln 4}{4!} - \frac{\ln 5}{5!} + \frac{\ln 6}{6!} + \cdots = -0 + \frac{\ln 2}{2} - \frac{\ln 3}{6} + \frac{\ln 4}{24} - \frac{\ln 5}{120} + \frac{\ln 6}{720} + \cdots$$
$$\approx 0.3466 - 0.1831 + 0.0578 - 0.0134 + 0.0025 = 0.2104$$

(B) This alternating series meets all of the conditions of Leibniz's theorem (Chapter 11): the terms alternate sign, the absolute values of the terms are nonincreasing, and the sequence of absolute values converges to zero. According to the alternating series estimation theorem, the error is less than the absolute value of the first unused term: $|a_7| = \frac{\ln 7}{7!} = \frac{\ln 7}{5040} \approx 0.0004$. The remainder, R_7 (which is the infinite sum minus the finite approximation) has the same sign as the 7th term, which is negative since $(-1)^7 = -1$. Since the infinite sum, S, minus the finite approximation, 0.2104, is negative, this means that the infinite sum is smaller than the finite approximation: $S < 0.2104$. Since the error is 0.0004, the infinite sum must lie in the range $0.21 < S < 0.2104$.

Example 6. Find the exact sum for the series below.

$$\sum_{n=0}^{\infty} \frac{1}{(2n+1)^2}$$

Solution: Write out the first handful of terms to see what this series looks like. As we've noted in previous chapters, $2n + 1$ makes an odd number.

$$\sum_{n=0}^{\infty} \frac{1}{(2n+1)^2} = 1 + \frac{1}{3^2} + \frac{1}{5^2} + \frac{1}{7^2} + \frac{1}{9^2} + \frac{1}{11^2} + \cdots$$

The denominators are odd numbers squared. This is similar to the series $\sum \frac{1}{n^2}$ that we showed has a sum of $\frac{\pi^2}{6}$; the only difference is that the even numbers are missing. Maybe we can use a similar method to find the sum of this series. When we found the sum for the series $\sum \frac{1}{n^2}$, we started with the Maclaurin series for $\sin x$. Review the common Maclaurin series from Chapter 14. When we used $\sin x$, we worked with the zeroes of the sine function, $\left(1-\frac{x}{\pi}\right)\left(1+\frac{x}{\pi}\right)\left(1-\frac{x}{2\pi}\right)\left(1+\frac{x}{2\pi}\right)\left(1-\frac{x}{3\pi}\right)\left(1+\frac{x}{3\pi}\right)\cdots$ We thought about multiplying this out and forming terms proportional to x^2. (It may help to review how we found the sum of $\frac{\pi^2}{6}$ in the text.) This gave rise to $\frac{1}{1}+\frac{1}{4}+\frac{1}{9}+\frac{1}{16}+\frac{1}{25}+\cdots$ We want to do something similar now, except not get $\frac{1}{4}, \frac{1}{16}, \frac{1}{36}$, etc. Think about the zeroes of the other trig functions. What are the zeroes of the $\cos x$? The cosine function is zero when $x=\pm\frac{\pi}{2}, \pm\frac{3\pi}{2}, \pm\frac{5\pi}{2}, \pm\frac{7\pi}{2}$, etc. Look at that; the numerators only have odd numbers. This suggests trying to do the same thing that we had done with the sine function, but using cosine instead. Begin with the Maclaurin series for cosine.

$$\cos x = 1-\frac{x^2}{2}+\frac{x^4}{24}-\frac{x^6}{720}+\cdots$$

With sine, we worked with the ratio $\frac{\sin x}{x}$ because the Maclaurin series for $\sin x$ began with x. In contrast, the Maclaurin series for $\cos x$ begins with 1 (not x), so there is no reason to divide $\cos x$ by x. The zeroes of $\cos x$ occur when $x=\pm\frac{\pi}{2}, \pm\frac{3\pi}{2}, \pm\frac{5\pi}{2}, \pm\frac{7\pi}{2}$, etc. The roots have the form $\left(1-\frac{2x}{\pi}\right), \left(1+\frac{2x}{\pi}\right), \left(1-\frac{2x}{3\pi}\right), \left(1+\frac{2x}{3\pi}\right), \left(1-\frac{2x}{5\pi}\right), \left(1+\frac{2x}{5\pi}\right)$, etc. For example, plug $x=\frac{5\pi}{2}$ into $\left(1-\frac{2x}{5\pi}\right)$ to see that this gives $1-\frac{2}{5\pi}\frac{5\pi}{2}=1-1=0$. Just as we had done for the sine function, we will multiply the roots of cosine on the left-hand side of the above equation.

$$\left(1-\frac{2x}{\pi}\right)\left(1+\frac{2x}{\pi}\right)\left(1-\frac{2x}{3\pi}\right)\left(1+\frac{2x}{3\pi}\right)\left(1-\frac{2x}{5\pi}\right)\left(1+\frac{2x}{5\pi}\right)\cdots = 1-\frac{x^2}{2}+\frac{x^4}{24}-\frac{x^6}{720}+\cdots$$

Once again, think about multiplying out the left-hand side, and just focus on how to get nonzero terms proportional to x^2. These terms will add up to $-\frac{x^2}{2}$ from the right-hand side. It's very similar to what we did before. We'll get:

$$-\frac{4x^2}{\pi^2}-\frac{4x^2}{9\pi^2}-\frac{4x^2}{25\pi^2}-\frac{4x^2}{49\pi^2}-\frac{4x^2}{81\pi^2}+\cdots=-\frac{x^2}{2}$$

Multiply by π^2 and divide by $-4x^2$ on both sides:

$$1+\frac{1}{3^2}+\frac{1}{5^2}+\frac{1}{7^2}+\frac{1}{9^2}+\frac{1}{11^2}+\cdots=\frac{\pi^2}{8}$$

Check the answer: Add up the first terms (enough terms so that we see that the terms are becoming relatively small) and see if this finite sum is near the exact value.

$$\approx 1 + 0.1111 + 0.04 + 0.0204 + 0.0123 + 0.0083 + 0.0059 + 0.0044 + 0.0035$$
$$+ 0.0028 + 0.0023 + 0.0019 + 0.0016 + 0.0014 + 0.0012 + 0.001$$
$$+ 0.0009 + 0.0008 + 0.0007 + 0.0007 + 0.0006 + 0.0005 + 0.0005$$
$$+ 0.0005 + 0.0004 + 0.0004 + 0.0004 + 0.0003 + 0.0003 + 0.0003$$
$$= 1.2254$$

As you can see this, series doesn't converge as rapidly as some others that we've seen. We carried this sum out to $\frac{1}{59^2}$. The more terms you keep, the better the approximation. Observe that 1.225 is near $\frac{\pi^2}{8} \approx 1.2337$. If you use a spreadsheet like Excel or write code, you can keep hundreds or thousands of terms and obtain better agreement.

Example 7. Find the exact sum for the series below.

$$\frac{1}{3} + \frac{1}{15} + \frac{1}{35} + \frac{1}{63} + \frac{1}{99} + \frac{1}{143} + \cdots$$

Solution: The denominators are products of consecutive odd numbers.

$$\frac{1}{3} + \frac{1}{15} + \frac{1}{35} + \frac{1}{63} + \frac{1}{99} + \frac{1}{143} + \cdots = \frac{1}{1(3)} + \frac{1}{3(5)} + \frac{1}{5(7)} + \frac{1}{7(9)} + \frac{1}{9(11)} + \frac{1}{11(13)} + \cdots$$

Try writing the series using summation notation. Since $2n + 1$ makes an odd number, the product $(2n + 1)(2n + 3)$ would form consecutive odd numbers, but that isn't the only choice. Observe that $(2n - 1)(2n + 1)$ also makes consecutive odd numbers. We'll choose the latter. With this choice, the sum needs to start from $n = 1$.

$$\sum_{n=1}^{\infty} \frac{1}{(2n-1)(2n+1)} = \frac{1}{1(3)} + \frac{1}{3(5)} + \frac{1}{5(7)} + \frac{1}{7(9)} + \frac{1}{9(11)} + \frac{1}{11(13)} + \cdots$$

This should seem somewhat familiar. It resembles the series where we applied the method of **partial fractions** in Chapter 4. Using the method of partial fractions, we get

$$\frac{1}{(2n-1)(2n+1)} = \frac{x}{2n-1} + \frac{y}{2n+1}$$

Multiply both sides by $(2n - 1)(2n + 1)$.

$$1 = x(2n + 1) + y(2n - 1) = 2nx + x + 2ny - y = x - y + 2n(x + y)$$

Both sides will be equal for any value of n if $x + y = 0$ (such that $y = -x$) and if $x - y = 1$. Plug $y = -x$ into $x - y = 1$ to get $x - (-x) = x + x = 2x = 1$, which shows that $x = \frac{1}{2}$ and $y = -\frac{1}{2}$. This means that the series may be expressed as follows (where we factor out the $\frac{1}{2}$ present in both x and y. (Incidentally, you can check your algebra by plugging in numbers. For example, for $n = 3$, the left-hand side gives $\frac{1}{[2(3)-1][2(3)+1]} = \frac{1}{5(7)}$ and the right-hand side gives $\frac{1}{2}\left(\frac{1}{2(3)-1} - \frac{1}{2(3)+1}\right) = \frac{1}{2}\left(\frac{1}{5} - \frac{1}{7}\right) = \frac{1}{2}\left(\frac{7}{35} - \frac{5}{35}\right) = \frac{2}{70} = \frac{1}{35} = \frac{1}{5(7)}$. Both sides equal $\frac{1}{5(7)}$ when $n = 3$.)

$$\sum_{n=1}^{\infty} \frac{1}{(2n-1)(2n+1)} = \frac{1}{2}\sum_{n=1}^{\infty}\left(\frac{1}{2n-1} - \frac{1}{2n+1}\right)$$

Write the right-hand side out longhand.

$$\frac{1}{2}\sum_{n=1}^{\infty}\left(\frac{1}{2n-1} - \frac{1}{2n+1}\right) = \frac{1}{2}\left(\frac{1}{1} - \frac{1}{3} + \frac{1}{3} - \frac{1}{5} + \frac{1}{5} - \frac{1}{7} + \frac{1}{7} - \frac{1}{9} + \frac{1}{9} - \frac{1}{11} + \cdots\right)$$

Every term cancels out except for the first and the last, and the last term approaches zero as n grows infinite, so only the first term remains.

$$\sum_{n=1}^{\infty} \frac{1}{(2n-1)(2n+1)} = \frac{1}{2}(1) = \frac{1}{2}$$

Check the answer: Add the first several terms of the given series and compare with $\frac{1}{2} = 0.5$.

$$\frac{1}{3} + \frac{1}{15} + \frac{1}{35} + \frac{1}{63} + \frac{1}{99} + \frac{1}{143} \approx 0.46$$

Chapter 15 Problems

Directions: Find the exact sum of each series for Exercise 1-16.

(1)

$$1 - \frac{1}{3} + \frac{1}{5} - \frac{1}{7} + \frac{1}{9} - \frac{1}{11} + \cdots$$

(2)

$$\sum_{n=1}^{\infty} \frac{6}{2^n}$$

(3)

$$\frac{1}{2}-\frac{1}{6}+\frac{1}{24}-\frac{1}{120}+\frac{1}{720}-\frac{1}{5040}+\cdots$$

(4)

$$\sum_{k=0}^{\infty}\frac{3^{2k}}{(2k)!}$$

(5)

$$\frac{1}{2}+\frac{1}{6}+\frac{1}{12}+\frac{1}{20}+\frac{1}{30}+\frac{1}{42}+\frac{1}{56}+\frac{1}{72}+\frac{1}{90}+\cdots$$

(6)

$$\sum_{i=1}^{\infty}\frac{i}{2^{i-1}}$$

(7)

$$\frac{1}{2(9)}-\frac{1}{6(27)}+\frac{1}{12(81)}-\frac{1}{20(243)}+\frac{1}{30(729)}-\frac{1}{42(2187)}+\cdots$$

(8)

$$\sum_{j=1}^{\infty}\frac{(-1)^{j+1}}{j(2^j)}$$

(9)

$$1+\frac{1}{3}+\frac{2}{15}+\frac{17}{315}+\cdots$$

(10)

$$\sum_{m=3}^{\infty}\frac{1}{(2m)!}$$

(11)

$$1 + \frac{1}{2} - \frac{1}{2(4)} + \frac{1(3)}{2(4)(6)} - \frac{1(3)(5)}{2(4)(6)(8)} + \cdots$$

(12)

$$\sum_{\ell=0}^{\infty} \frac{2^\ell}{\ell!}$$

(13)

$$9 - 3\sqrt{3} + 3 - \sqrt{3} + 1 - \frac{\sqrt{3}}{3} + \frac{1}{3} + \cdots$$

(14)

$$\sum_{k=4}^{\infty} \frac{2k+1}{k^2(k+1)^2}$$

(15)

$$\frac{1}{3}+\frac{1}{8}+\frac{1}{15}+\frac{1}{24}+\frac{1}{35}+\frac{1}{48}+\frac{1}{63}+\frac{1}{80}+\cdots$$

(16)

$$\frac{1}{2}+\frac{1}{3}-\frac{1}{3}\left(\frac{1}{8}+\frac{1}{27}\right)+\frac{1}{5}\left(\frac{1}{32}+\frac{1}{243}\right)-\frac{1}{7}\left(\frac{1}{128}+\frac{1}{2187}\right)+\frac{1}{9}\left(\frac{1}{512}+\frac{1}{19{,}683}\right)+\cdots$$

(17)

$$\frac{1}{4}+\frac{1}{16}+\frac{1}{36}+\frac{1}{64}+\frac{1}{100}+\frac{1}{144}+\frac{1}{196}+\cdots$$

(18)

$$\sum_{n=1}^{\infty}\frac{(-1)^{n+1}}{n^2}$$

(19) (A) Write out the Maclaurin series for e^{-x} with $x = \frac{1}{2}$.

(B) Find the sum of the first six terms of this series.

(C) Estimate the remainder.

(D) Estimate the complete (infinite) sum.

(E) Compare with your calculator's value for $e^{-1/2}$.

(20) (A) Find the sum of the first ten terms of the series below.

$$\sum_{k=1}^{\infty} \frac{k}{k^4 + 2k^2 + 1}$$

(B) What does the integral test tell you about the remainder?

(C) Estimate the complete (infinite) sum.

(21) A series begins with $a_1 = 1$ and its terms satisfy $a_{n+1} = \frac{n^2}{2n^2+3} a_n$.

(A) Find the sum of the first eight terms of the series.

(B) What does the ratio test tell you about the remainder?

(C) Estimate the complete (infinite) sum.

(22) (A) Find the sum of the first six terms of the series below.

$$\sum_{k=1}^{\infty} \frac{(-1)^k}{k\sqrt{k}}$$

(B) Estimate the error in this approximation.

(23) (A) Use the method prior to Example 1 that we used to show that $\sum \frac{1}{n^2} = \frac{\pi^2}{6}$ to derive the following double sum.

$$\sum_{m=1}^{\infty} \sum_{n=m+1}^{\infty} \frac{1}{m^2 n^2} = \frac{\pi^4}{120}$$

(B) Square both sides of $\sum \frac{1}{n^2} = \frac{\pi^2}{6}$.

(C) Combine the results from Parts A and B to show that $\sum \frac{1}{n^4} = \frac{\pi^4}{90}$ (starting from 1).

Chapter 1 Answers

(1) $-1, 2, 5, 8, 11$

Notes: $3(1) - 4 = -1, 3(2) - 4 = 2, 3(3) - 4 = 5, 3(4) - 4 = 8, 3(5) - 4 = 11$

(2) $\frac{1}{2}, \frac{2}{3}, \frac{3}{4}, \frac{4}{5}, \frac{5}{6}$

Notes: $\frac{1}{1+1} = \frac{1}{2}, \frac{2}{2+1} = \frac{2}{3}, \frac{3}{3+1} = \frac{3}{4}, \frac{4}{4+1} = \frac{4}{5}, \frac{5}{5+1} = \frac{5}{6}$

(3) $4, 20, 100, 500, 2500$

Notes: $a_2 = 5(4) = 20, a_3 = 5(20) = 100, a_4 = 5(100) = 500, a_5 = 5(500) = 2500$

(4) $-2, 4, -6, 8, -10$

Notes: $(-1)^1[2(1)] = (-1)(2) = -2, (-1)^2[2(2)] = (1)(4) = 4, (-1)^3[2(3)] = (-1)(6) = -6, (-1)^4[2(4)] = (1)(8) = 8, (-1)^5[2(5)] = (-1)(10) = -10$

(5) $-1, 0, 3, 8, 15$

Notes: $1(1 - 2) = -1, 2(2 - 2) = 0, 3(3 - 2) = 3, 4(4 - 2) = 8, 5(5 - 2) = 15$

(6) $1, 1, 2, 7, 29$

Notes: $b_3 = 5(1) - 3(1) = 2, b_4 = 5(2) - 3(1) = 7, b_5 = 5(7) - 3(2) = 29$

(7) $24, 12, 4, 1, \frac{1}{5}$

Notes: $\frac{24}{1!} = 24, \frac{24}{2!} = 12, \frac{24}{3!} = \frac{24}{6} = 4, \frac{24}{4!} = \frac{24}{24} = 1, \frac{24}{5!} = \frac{24}{120} = \frac{1}{5}$

The factorial function means to multiply successively smaller integers until you reach one. For example, $4! = 4(3)(2)(1) = 24$.

The recursion formula for this sequence is $a_{n+1} = \frac{a_n}{n+1}$ with $a_1 = 24$.

(8) $2, 2, 1, \frac{1}{3}, \frac{1}{12}$

Notes: $a_2 = a_{1+1} = \frac{a_1}{1} = \frac{2}{1} = 2, a_3 = a_{2+1} = \frac{a_2}{2} = \frac{2}{2} = 1, a_4 = a_{3+1} = \frac{a_3}{3} = \frac{1}{3}$

$a_5 = a_{4+1} = \frac{a_4}{4} = \frac{1/3}{4} = \frac{1}{3} \div 4 = \frac{1}{3} \times \frac{1}{4} = \frac{1}{12}$

To divide by 4, multiply by its reciprocal: $\frac{1/3}{4} = \frac{1}{3} \div 4 = \frac{1}{3} \div \frac{4}{1} = \frac{1}{3} \times \frac{1}{4} = \frac{1}{12}$.

(9) $3, 5, 9, 15, 23$

Notes: $c_2 = 3 + 2(1) = 5, c_3 = 5 + 2(2) = 9, c_4 = 9 + 2(3) = 15, c_5 = 15 + 2(4) = 23$

(10) $3, 9, 27, 81, 243$

Notes: $3^1 = 3, 3^2 = 9, 3^3 = 27, 3^4 = 81, 3^5 = 243$

(11) $\frac{1}{6}, \frac{10}{9}, \frac{25}{14}, \frac{46}{21}, \frac{73}{30}$

Notes: $\frac{3(1)^2-2}{1^2+5} = \frac{1}{6}, \frac{3(2)^2-2}{2^2+5} = \frac{10}{9}, \frac{3(3)^2-2}{3^2+5} = \frac{25}{14}, \frac{3(4)^2-2}{4^2+5} = \frac{46}{21}, \frac{3(5)^2-2}{5^2+5} = \frac{73}{30}$

(12) $2, \frac{1}{2}, \frac{8}{27}, \frac{1}{4}, \frac{32}{125}$

Notes: $\frac{2^1}{1^3} = \frac{2}{1} = 2, \frac{2^2}{2^3} = \frac{4}{8} = \frac{1}{2}, \frac{2^3}{3^3} = \frac{8}{27}, \frac{2^4}{4^3} = \frac{16}{64} = \frac{1}{4}, \frac{2^5}{5^3} = \frac{32}{125}$

(13) $0, 2, 0, 2, 0$

Notes: $1 + (-1)^1 = 1 - 1 = 0, 1 + (-1)^2 = 1 + 1 = 2, 1 + (-1)^3 = 1 - 1 = 0$

$1 + (-1)^4 = 1 + 1 = 2, 1 + (-1)^5 = 1 - 1 = 0$

(14) $2, 3, 6, 18, 108$

Notes: $a_3 = 2(3) = 6, a_4 = 3(6) = 18, a_5 = 6(18) = 108$

(15) $2, -4, 20, 788, 1{,}241{,}876$

Notes: $B_2 = 2(2)^2 - 12 = 2(4) - 12 = -4, B_3 = 2(-4)^2 - 12 = 2(16) - 12 = 20,$

$B_4 = 2(20)^2 - 12 = 2(400) - 12 = 788, B_5 = 2(788)^2 - 12 = 1{,}241{,}876$

(16) $\frac{1}{2}, -\frac{1}{3}, \frac{1}{4}, -\frac{1}{5}, \frac{1}{6}$ Examples: $\frac{(-1)^3}{3} = -\frac{1}{3}, \frac{(-1)^4}{4} = \frac{1}{4}$

(17) $-5, 25, -125, 625, -3125$

Notes: $(-5)^1 = -5, (-5)^2 = 25, (-5)^3 = -125, (-5)^4 = 625, (-5)^5 = -3125$

(18) $6, 12, 96, 8832, 77{,}968{,}896$

Notes: $c_2 = 6(6 - 4) = 6(2) = 12, c_3 = 12(12 - 4) = 12(8) = 96,$

$c_4 = 96(96 - 4) = 96(92) = 8832, c_5 = 8832(8832 - 4) = 8832(8828) = 77{,}968{,}896$

(19) $1, 0.2, 0.04, 0.008, 0.0016$ Alternate answer: $1, \frac{1}{5}, \frac{1}{25}, \frac{1}{125}, \frac{1}{625}$

Notes: $0.2^{1-1} = 0.2^0 = 1, 0.2^{2-1} = 0.2^1 = 0.2, 0.2^{3-1} = 0.2^2 = 0.04$

$0.2^{4-1} = 0.2^3 = 0.008, 0.2^{5-1} = 0.2^4 = 0.0016$

(20) $2, 3, \frac{6}{5}, \frac{6}{7}, \frac{1}{2}$

Notes: $\frac{1}{A_3} = \frac{1}{3} + \frac{1}{2} = \frac{3}{6} + \frac{2}{6} = \frac{5}{6}$, take the reciprocal of both sides of $\frac{1}{A_3} = \frac{5}{6}$ to get $A_3 = \frac{6}{5}$

$\frac{1}{A_4} = \frac{1}{6/5} + \frac{1}{3} = \frac{5}{6} + \frac{1}{3} = \frac{5}{6} + \frac{2}{6} = \frac{7}{6}$, take the reciprocal of both sides to get $A_4 = \frac{6}{7}$

$\frac{1}{A_5} = \frac{1}{6/7} + \frac{1}{6/5} = \frac{7}{6} + \frac{5}{6} = \frac{12}{6} = 2$, take the reciprocal of both sides to get $A_5 = \frac{1}{2}$

These fractions can be tricky. Note that $\frac{1}{A_{k+2}} = \frac{1}{A_{k+1}} + \frac{1}{A_k}$ solves for $\frac{1}{A_{k+2}}$, not A_{k+2}, so you need to find the reciprocal of both sides to solve for A_{k+2}.

(21) $a_{n+1} = a_n + 3$ is the recursion formula, whereas $3n - 5$ is the formula for n.
The next two terms are 16 and 19.
Checks: $a_5 = a_4 + 3 = 7 + 3 = 10, 3(5) - 5 = 15 - 5 = 10$
Notes: $3n$ makes 3, 6, 9, 12, 15, 18, ... Subtract 5 from this to get the given sequence.
The way sequences were defined, they start with $n = 1$.
The next two terms are $a_7 = 3(7) - 5 = 16$ and $a_8 = 3(8) - 5 = 19$.

(22) $\frac{n+2}{n+4}$ is the formula for n. The next two terms are $\frac{9}{11}$ and $\frac{5}{6}$.
Note: According to the instructions, you need the recursion formula **or** the formula for n; you don't need both. If you can get both, that's even better, but in some cases the alternative answer is challenging to get. The answer key will give at least one of the two formulas (the recursion formula or the formula for n) for each exercise.
Checks: $\frac{1+2}{1+4} = \frac{3}{5}, \frac{2+2}{2+4} = \frac{4}{6} = \frac{2}{3}, \frac{3+2}{3+4} = \frac{5}{7}, \frac{4+2}{4+4} = \frac{6}{8} = \frac{3}{4}, \frac{5+2}{5+4} = \frac{7}{9}, \frac{6+2}{6+4} = \frac{8}{10} = \frac{4}{5}$
Notes: The 'trick' is to realize that $\frac{2}{3} = \frac{4}{6}, \frac{3}{4} = \frac{6}{8}$, and $\frac{4}{5} = \frac{8}{10}$.
The next two terms are $a_7 = \frac{7+2}{7+4} = \frac{9}{11}$ and $a_8 = \frac{8+2}{8+4} = \frac{10}{12} = \frac{5}{6}$.

(23) $a_{n+2} = a_n + a_{n+1}$ is the recursion formula. The next two terms are 47 and 76.
Checks: $3 + 4 = 7, 4 + 7 = 11, 7 + 11 = 18, 11 + 18 = 29$
Notes: This is like the Fibonacci sequence except for changing the starting values.
The next two terms are $18 + 29 = 47$ and $29 + 47 = 76$.

(24) $(n + 1)^3 - 1$ is the formula for n; it's equivalent to $n^3 + 3n^2 + 3n$.
The next two terms are 511 and 728.
Checks: $(1 + 1)^3 - 1 = 8 - 1 = 7, (2 + 1)^3 - 1 = 27 - 1 = 26, (3 + 1)^3 - 1 = 64 - 1 = 63, (4 + 1)^3 - 1 = 125 - 1 = 124, (5 + 1)^3 - 1 = 216 - 1 = 215$
Notes: The sequence 8, 27, 64, 125, 216, 343 is made of cubes starting with 2^3. The formula has $n + 1$ so that the first base is 2 when $n = 1$, and it subtracts 1 because the actual sequence (7, 26, 63, 124, 215, 342, ...) is 1 less than the cube sequence.
The next two terms are $a_7 = (7 + 1)^3 - 1 = 8^3 - 1 = 512 - 1 = 511$ and $a_8 = (8 + 1)^3 - 1 = 9^3 - 1 = 729 - 1 = 728$.

(25) $\frac{1}{(-3)^{n-1}}$ is the formula for n; the recursion formula is $a_{n+1} = -\frac{a_n}{3}$.
The next two terms are $\frac{1}{729}$ and $-\frac{1}{2187}$.

Checks: $\frac{1}{(-3)^{1-1}} = \frac{1}{(-3)^0} = \frac{1}{1} = 1$ (because $3^0 = 1$), $\frac{1}{(-3)^{2-1}} = \frac{1}{(-3)^1} = -\frac{1}{3}$, $a_2 = -\frac{1}{3}$

$\frac{1}{(-3)^{3-1}} = \frac{1}{(-3)^2} = \frac{1}{9}$, $a_3 = -\frac{(-1/3)}{3} = \frac{1}{3} \div 3 = \frac{1}{3} \div \frac{3}{1} = \frac{1}{3} \times \frac{1}{3} = \frac{1}{9}$

$\frac{1}{(-3)^{4-1}} = \frac{1}{(-3)^3} = -\frac{1}{27}$, $a_4 = -\frac{1/9}{3} = -\frac{1}{9} \div 3 = -\frac{1}{9} \div \frac{3}{1} = -\frac{1}{9} \times \frac{1}{3} = -\frac{1}{27}$

Note: The next two terms of the sequence are $a_7 = \frac{1}{(-3)^{7-1}} = \frac{1}{(-3)^6} = \frac{1}{729}$ and

$a_8 = \frac{1}{(-3)^{8-1}} = \frac{1}{(-3)^7} = -\frac{1}{2187}$.

(26) $a_{n+1} = a_n + \frac{9}{10^{n+1}}$ is the recursion formula; $1 - 0.1^n = 1 - \frac{1}{10^n}$ is the formula for n.

The next two terms are 0.999999 (with 6 nines) and 0.9999999 (with 7 nines).

Checks: $a_2 = 0.9 + \frac{9}{10^{1+1}} = 0.9 + \frac{9}{10^2} = 0.9 + \frac{9}{100} = 0.9 + 0.09 = 0.99$

$a_3 = 0.99 + \frac{9}{10^{2+1}} = 0.99 + \frac{9}{10^3} = 0.99 + \frac{9}{1000} = 0.99 + 0.009 = 0.999$

$a_4 = 0.999 + \frac{9}{10^{3+1}} = 0.999 + \frac{9}{10^4} = 0.999 + \frac{9}{10{,}000} = 0.999 + 0.0009 = 0.9999$

$a_5 = 0.9999 + \frac{9}{10^{4+1}} = 0.9999 + \frac{9}{10^5} = 0.9999 + \frac{9}{100{,}000} = 0.99999$

Note: The next two terms are $a_6 = 0.99999 + \frac{9}{10^{5+1}} = 0.99999 + \frac{9}{1{,}000{,}000} = 0.999999$

and $a_7 = 0.999999 + \frac{9}{10^{6+1}} = 0.999999 + \frac{9}{10{,}000{,}000} = 0.9999999$.

(27) $a_{n+1} = (6|a_n| - 2)(-1)^n$ is the recursion formula.

The next two terms are −167,962 and 1,007,770.

Checks: $a_2 = (6|-4| - 2)(-1)^2 = [6(4) - 2](1) = 24 - 2 = 22$

$a_3 = (6|22| - 2)(-1)^3 = [6(22) - 2](-1) = (132 - 2)(-1) = 130(-1) = -130$

$a_4 = (6|-130| - 2)(-1)^4 = [6(130) - 2](1) = 780 - 2 = 778$

$a_5 = (6|778| - 2)(-1)^5 = (4668 - 2)(-1) = 4666(-1) = -4666$

$a_6 = (6|-4666| - 2)(-1)^6 = [6(4666) - 2](1) = (27{,}996 - 2)(1) = 27{,}994$

Notes: $|a_n|$ is the absolute value of a_n, meaning to disregard any minus sign.

The 'trick' is to realize that each term is nearly 6 times as large as the previous term (if you ignore the signs); you just need to subtract 2 to make it work. The way to ignore the signs in the multiplication is to use absolute values. Then use $(-1)^n$ as usual to make an alternating sequence where the $n = 2$ term is positive.

The next two terms are $a_7 = (6|27{,}994| - 2)(-1)^7 = (167{,}964 - 2)(-1) = -167{,}962$ and $a_8 = (6|-167{,}962| - 2)(-1)^6 = [6(167{,}962) - 2](1) = 1{,}007{,}770$.

(28) $\frac{1}{(2n-5)^2}$ is the formula for n. The next two terms are $\frac{1}{169}$ and $\frac{1}{225}$.

Checks: $\frac{1}{[2(1)-5]^2} = \frac{1}{(-3)^2} = \frac{1}{9}, \frac{1}{[2(2)-5]^2} = \frac{1}{(-1)^2} = \frac{1}{1} = 1, \frac{1}{[2(3)-5]^2} = \frac{1}{1^2} = \frac{1}{1} = 1$

$\frac{1}{[2(4)-5]^2} = \frac{1}{3^2} = \frac{1}{9}, \frac{1}{[2(5)-5]^2} = \frac{1}{5^2} = \frac{1}{25}, \frac{1}{[2(6)-5]^2} = \frac{1}{7^2} = \frac{1}{49}, \frac{1}{[2(7)-5]^2} = \frac{1}{9^2} = \frac{1}{81}$

Notes: The 'tricks' are to realize that each term involves the reciprocal of a square of an odd number and that the first two terms involve the square of a negative odd number. Note that $2n$ makes the even numbers; when we add or subtract an odd number from $2n$, that makes the odd numbers; $2n - 5$ makes the odd numbers starting with $2(1) - 5 = 2 - 5 = -3$.

The next two terms are $a_9 = \frac{1}{[2(9)-5]^2} = \frac{1}{13^2} = \frac{1}{169}$ and $a_{10} = \frac{1}{[2(10)-5]^2} = \frac{1}{15^2} = \frac{1}{225}$.

(29) $2\sqrt{n+2}$ is the formula for n; the recursion formula is $a_{n+1} = a_n \frac{\sqrt{n+3}}{\sqrt{n+2}}$.

The next two terms are 6 and $2\sqrt{10}$.

Checks: $2\sqrt{1+2} = 2\sqrt{3}$

$2\sqrt{2+2} = 2\sqrt{4} = 2(2) = 4, a_2 = 2\sqrt{3}\frac{\sqrt{1+3}}{\sqrt{1+2}} = 2\sqrt{3}\frac{\sqrt{4}}{\sqrt{3}} = 2\sqrt{4} = 2(2) = 4$

$2\sqrt{3+2} = 2\sqrt{5}, a_3 = 4\frac{\sqrt{2+3}}{\sqrt{2+2}} = 4\frac{\sqrt{5}}{\sqrt{4}} = 4\frac{\sqrt{5}}{2} = 2\sqrt{5}$

$2\sqrt{4+2} = 2\sqrt{6}, a_4 = 2\sqrt{5}\frac{\sqrt{3+3}}{\sqrt{3+2}} = 2\sqrt{5}\frac{\sqrt{6}}{\sqrt{5}} = 2\sqrt{6}$

$2\sqrt{5+2} = 2\sqrt{7}, a_5 = 2\sqrt{6}\frac{\sqrt{4+3}}{\sqrt{4+2}} = 2\sqrt{6}\frac{\sqrt{7}}{\sqrt{6}} = 2\sqrt{7}$

$2\sqrt{6+2} = 2\sqrt{8} = 2\sqrt{4(2)} = 2\sqrt{4}\sqrt{2} = 2(2)\sqrt{2} = 4\sqrt{2},$

$a_5 = 2\sqrt{7}\frac{\sqrt{1+7}}{\sqrt{1+6}} = 2\sqrt{7}\frac{\sqrt{8}}{\sqrt{7}} = 2\sqrt{8} = 2\sqrt{4(2)} = 2\sqrt{4}\sqrt{2} = 2(2)\sqrt{2} = 4\sqrt{2}$

Notes: One 'trick' is to realize that $2\sqrt{4} = 4$ and $2\sqrt{8} = 4\sqrt{2}$. Once you realize this, the sequence looks like $2\sqrt{3}, 2\sqrt{4}, 2\sqrt{5}, 2\sqrt{6}, 2\sqrt{7}, 2\sqrt{8}, \ldots$

The next two terms are $a_7 = 2\sqrt{7+2} = 2\sqrt{9} = 2(3) = 6$ and $a_8 = 2\sqrt{8+2} = 2\sqrt{10}$.

(30) $a_{n+1} = a_n - n - 3$ is the recursion formula; the formula for n is $-\frac{(n+2)(n+3)}{2} - 1$

The next two terms are -46 and -56.

Checks: $a_1 = -\frac{(1+2)(1+3)}{2} - 1 = -\frac{3(4)}{2} - 1 = -6 - 1 = -7$

$a_2 = -7 - 1 - 3 = -11, a_2 = -\frac{(2+2)(2+3)}{2} - 1 = -\frac{4(5)}{2} - 1 = -10 - 1 = -11$

$a_3 = -11 - 2 - 3 = -16, a_3 = -\frac{(3+2)(3+3)}{2} - 1 = -\frac{5(6)}{2} - 1 = -15 - 1 = -16$

$a_4 = -16 - 3 - 3 = -22, a_4 = -\frac{(4+2)(4+3)}{2} - 1 = -\frac{6(7)}{2} - 1 = -21 - 1 = -22$

$a_5 = -22 - 4 - 3 = -29, a_3 = -\frac{(5+2)(5+3)}{2} - 1 = -\frac{7(8)}{2} - 1 = -28 - 1 = -29$

$a_6 = -29 - 5 - 3 = -37, a_6 = -\frac{(6+2)(6+3)}{2} - 1 = -\frac{8(9)}{2} - 1 = -36 - 1 = -37$

Notes: The recursion formula is easier. The differences are 4, 5, 6, 7, 8, ..., which suggests including $n + 3$ to make the recursion formula. If you subtract $n + 3$ from a_n, you get $a_{n+1} = a_n - (n + 3) = a_n - n - 3$.

The formula in terms of n is trickier. First ignore the minus signs and look at 7, 11, 16, 22, 29, 37, ... Try adding or subtracting values to this sequence. If you subtract 1 from each term, you get 6, 10, 15, 21, 28, 36. This is actually a common number pattern. The formula $\frac{n(n+1)}{2}$ makes the sequence 1, 3, 6, 10, 15, 21, 28, 36, ... These are called **triangular numbers** because they can be formed geometrically by arranging objects in equilateral triangles. (Example: Put 1 object on a vertex, 2 objects below it, 3 objects below that row, and 4 objects on the edge opposite to the starting vertex. This makes the number 10.) We'll see triangular numbers again in Chapter 3. Since our pattern starts with 6, the formula is $\frac{(n+2)(n+3)}{2}$; this skips the 1 and 3; when $n = 1$, $\frac{(n+2)(n+3)}{2}$ equals 6. Now add 1 to $\frac{(n+2)(n+3)}{2}$ to shift the sequence to 7, 11, 16, 22, 29, 37, ... This makes $\frac{(n+2)(n+3)}{2} + 1$. Finally, multiply by negative one to insert the minus signs: $-\frac{(n+2)(n+3)}{2} - 1$.

The next two terms are $a_7 = -37 - 6 - 3 = -46$ and $a_8 = -46 - 7 - 3 = -56$.

Chapter 2 Answers

(1) $\left\{\frac{500}{n^5}\right\}$ converges to $\lim_{n\to\infty} \frac{500}{n^5} = 0$. Notes: $\left\{\frac{1}{n^c}\right\}$ converges to zero if $c > 0$. $\lim_{n\to\infty} (ca_n) = cL_a$. (The two c's aren't the same; one is 5, the other is 500.)

(2) $\left\{-\frac{m^{10}}{1{,}000{,}000}\right\}$ diverges. Notes: $\{m^c\}$ diverges if $c > 0$. Dividing by $-1{,}000{,}000$ doesn't change that. (It still diverges to minus infinity.)

(3) $\left\{\frac{(-1)^k}{k}\right\}$ converges to $\lim_{k\to\infty} \frac{(-1)^k}{k} = 0$. Notes: $\left\{\frac{1}{k}\right\}$ converges to zero. Let $\{c_k\} = \left\{\frac{1}{k}\right\}$. $\{(-1)^k c_k\}$ converges to zero if $\lim_{k\to\infty} c_k = 0$ according to the sandwich theorem. (See the note in the text concerning alternating sequences and the sandwich theorem.)

(4) $\left\{\frac{(-1)^{i+1}}{2i+3}\right\}$ converges to $\lim_{i\to\infty} \frac{(-1)^{i+1}}{2i+3} = 0$. Notes: $\left\{\frac{1}{2i+3}\right\}$ converges to zero. One way to see this is to divide the numerator and denominator each by i to get $\lim_{i\to\infty} \frac{1}{2i+3} = \lim_{i\to\infty} \frac{\frac{1}{i}}{2+\frac{3}{i}} = \frac{0}{2+0} = \frac{0}{2} = 0$; recall the tip following Example 10. Let $\{c_i\} = \left\{\frac{1}{2i+3}\right\}$. $\{(-1)^{i+1} c_i\}$ converges to zero if $\lim_{i\to\infty} c_i = 0$ according to the sandwich theorem. (See the note in the text concerning alternating sequences and the sandwich theorem.)

(5) $a_n = 5$ converges to $\lim_{n\to\infty} 5 = 5$. Note: This sequence is 5, 5, 5, 5, 5, 5, 5, 5, 5, ...

(6) $\left\{\frac{\ell-4}{\ell}\right\}$ converges to $\lim_{\ell\to\infty} \ell = 1$. Notes: Divide the numerator and denominator each by ℓ to get $\lim_{\ell\to\infty} \frac{\ell-4}{\ell} = \lim_{\ell\to\infty} \frac{1-\frac{4}{\ell}}{1} = \frac{1-0}{1} = \frac{1}{1} = 1$. Alternatively, use l'Hôpital's rule. Recall the tip following Example 10.

(7) $\left\{\frac{j-4}{j}(-1)^j\right\}$ diverges. Notes: With Exercise 6 in mind, the end of this sequence will have values that are alternately close to 1 and -1; the tail of the sequence is nearly $1, -1, 1, -1, 1, -1, 1, -1, 1, -1, \ldots$ which diverges because it doesn't approach a single value. To show this formally, use the formal definition of a limiting value and choose ϵ less than one (such as $\epsilon = 0.5$). See the paragraph regarding uniqueness on the second page of this chapter; similar reasoning works to show that this sequence diverges.

(8) $a_k = \frac{a_{k-1}}{4}$ with $a_1 = 8$ converges to $\lim_{k\to\infty} \frac{8}{4^k} = 0$. Notes: The sequence is $8, 2, \frac{1}{2}, \frac{1}{8}, \frac{1}{32}, \frac{1}{128}, \ldots$ Factor out the 8 to write this as $8\left(1, \frac{1}{4}, \frac{1}{16}, \frac{1}{64}, \frac{1}{256}, \frac{1}{1024}\right), \ldots$ This is the same as $\left\{\frac{32}{4^k}\right\}$, which converges to $\lim_{k\to\infty} \frac{32}{4^k} = 0$. $\left\{\frac{1}{c^k}\right\}$ converges to zero if $|c| > 1$. $\lim_{k\to\infty}(ca_k) = cL_a$. (The two c's aren't the same; one is 4, the other is 32.)

(9) $\left\{\frac{9}{3^m}\right\}$ converges to $\lim_{m\to\infty} \frac{9}{3^m} = 0$. Notes: $\left\{\frac{1}{c^m}\right\}$ converges to zero if $|c| > 1$. $\lim_{m\to\infty}(ca_m) = cL_a$. (The two c's aren't the same; one is 3, the other is 9.).

(10) $\left\{\frac{3^j}{2^j}\right\}$ diverges. Notes: $\frac{3^j}{2^j} = \left(\frac{3}{2}\right)^j = 1.5^j$. $\{c^n\}$ diverges if $|c| > 1$. Here, $c = 1.5$.

(11) $\left\{\frac{2\ell-5}{8\ell+3}\right\}$ converges to $\lim_{\ell\to\infty} \frac{2\ell-5}{8\ell+3} = \frac{1}{4}$. Notes: Divide the numerator and denominator each by ℓ to get $\lim_{\ell\to\infty} \frac{2\ell-5}{8\ell+3} = \lim_{\ell\to\infty} \frac{2-\frac{5}{\ell}}{8+\frac{3}{\ell}} = \frac{2-0}{8+0} = \frac{2}{8} = \frac{1}{4}$. Alternatively, use l'Hôpital's rule. Recall the tip following Example 10.

(12) $\left\{\frac{n-3n^2}{6-n}\right\}$ diverges. Notes: Divide the numerator and denominator each by n^2 to get $\lim_{n\to\infty} \frac{n-3n^2}{6-n} = \lim_{n\to\infty} \frac{\frac{1}{n}-3}{\frac{6}{n^2}-\frac{1}{n}} = \frac{0-3}{0-0} = \frac{-3}{0} = \infty$. Alternatively, use l'Hôpital's rule. Recall the tip following Example 10.

(13) $\left\{\frac{m}{m^2+2m+4}\right\}$ converges to $\lim_{m\to\infty} \frac{m}{m^2+2m+4} = 0$. Notes: Divide the numerator and denominator each by m^2 to get $\lim_{m\to\infty} \frac{m}{m^2+2m+4} = \lim_{m\to\infty} \frac{\frac{1}{m}}{1+\frac{2}{m}+\frac{4}{m^2}} = \frac{0}{1+0+0} = \frac{0}{1} = 0$. Alternatively, use l'Hôpital's rule. Recall the tip following Example 10.

(14) $\left\{\frac{5+9i-2i^2}{4-3i+6i^2}\right\}$ converges to $\lim_{i\to\infty} \frac{5+9i-2i^2}{4-3i+6i^2} = -\frac{1}{3}$. Notes: Divide the numerator and denominator each by i^2 to get $\lim_{i\to\infty} \frac{5+9i-2i^2}{4-3i+6i^2} = \lim_{i\to\infty} \frac{\frac{5}{i^2}+\frac{9}{i}-2}{\frac{4}{i^2}-\frac{3}{i}+6} = \frac{0+0-2}{0-0+6} = \frac{-2}{6} = -\frac{1}{3}$. Alternatively, use l'Hôpital's rule. Recall the tip following Example 10.

(15) $\left\{7 - \frac{2}{k}\right\}$ converges to $\lim_{k\to\infty}\left(7 - \frac{2}{k}\right) = 7$. Notes: Let $\{a_k\} = \{7\}$ and $\{b_k\} = \left\{-\frac{2}{k}\right\}$. Then $\lim_{k\to\infty}(a_k + b_k) = L_a + L_b = 7 + 0$ because $L_a = \lim_{k\to\infty} 7 = 7$ and $L_b = \lim_{k\to\infty}\left(-\frac{2}{k}\right) = -2\lim_{k\to\infty}\left(\frac{1}{k}\right) = 0$.

Tip: It often helps to write out the first terms of the sequence. For example, in Exercise 15, $\left\{7 - \frac{2}{k}\right\} \approx 5, 6, 6.333, 6.5, 6.6, 6.667, 6.714, 6.75, 6.778, 6.8, \ldots$

(16) $\left\{\left(3 + \frac{1}{j}\right)\left(2 - \frac{1}{j}\right)\right\}$ converges to $\lim_{j\to\infty}\left[\left(3 + \frac{1}{j}\right)\left(2 - \frac{1}{j}\right)\right] = 6$. Notes: Let $\{a_j\} = \left\{3 + \frac{1}{j}\right\}$ and $\{b_j\} = \left\{2 - \frac{1}{j}\right\}$. Then $\lim_{j\to\infty} a_j b_j = L_a L_b = (3)(2) = 6$ using the results of Exercise 15. Using the above tip, $\left\{\left(3 + \frac{1}{j}\right)\left(2 - \frac{1}{j}\right)\right\} = 4, 5.25, 5.556, 5.688, 5.76, \ldots$

(17) $\{\sqrt{n}\}$ diverges. Notes: $\{n\}$ diverges. Taking the squareroot of each value won't change this since $\{n\}$ grows to infinity.

(18) $\left\{\frac{1}{\sqrt{m+6}}\right\}$ converges to $\lim_{m\to\infty}\frac{1}{\sqrt{m+6}} = 0$. Notes: $\left\{\frac{1}{m}\right\}$ converges to zero; $\left\{\frac{1}{m+6}\right\}$ also converges to zero since it merely shifts the index by six ($m + 6$ and m each grow to infinity, while their reciprocals shrink to zero). Let $f(a_m) = \sqrt{a_m}$ and $a_m = \frac{1}{m+6}$. Note that $\frac{1}{\sqrt{m+6}} = \sqrt{\frac{1}{m+6}}$. Since $\left\{\frac{1}{m+6}\right\}$ converges to $L = 0$ and since $f(a_m) = \sqrt{a_m}$ is continuous at $f(L) = \sqrt{0} = 0$, $\{f(a_m)\} = \sqrt{a_m}$ converges to $f(L) = \sqrt{0} = 0$.

(19) $\left\{\sqrt{\frac{4i^2+3}{9i^2-8}}\right\}$ converges to $\lim_{i\to\infty}\sqrt{\frac{4i^2+3}{9i^2-8}} = \frac{2}{3}$. Notes: Divide the numerator and denominator each by i^2 to get $\lim_{i\to\infty}\frac{4i^2+3}{9i^2-8} = \lim_{i\to\infty}\frac{4+\frac{3}{i^2}}{9-\frac{8}{i^2}} = \frac{4+0}{9-0} = \frac{4}{9}$. Alternatively, use l'Hôpital's rule. Recall the tip following Example 10. Now let $f(a_i) = \sqrt{a_i}$ and $a_i = \frac{4i^2+3}{9i^2-8}$. Since $\left\{\frac{4i^2+3}{9i^2-8}\right\}$ converges to $L = \frac{4}{9}$ and since $f(a_i) = \sqrt{a_i}$ is continuous at $f(L) = \sqrt{\frac{4}{9}} = \frac{2}{3}$, $\{f(a_i)\} = \sqrt{a_i}$ converges to $f(L) = \sqrt{\frac{4}{9}} = \frac{2}{3}$.

(20) $\left\{\sqrt{\frac{\ell^3-1}{\ell^2+5}}\right\}$ diverges. Notes: Divide the numerator and denominator each by ℓ^3 to get $\lim_{\ell\to\infty}\frac{\ell^3-1}{\ell^2+5} = \lim_{\ell\to\infty}\frac{1-\frac{1}{\ell^3}}{\frac{1}{\ell}+\frac{5}{\ell^3}} = \frac{1-0}{0+0} = \frac{1}{0} = \infty$. Alternatively, use l'Hôpital's rule. Recall the tip following Example 10. Taking the square root of each term of a sequence that grows to infinity won't stop it from diverging. Recall the solution to Exercise 17.

(21) $\left\{\frac{\sqrt{j}}{j^2}\right\}$ converges to $\lim_{j\to\infty}\frac{\sqrt{j}}{j^2} = 0$. Notes: $\frac{\sqrt{j}}{j^2} = \frac{j^{1/2}}{j^2} = j^{\frac{1}{2}-2} = j^{-3/2} = \frac{1}{j^{3/2}} = \frac{1}{j^{1.5}}$ according to the rules $x^{1/2} = \sqrt{x}$ and $\frac{x^p}{x^q} = x^{p-q}$. $\left\{\frac{1}{j^c}\right\}$ converges to zero if $c > 0$.

(22) $\{\sqrt{\cos n}\}$ diverges. Notes: $\{\cos n\}$ diverges (similar to Example 3) because a periodic function forms a repeating pattern; it doesn't converge to any single value. Taking the square root of each term won't change this.

(23) $a_{m+2} = \frac{a_{m+1}}{a_m}$ with $a_1 = 2$ and $a_2 = 8$ diverges. Notes: This recursion formula makes the repeating pattern $2, 8, 4, \frac{1}{2}, \frac{1}{8}, \frac{1}{4}, 2, 8, \ldots$ since $\frac{8}{2} = 4$, $\frac{4}{8} = \frac{1}{2}$, $\frac{1/2}{4} = \frac{1}{8}$, $\frac{1/8}{1/2} = \frac{1}{4}$, $\frac{1/4}{1/8} = 2$, $\frac{2}{1/4} = 8, \ldots$ This sequence repeats forever. Since it never converges to a single value, it diverges.

(24) $\{\tan(2k)\}$ diverges. Notes: tangent diverges (similar to Exercise 22 and Example 3) because a periodic function repeats; it doesn't converge to any single value.

(25) $\left\{\tan\frac{1}{2\ell}\right\}$ converges to $\lim_{\ell\to\infty}\tan\frac{1}{2\ell} = 0$. Notes: This problem is similar to Example 4, whereas the previous problem is similar to Example 3; compare Examples 3-4. In this problem, the angle approaches zero, whereas in the previous problem, the angle grows to infinity. Let $f(a_\ell) = \tan a_\ell$ and $a_\ell = \frac{1}{2\ell}$. We know that $\{a_\ell\} = \left\{\frac{1}{2\ell}\right\}$ converges to $L = 0$ since $\left\{\frac{1}{\ell}\right\}$ converges to zero, according to the rule $\lim_{\ell\to\infty}(ca_\ell) = cL_a$. Since $f(a_\ell) = \tan a_\ell$ is continuous at $f(L) = \tan 0$, $\{f(a_\ell)\} = \{\tan a_\ell\} = \left\{\tan\left(\frac{1}{2\ell}\right)\right\}$ converges to $f(L) = \tan 0 = 0$.

(26) $\left\{\sin\frac{1}{n} + \cos\frac{1}{n}\right\}$ converges to $\lim_{n\to\infty}\sin\frac{1}{n} + \cos\frac{1}{n} = 1$. Notes: $\left\{\frac{1}{n}\right\}$ converges to zero. Let $f(a_n) = \sin a_n$, $g(a_n) = \cos a_n$, and $a_n = \frac{1}{n}$. Since $f(a_n) = \sin a_n$ and $g(a_n) = \cos a_n$ are continuous at $f(L_1) = \sin 0$ and $g(L_2) = \cos 0$, $\{f(a_n)\} = \{\sin a_n\} = \left\{\sin\frac{1}{n}\right\}$ and $\{g(a_n)\} = \{\cos a_n\} = \left\{\cos\frac{1}{n}\right\}$ converge to $f(L_1) = \sin 0 = 0$ and $g(L_2) = \cos 0 = 1$, respectively. Apply $\lim_{n\to\infty}\big(f(a_n) + g(a_n)\big) = f(L_1) + g(L_2) = 0 + 1 = 1$. (This is the first practical limit theorem stated in the chapter, adapted here since these sequences were formed with functions.)

(27) $\left\{\frac{120}{k!}\right\}$ converges to $\lim_{k\to\infty}\frac{120}{k!} = 0$. Notes: $\left\{\frac{c^n}{n!}\right\}$ converges to zero for all c. Let $c = 1$ to see that $\left\{\frac{1}{n!}\right\}$ converges to zero. Also, $\lim_{n\to\infty}(120a_n) = 120L_a$.

(28) $\left\{\frac{1}{(j+2)!}\right\}$ converges to $\lim_{j\to\infty}\frac{1}{(j+2)!} = 0$. Notes: $\left\{\frac{1}{j!}\right\}$ converges to zero (see the solution to Exercise 27). Here, the $+2$ simply shifts the index by 2; as j grows to infinity, $j!$ and $(j+2)!$ both grow to infinity, while their reciprocals shrink to zero.

(29) $\left\{-\frac{n!}{3^n}\right\}$ diverges. Notes: $\left\{\frac{n!}{c^n}\right\}$ diverges for all c. The minus sign makes it diverge to $-\infty$ instead of ∞.

(30) $\left\{\frac{3^m}{m!}\right\}$ converges to $\lim_{m\to\infty}\frac{3^m}{m!} = 0$. Notes: $\left\{\frac{c^m}{m!}\right\}$ converges to zero for all c.

(31) $\{e^k\}$ diverges. Notes: $\{c^k\}$ diverges if $|c| > 1$. Here, $e \approx 2.71828$.

(32) $\{e^{-n}\}$ converges to $\lim_{n\to\infty} e^{-n} = 0$. Notes: $\left\{\frac{1}{c^m}\right\}$ converges to zero if $|c| > 1$. Here, $e \approx 2.71828$. Note that $e^{-n} = \frac{1}{e^n}$.

(33) $\left\{\frac{m^2}{2^m}\right\}$ converges to $\lim_{m\to\infty}\frac{m^2}{2^m} = 0$. Notes: Apply l'Hôpital's rule to the function $f(x) = \frac{x^2}{2^x}$. Recall from calculus that $\frac{d}{dx}2^x = 2^x \ln 2$.

$$\lim_{x\to\infty}\frac{x^2}{2^x} = \frac{\left.\frac{d}{dx}x^2\right|_{x\to\infty}}{\left.\frac{d}{dx}2^x\right|_{x\to\infty}} = \frac{\lim\limits_{x\to\infty} 2x}{\lim\limits_{x\to\infty} 2^x \ln 2} = \frac{\infty}{\infty} \text{ (indeterminate)}$$

Since the result is again indeterminate, apply l'Hôpital's rule again, this time to the function $g(x) = \frac{2x}{2^x \ln 2}$ obtained above. (Note that ln 2 is a constant.)

$$\lim_{x\to\infty}\frac{2x}{2^x \ln 2} = \frac{\left.\frac{d}{dx}2x\right|_{x\to\infty}}{\left.\frac{d}{dx}(2^x \ln 2)\right|_{x\to\infty}} = \frac{\lim\limits_{x\to\infty} 2}{\lim\limits_{x\to\infty} 2^x (\ln 2)^2} = \frac{2}{\infty} = 0$$

Since $\lim_{x\to\infty}\frac{x^2}{2^x} = \lim_{x\to\infty}\frac{2x}{2^x \ln 2} = 0$ and $\frac{x^2}{2^x}$ and $\frac{2x}{2^x \ln 2}$ are defined for all x, it follows that $\lim_{m\to\infty}\frac{m^2}{2^m} = 0$.

(34) $\left\{\frac{\ell}{\ln \ell}\right\}$ diverges. Notes: Apply l'Hôpital's rule to the function $f(x) = \frac{x}{\ln x}$. Recall from calculus that $\frac{d}{dx}\ln x = \frac{1}{x}$.

$$\lim_{x\to\infty}\frac{x}{\ln x}=\frac{\left.\frac{d}{dx}x\right|_{x\to\infty}}{\left.\frac{d}{dx}\ln x\right|_{x\to\infty}}=\frac{\lim_{x\to\infty}1}{\lim_{x\to\infty}\frac{1}{x}}=\frac{1}{0}=\infty$$

(35) $\left\{\frac{\ln(3k)}{\ln(2k)}\right\}$ converges to $\lim_{k\to\infty}\frac{\ln(3k)}{\ln(2k)}=1$ (it isn't 3/2; many students are surprised by this, but read the notes carefully and thoroughly). Notes: Apply l'Hôpital's rule to the function $f(x)=\frac{\ln(3x)}{\ln(2x)}$. Did you notice and remember in first-year calculus that $\frac{d}{dx}\ln(cx)=\frac{1}{x}$. Most students are surprised that the constant c cancels out in this derivative. There are two ways to see this. One way is to remember the product rule of logarithms: $\ln(pq)=\ln p+\ln q$ such that $\frac{d}{dx}\ln(cx)=\frac{d}{dx}(\ln c+\ln x)=0+\frac{1}{x}$; see, the constant c vanishes because $\ln c$ is a constant. A second way is to apply the chain rule carefully; after you take a derivative of $\ln(cx)$ with respect to x, then you take a derivative of what's inside (the cx) with respect to x: $\frac{d}{dx}\ln(cx)=\frac{1}{cx}\frac{d}{dx}(cx)=\frac{1}{cx}c=\frac{1}{x}$. If you do this correctly, the c cancels out (to the surprise of many students). If you still don't believe it, we'll present a clear and convincing way to check the answer at the end of the solution.

$$\lim_{x\to\infty}\frac{\ln(3x)}{\ln(2x)}=\frac{\left.\frac{d}{dx}\ln(3x)\right|_{x\to\infty}}{\left.\frac{d}{dx}\ln(2x)\right|_{x\to\infty}}=\lim_{x\to\infty}\frac{1/x}{1/x}=\lim_{x\to\infty}1=1$$

After taking derivatives, the numerator and denominator are identical. Since $\lim_{x\to\infty}\frac{\ln(3x)}{\ln(2x)}=1$ and $\frac{\ln(3x)}{\ln(2x)}$ is defined for all $x\geq 1$ (as noted in the text, it's defined for $x>0$, but this is all we need), it follows that $\lim_{k\to\infty}\frac{\ln(3k)}{\ln(2k)}=1$. (Again, it's 1 and not 3/2. Now we'll check the answer so you can see firsthand that 3/2 is incorrect.)

Check the answer: There is a simple way to check all of the answers in this chapter. We've mentioned it before as a "tip." Plug large values of the index into your calculator and see what happens. Here, the sequence is $\left\{\frac{\ln(3k)}{\ln(2k)}\right\}$. When $k=100$, we get $\frac{\ln 300}{\ln 200}\approx 1.0765$. When $k=1000$, we get $\frac{\ln 3000}{\ln 2000}\approx 1.0533$. When k is 1,000,000, we get $\frac{\ln 3{,}000{,}000}{\ln 2{,}000{,}000}\approx 1.0279$. See! As k increases, the ratio clearly gets closer to 1.

(36) $\left\{\frac{n^2}{e^n}\right\}$ converges to $\lim\limits_{n\to\infty}\frac{n^2}{e^n} = 0$. Notes: Apply l'Hôpital's rule to the function $f(x) = \frac{x^2}{e^x}$. Recall from calculus that $\frac{d}{dx}e^x = e^x$. (The exponential function, e^x, is the function whose value equals its own slope.)

$$\lim_{x\to\infty}\frac{x^2}{e^x} = \frac{\left.\frac{d}{dx}x^2\right|_{x\to\infty}}{\left.\frac{d}{dx}e^x\right|_{x\to\infty}} = \frac{\lim\limits_{x\to\infty}2x}{\lim\limits_{x\to\infty}e^x} = \frac{2}{\infty} = 0$$

Since $\lim\limits_{x\to\infty}\frac{x^2}{e^x} = 0$ and $\frac{x^2}{e^x}$ is defined for all x, it follows that $\lim\limits_{n\to\infty}\frac{n^2}{e^n} = 0$.

(37) $\left\{\sqrt{\frac{1}{j!}}\right\}$ converges to $\lim\limits_{j\to\infty}\sqrt{\frac{1}{j!}} = 0$. Notes: $\left\{\frac{1}{j!}\right\}$ converges to zero (recall the solution to Exercise 27). Let $f(a_j) = \sqrt{a_j}$ and $a_j = \frac{1}{j!}$. Since $\left\{\frac{1}{j!}\right\}$ converges to $L = 0$ and since $f(a_j) = \sqrt{a_j}$ is continuous at $f(L) = \sqrt{0} = 0$, $\{f(a_j)\} = \sqrt{a_j}$ converges to $f(L) = \sqrt{0} = 0$.

(38) $\left\{\frac{1}{3^{1/i}}\right\}$ converges to $\lim\limits_{i\to\infty}\frac{1}{3^{1/i}} = 1$. Notes: $\{3^{1/i}\}$ converges to one (see Example 15), which means that its reciprocal, $\left\{\frac{1}{3^{1/i}}\right\}$, converges to $\frac{1}{1} = 1$.

(39) $\{n^{1/n}\}$ converges to $\lim\limits_{n\to\infty}n^{1/n} = 1$. Notes: The tip from Example 15 suggests taking the natural logarithm of both sides of $a_n = n^{1/n}$ to get $\ln a_n = \ln n^{1/n}$. Apply the rule $\ln p^q = q\ln p$ to get $\ln a_n = \frac{1}{n}\ln n$. Apply l'Hôpital's rule to the function $f(x) = \frac{\ln x}{x}$. Recall from calculus that $\frac{d}{dx}\ln x = \frac{1}{x}$.

$$\lim_{x\to\infty}\frac{\ln x}{x} = \frac{\left.\frac{d}{dx}\ln x\right|_{x\to\infty}}{\left.\frac{d}{dx}x\right|_{x\to\infty}} = \frac{\lim\limits_{x\to\infty}\frac{1}{x}}{\lim\limits_{x\to\infty}1} = \frac{0}{1} = 0$$

Since $\lim\limits_{x\to\infty}\frac{\ln x}{x} = 0$ and $\frac{\ln x}{x}$ is defined for all $x \geq 1$ (as noted in the text, it's defined for $x > 0$, but this is all we need), it follows that $\{\ln a_n\} = \left\{\frac{1}{n}\ln n\right\}$ converges to zero. Now use the property of logarithms that $e^{\ln a_n} = a_n$, which equals $n^{1/n}$. Let $f(b_n) = e^{b_n}$ and $b_n = \ln a_n$. We already know that $\{b_n\} = \{\ln a_n\}$ converges to $L = 0$. Since $f(b_n) = e^{b_n}$ is continuous at $f(L) = e^0 = 1$, $\{f(b_n)\} = \{e^{b_n}\} = \{e^{\ln a_n}\} = \{a_n\}$

$= \{n^{1/n}\}$ converges to $f(L) = e^0 = 1$. The main idea is that the exponent of $n^{1/n}$ shrinks to zero, and any (nonzero) number raised to the power of zero equals one.

(40) $\left\{\left(1+\frac{3}{k}\right)^k\right\}$ converges to $\lim_{k\to\infty}\left(1+\frac{3}{k}\right)^k = e^3$. Notes: Use the tip from Example 15. Take the natural logarithm of both sides of $a_k = \left(1+\frac{3}{k}\right)^k$ to get $\ln a_k = \ln\left(1+\frac{3}{k}\right)^k$. Apply the rule $\ln p^q = q\ln p$ to get $\ln a_k = k\ln\left(1+\frac{3}{k}\right)$. We could use l'Hôpital's rule if we could write this as a ratio. Observe that $k\ln\left(1+\frac{3}{k}\right) = \frac{\ln\left(1+\frac{3}{k}\right)}{1/k}$. Apply l'Hôpital's rule to the function $f(x) = \frac{\ln\left(1+\frac{3}{x}\right)}{1/x}$. Recall from calculus that $\frac{d}{dx}\ln x = \frac{1}{x}$. Use the chain rule; after taking a derivative of the natural logarithm with respect to x, take a derivative of the inside (which is $1+\frac{3}{x}$) with respect to x.

$$\lim_{x\to\infty}\frac{\ln\left(1+\frac{3}{x}\right)}{1/x} = \frac{\frac{d}{dx}\ln\left(1+\frac{3}{x}\right)\Big|_{x\to\infty}}{\frac{d}{dx}\frac{1}{x}\Big|_{x\to\infty}}$$

The derivative in the numerator is $\frac{d}{dx}\ln\left(1+\frac{3}{x}\right) = \frac{1}{1+3/x}\left(-\frac{3}{x^2}\right)$, where we applied the chain rule from first-semester calculus (as previously mentioned). The derivative in the denominator is $-\frac{1}{x^2}$. When we divide this, the minus signs and the x^2's both cancel out: $\frac{3}{x^2} \div \frac{1}{x^2} = \frac{3}{x^2}x^2 = 3$ (since the way to divide by a fraction is to multiply by its reciprocal). That means the fraction simplifies to:

$$\lim_{x\to\infty}\left[\left(1+\frac{3}{x}\right)(3)\right] = \lim_{x\to\infty}\left(3+\frac{9}{x}\right) = 3+0 = 3$$

Since $\lim_{x\to\infty}\frac{\ln\left(1+\frac{3}{x}\right)}{1/x} = 3$ and $\frac{\ln\left(1+\frac{3}{x}\right)}{1/x}$ is defined for all $x \geq 1$ (as noted in the text, it's defined for $x > 0$, but this is all we need), it follows that $\{\ln a_k\} = \left\{k\ln\left(1+\frac{3}{k}\right)\right\}$ converges to 3. Now use the property of logarithms that $e^{\ln a_k} = a_k$,which equals $\left(1+\frac{3}{k}\right)^k$. Let $f(b_k) = e^{b_k}$ and $b_k = \ln a_k$. We already know that $\{b_k\} = \{\ln a_k\}$ converges to $L = 3$. Since $f(b_k) = e^{b_k}$ is continuous at $f(L) = e^3$, $\{f(b_k)\} = \{e^{b_k}\} = \{e^{\ln a_k}\} = \{a_k\} = \left\{\left(1+\frac{3}{k}\right)^k\right\}$ converges to $f(L) = e^3$.

Chapter 3 Answers

(1)

$$\sum_{n=1}^{4}(-1)^n n^4 = (-1)^1 1^4 + (-1)^2 2^4 + (-1)^3 3^4 + (-1)^4 4^4$$

$$= -1 + 16 - 81 + 256 = -82 + 272 = 190$$

(2)

$$\sum_{k=1}^{5}\frac{(-1)^{k+1}}{k(k+1)} = \frac{(-1)^2}{1(2)} + \frac{(-1)^3}{2(3)} + \frac{(-1)^4}{3(4)} + \frac{(-1)^5}{4(5)} + \frac{(-1)^6}{5(6)}$$

$$= \frac{1}{2} - \frac{1}{6} + \frac{1}{12} - \frac{1}{20} + \frac{1}{30} = \frac{30}{60} - \frac{10}{60} + \frac{5}{60} - \frac{3}{60} + \frac{2}{60} = \frac{37-13}{60} = \frac{24}{60} = \frac{2}{5}$$

(3)

$$\sum_{j=4}^{10}(3j^2 - 12) = 3(4)^2 - 12 + 3(5)^2 - 12 + 3(6)^2 - 12 + 3(7)^2 - 12 + 3(8)^2 - 12$$

$$+ 3(9)^2 - 12 + 3(10)^2 - 12$$

$$= 48 + 75 + 108 + 147 + 192 + 243 + 300 - 7(12) = 1113 - 84 = 1029$$

(4) Note: There are $23 - 14 + 1 = 9 + 1 = 10$ terms. Each term is a 7. It's a **common mistake** for students to think that there are $23 - 14 = 9$ terms and to multiply $9(7)$ to make 63 (which is an incorrect answer). To demonstrate that there are 10 sevens (and not 9 sevens), let's count the numbers fourteen thru twenty-three: 14, 15, 16, 17, 18, 19, 20, 21, 22, 23. See! There are 10 terms (not 9). Imagine that you read pages 3-4 in a book (starting at the top of page 3 and finishing at the bottom of page 4). It would be incorrect to subtract $4 - 3 = 1$ and say that you read just 1 page. You read page 3 and page 4, so that's 2 pages. You have to add 1 after subtracting in situations like these. Not doing so is a common mistake that students make.

$$\sum_{m=14}^{23} 7 = 7 + 7 + 7 + 7 + 7 + 7 + 7 + 7 + 7 + 7 = 10(7) = 70$$

(5) Note: The argument is in radians. If you prefer degrees, multiply by 180 and divide by π. For example, $\frac{\pi}{3}$ radians $= \frac{\pi}{3}\frac{180}{\pi} = 60$ degrees.

$$\sum_{\ell=3}^{9}\cos\left(\frac{\ell\pi}{3}\right) = \cos\frac{3\pi}{3} + \cos\frac{4\pi}{3} + \cos\frac{5\pi}{3} + \cos\frac{6\pi}{3} + \cos\frac{7\pi}{3} + \cos\frac{8\pi}{3} + \cos\frac{9\pi}{3}$$

$$= \cos\pi + \cos\frac{4\pi}{3} + \cos\frac{5\pi}{3} + \cos 2\pi + \cos\frac{7\pi}{3} + \cos\frac{8\pi}{3} + \cos 3\pi$$

$$= -1 - \frac{1}{2} + \frac{1}{2} + 1 + \frac{1}{2} - \frac{1}{2} - 1 = 2 - 3 = -1$$

Note: In degrees, it's $\cos 180° + \cos 240° + \cos 300° + \cos 360° + \cos 420° + \cos 480° + \cos 540° = -1 - \frac{1}{2} + \frac{1}{2} + 1 + \frac{1}{2} - \frac{1}{2} - 1 = 2 - 3 = -1$.

(6) Here and elsewhere in this book, i is an index like k or n. (It's **not** imaginary.)

$$\sum_{i=5}^{11}\frac{6i(-1)^i}{(i-2)^2} = \frac{6(5)(-1)^5}{(5-2)^2} + \frac{6(6)(-1)^6}{(6-2)^2} + \frac{6(7)(-1)^7}{(7-2)^2} + \frac{6(8)(-1)^8}{(8-2)^2} + \frac{6(9)(-1)^9}{(9-2)^2}$$

$$+ \frac{6(10)(-1)^{10}}{(10-2)^2} + \frac{6(11)(-1)^{11}}{(11-2)^2} = -\frac{30}{3^2} + \frac{36}{4^2} - \frac{42}{5^2} + \frac{48}{6^2} - \frac{54}{7^2} + \frac{60}{8^2} - \frac{66}{9^2}$$

$$= -\frac{10}{3} + \frac{9}{4} - \frac{42}{25} + \frac{4}{3} - \frac{54}{49} + \frac{15}{16} - \frac{22}{27}$$

$$= \frac{-1{,}764{,}000 + 1{,}190{,}700 - 889{,}056 + 705{,}600 - 583{,}200 + 496{,}125 - 431{,}200}{529{,}200}$$

$$= \frac{-1{,}275{,}031}{529{,}200} \approx -2.41$$

(7)

$$\sum_{m=1}^{3}\sum_{n=1}^{3}\frac{m^2 n}{2} = \frac{1^2(1)}{2} + \frac{2^2(1)}{2} + \frac{3^2(1)}{2} + \frac{1^2(2)}{2} + \frac{2^2(2)}{2} + \frac{3^2(2)}{2} + \frac{1^2(3)}{2} + \frac{2^2(3)}{2} + \frac{3^2(3)}{2}$$

$$= \frac{1}{2} + 2 + \frac{9}{2} + 1 + 4 + 9 + \frac{3}{2} + 6 + \frac{27}{2} = 22 + \frac{40}{2} = 22 + 20 = 42$$

(8)

$$\sum_{i=1}^{3}\sum_{j=1}^{3}(3i - j)$$

$$= 3(1) - 1 + 3(2) - 1 + 3(3) - 1 + 3(1) - 2 + 3(2) - 2 + 3(3) - 2$$
$$+ 3(1) - 3 + 3(2) - 3 + 3(3) - 3$$

$$= 3 - 1 + 6 - 1 + 9 - 1 + 3 - 2 + 6 - 2 + 9 - 2 + 3 - 3 + 6 - 3 + 9 - 3$$

$$= 2 + 5 + 8 + 1 + 4 + 7 + 0 + 3 + 6 = 36$$

(9)

$$\sum_{k=1}^{3}\sum_{\ell=k}^{3}\frac{6\ell}{k} = \frac{6(1)}{1} + \frac{6(1)}{2} + \frac{6(1)}{3} + \frac{6(2)}{1} + \frac{6(2)}{2} + \frac{6(2)}{3} + \frac{6(3)}{1} + \frac{6(3)}{2} + \frac{6(3)}{3}$$

$$= 6 + 3 + 2 + 12 + 6 + 4 + 18 + 9 + 6 = 66$$

(10) Note: There are $3(3)(3) = 27$ pairs of terms (for 54 terms) all together.

$$\sum_{i=1}^{3}\sum_{j=1}^{3}\sum_{k=1}^{3}(k^2 - ij)$$

$$= 1^2 - 1(1) + 1^2 - 2(1) + 1^2 - 3(1) + 1^2 - 1(2) + 1^2 - 2(2) + 1^2$$
$$- 3(2) + 1^2 - 1(3) + 1^2 - 2(3) + 1^2 - 3(3) + 2^2 - 1(1) + 2^2 - 2(1)$$
$$+ 2^2 - 3(1) + 2^2 - 1(2) + 2^2 - 2(2) + 2^2 - 3(2) + 2^2 - 1(3) + 2^2$$
$$- 2(3) + 2^2 - 3(3) + 3^2 - 1(1) + 3^2 - 2(1) + 3^2 - 3(1) + 3^2 - 1(2)$$
$$+ 3^2 - 2(2) + 3^2 - 3(2) + 3^2 - 1(3) + 3^2 - 2(3) + 3^2 - 3(3)$$

$$= 1 - 1 + 1 - 2 + 1 - 3 + 1 - 2 + 1 - 4 + 1 - 6 + 1 - 3 + 1 - 6 + 1 - 9$$
$$+ 4 - 1 + 4 - 2 + 4 - 3 + 4 - 2 + 4 - 4 + 4 - 6 + 4 - 3 + 4 - 6 + 4 - 9$$
$$+ 9 - 1 + 9 - 2 + 9 - 3 + 9 - 2 + 9 - 4 + 9 - 6 + 9 - 3 + 9 - 6 + 9 - 9$$

$$= 14 - 3 + 14 - 6 + 14 - 9 + 14 - 6 + 14 - 12 + 14 - 18 + 14 - 9 + 14 - 18 + 14 - 27$$

$$= 11 + 8 + 5 + 8 + 2 - 4 + 5 - 4 - 13 = 39 - 21 = 18$$

(11) Note: Two possible answers are given below. Observe that $k = n + 10$.

$$\sum_{n=1}^{6}\frac{2}{n+10} = \sum_{k=11}^{16}\frac{2}{k}$$

Note: The 'trick' is to realize that $\frac{2}{11} + \frac{1}{6} + \frac{2}{13} + \frac{1}{7} + \frac{2}{15} + \frac{1}{8} = \frac{2}{11} + \frac{2}{12} + \frac{2}{13} + \frac{2}{14} + \frac{2}{15} + \frac{2}{16}$.

Exercises 11-22 are about rewriting series using summation notation. The instructions for these problems didn't ask you to find the sum.

(12) Note: The two answers below are related by $k = n + 6$.

$$\sum_{n=1}^{9}(-1)^{n+1}(2n + 12) = \sum_{k=7}^{15} 2k(-1)^{k+1}$$

Example: $(-1)^{9+1}[2(9) + 12] = (-1)^{10}(30) = 30$.

(13) Example: $(-1)^8[4(8)-2]=(1)(30)=30$

$$\sum_{n=1}^{8}(-1)^n(4n-2)=\sum_{n=1}^{8}(-1)^n 2(2n-1)$$

(14) Note: In the answers below, $k=n+2=i-1$.

$$\sum_{n=1}^{7}\frac{n+2}{(n+3)^2}=\sum_{k=3}^{9}\frac{k}{(k+1)^2}=\sum_{i=4}^{10}\frac{i-1}{i^2}$$

Example: $\frac{3}{(3+1)^2}=\frac{3}{4^2}=\frac{3}{16}$.

(15) Notes: In the answers below, $n=k+1$ such that $9-2n=9-2(k+1)=9-2k-2=7-2k$. (Of course, $9-2n=-2n+9$ and $7-2k=-2k+7$.)

$$\sum_{n=1}^{8}(9-2n)=\sum_{k=0}^{7}(7-2k)$$

(16) Examples: $2+(-1)^1=2-1=1$ and $2+(-1)^2=2+1=3$.

$$\sum_{n=1}^{\infty}[2+(-1)^n]$$

(17) Examples: $7(-3)^{1-1}=7(-3)^0=7(1)=7$ and $7(-3)^{2-1}=7(-3)^1=-21$.

$$\sum_{n=1}^{5}7(-3)^{n-1}=\sum_{n=0}^{4}7(-3)^k$$

(18) Note: In the answers below, $k=n+6$.

$$\sum_{n=1}^{7}25(n+6)=\sum_{k=7}^{13}25k$$

Example: $25(7+6)=25(13)=325$.

(19) Note: The 'trick' is to realize that $\frac{1}{6}+\frac{1}{3}+\frac{1}{2}+\frac{2}{3}+\frac{5}{6}=\frac{1}{6}+\frac{2}{6}+\frac{3}{6}+\frac{4}{6}+\frac{5}{6}$.

$$\sum_{n=1}^{5}\frac{n}{6}$$

(20) Examples: $\frac{32}{4^{8-1}}=\frac{32}{4^7}=\frac{32}{16{,}384}=\frac{1}{512}$ and $\frac{32}{4^{1-1}}=\frac{32}{4^0}=\frac{32}{1}=32$.

$$\sum_{n=1}^{8}\frac{32}{4^{n-1}}=\sum_{k=0}^{7}\frac{32}{4^k}=\sum_{n=1}^{8}\frac{128}{4^n}=\sum_{k=0}^{7}\frac{128}{4^{k+1}}$$

(21) Note: In the answers below, $n+1=k$ such that $n=k-1$. Observe that $(n+1)^2+n-1=k^2+(k-1)-1=k^2+k-2$.

$$\sum_{n=1}^{7}\frac{2n+7}{(n+1)^2+n-1}=\sum_{n=1}^{7}\frac{2n+7}{n^2+3n}=\sum_{n=1}^{7}\frac{2n+7}{n(n+3)}$$

$$=\sum_{k=2}^{8}\frac{2k+5}{k^2+k-2}=\sum_{k=2}^{8}\frac{2k+5}{(k+2)(k-1)}=\sum_{i=4}^{10}\frac{2i+1}{i(i-3)}$$

Examples: $\frac{2(1)+7}{(1+1)^2+1-1}=\frac{2+7}{2^2+0}=\frac{9}{4}$ agrees with $\frac{2(1)+7}{1^2+3(1)}=\frac{2+7}{1+3}=\frac{9}{4}$ and $\frac{2(1)+7}{1(1+3)}=\frac{2+7}{1(4)}=\frac{9}{4}$, as well as $\frac{2(2)+5}{2^2+2-2}=\frac{4+5}{4+2-2}=\frac{9}{4}$ and $\frac{2(2)+5}{(2+2)(2-1)}=\frac{4+5}{4(1)}=\frac{9}{4}$.

$\frac{2(2)+7}{(2+1)^2+2-1}=\frac{4+7}{3^2+1}=\frac{11}{10}$ agrees with $\frac{2(2)+7}{2^2+3(2)}=\frac{4+7}{4+6}=\frac{11}{10}$ as well as $\frac{2(3)+5}{3^2+3-2}=\frac{6+5}{9+3-2}=\frac{11}{10}$.

(22) Notes: The exponent, $(-1)^n$, alternates between 1 and -1; when n is odd, the exponent is -1, which makes the reciprocal of the fraction since $x^{-1}=\frac{1}{x}$ and $\left(\frac{x}{y}\right)^{-1}=\frac{y}{x}$. This series is $\frac{4}{1}+\frac{8}{5}+\frac{6}{27}+\frac{64}{7}+\frac{8}{125}+\frac{216}{9}+\frac{10}{343}+\frac{512}{11}+\frac{12}{729}+\frac{1000}{13}+\frac{14}{1331}$, which reduces to $4+\frac{8}{5}+\frac{2}{9}+\frac{64}{7}+\frac{8}{125}+24+\frac{10}{343}+\frac{512}{11}+\frac{4}{243}+\frac{1000}{13}+\frac{14}{1331}$.

$$\sum_{n=1}^{11}\left(\frac{n^3}{n+3}\right)^{(-1)^n}$$

(23) Notes: $k=n-2$ leads to $k+2=n$. Replace n with $k+2$ to get $4n-2=4(k+2)-2=4k+8-2=4k+6$. When $n=3$, $k=1$; compare $4(3)-2=12-2=10$ with $4(1)+6=4+6=10$. When $n=11$, $k=9$; compare $4(11)-2=44-2=42$ with $4(9)+6=36+6=42$.

$$\sum_{n=3}^{11}(4n-2)=\sum_{k=1}^{9}(4k+6)=\sum_{k=1}^{9}2(2k+3)$$

(24) Notes: $j=k+1$ leads to $j-1=k$. Replace k with $j-1$ to get $k-2=j-1-2=j-3$ and to get $k+2=j-1+2=j+1$. When $k=0, j=1$; compare $\frac{0-2}{0+2}=-\frac{2}{2}=-1$ with $\frac{1-3}{1+1}=-\frac{2}{2}=-1$. When $k=8, j=9$; compare $\frac{8-2}{8+2}=\frac{6}{10}=\frac{3}{5}$ with $\frac{9-3}{9+1}=\frac{6}{10}=\frac{3}{5}$.

$$\sum_{k=0}^{8}\frac{k-2}{k+2}=\sum_{j=1}^{9}\frac{j-3}{j+1}$$

(25) Notes: Since $\ell = -m$, replace m with $-\ell$ to get $m+1 = -\ell+1 = 1-\ell$. When $m = 0, \ell = 0$; compare $\frac{0^2}{0+1} = \frac{0}{1} = 0$ with $\frac{0^2}{1-0} = \frac{0}{1} = 0$. When $m = 4, \ell = -4$; compare $\frac{4^2}{4+1} = \frac{16}{5}$ with $\frac{(-4)^2}{1-(-4)} = \frac{16}{1+4} = \frac{16}{5}$. The values of m includes 0, 1, 2, 3, and 4; the values of ℓ include $-4, -3, -2, -1$, and 0. Since a sum goes from the least value to the greatest value, the lower limit is -4 and the upper limit is 0.

$$\sum_{m=0}^{4} \frac{m^2}{m+1} = \sum_{\ell=-4}^{0} \frac{\ell^2}{1-\ell}$$

(26) Notes: $n = i+3$ leads to $n-3 = i$. Replace i with $n-3$ to get $i^2+6i+3 = (n-3)^2+6(n-3)+3 = n^2-6n+9+6n-18+3 = n^2-6$. When $i = 1, n = 4$; compare $\frac{1^2+6(1)+3}{1+3} = \frac{1+6+3}{4} = \frac{10}{4} = \frac{5}{2}$ with $\frac{4^2-6}{4} = \frac{16-6}{4} = \frac{10}{4} = \frac{5}{2}$. When $i = 7, n = 10$; compare $\frac{7^2+6(7)+3}{7+3} = \frac{49+42+3}{10} = \frac{94}{10} = \frac{47}{5}$ with $\frac{10^2-6}{10} = \frac{100-6}{10} = \frac{94}{10} = \frac{47}{5}$.

$$\sum_{i=1}^{7} \frac{i^2+6i+3}{i+3} = \sum_{n=4}^{10} \frac{n^2-6}{n} = \sum_{n=4}^{10} \left(n - \frac{6}{n}\right)$$

(27) Note: Use the triangular number formula with $N = 18$.

$$\frac{N(N+1)}{2} = \frac{18(19)}{2} = 171$$

(28) Notes: Use the triangular number formula with $N = 144$. This series is $1+2+3+4+5+6+\cdots+144$.

$$\frac{N(N+1)}{2} = \frac{144(145)}{2} = 10{,}440$$

(29) Notes: You can factor out the 4. The series is $4(25)+4(26)+4(27)+4(28)+4(29)+\cdots+4(75) = 4(25+26+27+28+29+\cdots 75)$. Since this sum doesn't begin at 1, you need to find the sum from 1 to 75 and then subtract the sum from 1 to 24. This will leave 25 thru 75. We used this same concept in Example 9.

$$\sum_{i=25}^{75} 4i = 4\sum_{i=25}^{75} i = 4\left(\sum_{i=1}^{75} i - \sum_{i=1}^{24} i\right)$$

$$= 4\left[\frac{75(76)}{2} - \frac{24(25)}{2}\right] = 4(2850-300) = 4(2550) = 10{,}200$$

(30) Notes: The difference between any two consecutive terms is 8. This is an arithmetic series with $N = 15$, $d = 8$, $a = 8$, and $b = 120$.

$$\frac{N(a+b)}{2} = \frac{15(8+120)}{2} = \frac{15(128)}{2} = 960$$

$$Na + \frac{N(N-1)d}{2} = 15(8) + \frac{15(14)8}{2} = 120 + 840 = 960$$

Alternate solution: Factor out the 8 to get $8(1 + 2 + 3 + \cdots + 15)$.

$$8\frac{15(16)}{2} = 960$$

Note: You can simply add the given terms together to **check your answer**, but for your solution, the instructions ask you to **use a formula from this chapter**.

(31) Notes: The difference between any two consecutive terms is 7. This is an arithmetic series with $N = 12$, $d = 7$, $a = 18$, and $b = 95$.

$$\frac{N(a+b)}{2} = \frac{12(18+95)}{2} = \frac{12(113)}{2} = 678$$

$$Na + \frac{N(N-1)d}{2} = 12(18) + \frac{12(11)7}{2} = 216 + 462 = 678$$

(32) Notes: This is an arithmetic series. Compare the general form $a + (m-1)d$ with $6 + 4(m-1)$ to see that $a = 6$ and $d = 4$. Looking at the sum, $N = 50$.

$$Na + \frac{N(N-1)d}{2} = 50(6) + \frac{50(49)4}{2} = 300 + 4900 = 5200$$

Alternatively, first find the last term: $b = a + (N-1)d = 6 + (50-1)4 = 202$.

$$\frac{N(a+b)}{2} = \frac{50(6+202)}{2} = \frac{50(208)}{2} = 5200$$

(33) Notes: This is an arithmetic series. Before identifying the constants, we need to put $5\ell - 2$ in the standard form $a + (\ell - 1)d$. Somehow, we need to write $(\ell - 1)d$. The 'trick' is to subtract and add five: $5\ell - 2 = 5\ell - 5 + 5 - 2 = 5(\ell - 1) + 3$, which we can rewrite as $3 + (\ell - 1)5$. Since it's easy to make a mistake with this, let's write out the first few terms for each expression to check our work thus far:

- $3 + (\ell - 1)5 \to 3 + (3 + 5) + [3 + 2(5)] + \cdots = 3 + 8 + 13 + \cdots$
- $5\ell - 2 \to (5 - 2) + (10 - 2) + (15 - 2) + \cdots = 3 + 8 + 13 + \cdots$

Now compare $3 + (\ell - 1)5$ with $a + (\ell - 1)d$ to see that $a = 3$ and $d = 5$. Looking at the sum, $N = 15$.

$$Na + \frac{N(N-1)d}{2} = 15(3) + \frac{15(14)5}{2} = 45 + 525 = 570$$

Alternatively, first find the last term: $b = a + (N-1)d = 3 + (15-1)5 = 73$.

$$\frac{N(a+b)}{2} = \frac{15(3+73)}{2} = \frac{15(76)}{2} = 570$$

(34) Notes: The difference between any two consecutive terms is 11. This is an arithmetic series with $N = 9, d = 11, a = -18$, and $b = 70$.

$$\frac{N(a+b)}{2} = \frac{9(-18+70)}{2} = \frac{9(52)}{2} = 234$$

$$Na + \frac{N(N-1)d}{2} = 9(-18) + \frac{9(8)11}{2} = -162 + 396 = 234$$

(35) Notes: The difference between any two consecutive terms is -10. This is an arithmetic series with $N = 12, d = -10, a = 43$, and $b = -67$. The reason that $d = -10$ is **negative** is that each term gets smaller (not bigger).

$$\frac{N(a+b)}{2} = \frac{12(43-67)}{2} = \frac{12(-24)}{2} = -144$$

$$Na + \frac{N(N-1)d}{2} = 12(43) + \frac{12(11)(-10)}{2} = 516 - 660 = -144$$

Alternate solution: Think of the series backwards. In reverse order, $N = 12, d = 10$, $a = -67$, and $b = 43$. If you go **backward**, then d is positive (not negative).

$$-67 - 57 - 47 - 37 - 27 - 17 - 7 + 3 + 13 + 23 + 33 + 43$$

$$\frac{N(a+b)}{2} = \frac{12(-67+43)}{2} = \frac{12(-24)}{2} = -144$$

$$Na + \frac{N(N-1)d}{2} = 12(-67) + \frac{12(11)(10)}{2} = -804 + 660 = -144$$

(36) Notes: Since each term is 3 times the previous term, this is a geometric series with $N = 12, r = \frac{b_{n+1}}{b_n} = 3, a = 1$, and $b = 177{,}147$.

$$\frac{a(1-r^N)}{1-r} = \frac{1[1-(3)^{12}]}{1-3} = \frac{1-531{,}441}{-2} = \frac{-531{,}440}{-2} = 265{,}720$$

$$\frac{a-br}{1-r} = \frac{1-177{,}147(3)}{1-3} = \frac{1-531{,}441}{-2} = \frac{-531{,}440}{-2} = 265{,}720$$

(37) Notes: Since each term is 4 times the previous term, this is a geometric series with $N = 9, r = \frac{b_{n+1}}{b_n} = 4, a = 12$, and $b = 768{,}432$.

$$\frac{a(1-r^N)}{1-r} = \frac{12[1-(4)^9]}{1-4} = \frac{12(1-262{,}144)}{-3} = \frac{-3{,}145{,}716}{-3} = 1{,}048{,}572$$

$$\frac{a-br}{1-r} = \frac{12-768{,}432(4)}{1-4} = \frac{12-3{,}073{,}728}{-3} = \frac{-3{,}145{,}716}{-3} = 1{,}048{,}572$$

(38) Notes: This is a geometric series. Compare the general form ar^{j-1} with 5^{j-1} to see that $a = 1$ and $r = 5$. Note that $(1)5^{j-1} = 5^{j-1}$. Looking at the sum, $N = 12$.

$$\frac{a(1-r^N)}{1-r} = \frac{1[1-(5)^{12}]}{1-5} = \frac{1-244{,}140{,}625}{-4} = \frac{-244{,}140{,}624}{-4} = 61{,}035{,}156$$

Alternatively, first find the last term: $b = ar^{N-1} = (1)5^{12-1} = 5^{11} = 48{,}828{,}125$.

$$\frac{a-br}{1-r} = \frac{1-48{,}828{,}125(5)}{1-5} = \frac{-244{,}140{,}624}{-4} = 61{,}035{,}156$$

(39) Notes: This is a geometric series. Before we identify the constants, we need to write 4^n in the standard form ar^{n-1}. We will show two ways to solve this problem; after the first solution, we will present an **alternate solution** (see the next page). Also, after the alternate solution, we will demonstrate how to **check the answer**.

One way is to note that $4^n = 4(4^{n-1})$. Now we have another problem; this series doesn't start from 1, which the formula requires. We encountered a similar issue in Problem 29. We'll find the sum from 1 to 10, then find the sum from 1 to 4, and then subtract. We define the following:

$$S_1 = \sum_{n=1}^{10} 4^n = \sum_{n=1}^{10} 4(4)^{n-1} \quad , \quad S_2 = \sum_{n=1}^{4} 4^n = \sum_{n=1}^{4} 4(4)^{n-1} \quad , \quad S_3 = S_1 - S_2$$

To find S_1, use $N = 10$, $r = \frac{b_{n+1}}{b_n} = 4$, and $a = 4$.

$$S_1 = \frac{a(1-r^N)}{1-r} = \frac{4[1-(4)^{10}]}{1-4} = \frac{4(1-1{,}048{,}576)}{-3} = \frac{-4{,}194{,}300}{-3} = 1{,}398{,}100$$

To find S_2, use $N = 4$, $r = \frac{b_{n+1}}{b_n} = 4$, and $a = 4$.

$$S_1 = \frac{a(1-r^N)}{1-r} = \frac{4[1-(4)^4]}{1-4} = \frac{4(1-256)}{-3} = \frac{-1020}{-3} = 340$$

Now subtract: $S_3 = 1{,}398{,}100 - 340 = 1{,}397{,}760$ is our **final answer**.

Alternate solution: Shift the series using the method from Chapter 3 to the new index $p = n - 4$ (such that $n = p + 4$). The new series will run from $5 - 4 = 1$ to $10 - 4 = 6$. Note that $4^n = 4^{p+4} = 4^{p-1+5} = 4^{p-1}4^5 = (1024)4^{p-1}$.

$$\sum_{n=5}^{10} 4^n = \sum_{p=1}^{6} 4^{p+4} = \sum_{p=1}^{6} 4^{p-1+5} = \sum_{p=1}^{6} 4^{p-1}4^5 = \sum_{p=1}^{6} (1024)4^{p-1}$$

Use the formula for a geometric series with $a = 1024$, $r = 4$, and $N = 6$.

$$\frac{a(1-r^N)}{1-r} = \frac{1024(1-4^6)}{1-4} = \frac{1024(1-4096)}{-3} = \frac{1024(-4095)}{-3} = 1{,}397{,}760$$

Check the answer: Write the series out longhand and add.

$$\sum_{n=5}^{10} 4^n = 4^5 + 4^6 + 4^7 + 4^8 + 4^9 + 4^{10}$$

$$= 1024 + 4096 + 16{,}384 + 65{,}536 + 262{,}144 + 1{,}048{,}576 = 1{,}397{,}760$$

(40) Notes: Since each term is -6 times the previous term, this is a geometric series with $N = 9$, $r = \frac{b_{n+1}}{b_n} = -6$, $a = 5$, and $b = 8{,}398{,}080$.

$$\frac{a(1-r^N)}{1-r} = \frac{5[1-(-6)^9]}{1-(-6)} = \frac{5[1-(-10{,}077{,}696)]}{1+6} = \frac{5(1+10{,}077{,}696)}{7} = 7{,}198{,}355$$

$$\frac{a-br}{1-r} = \frac{5-8{,}398{,}080(-6)}{1-(-6)} = \frac{5+50{,}388{,}480}{1+6} = \frac{50{,}388{,}485}{7} = 7{,}198{,}355$$

(41) Notes: The difference between any two consecutive terms is $\frac{1}{6}$. This is an arithmetic series with $N = 16$, $d = \frac{1}{6}$, $a = \frac{1}{2}$, and $b = 3$. The series can be rewritten as $\frac{3}{6} + \frac{4}{6} + \frac{5}{6} + \frac{6}{6} + \frac{7}{6} + \frac{8}{6} + \frac{9}{6} + \frac{10}{6} + \frac{11}{6} + \frac{12}{6} + \frac{13}{6} + \frac{14}{6} + \frac{15}{6} + \frac{16}{6} + \frac{17}{6} + \frac{18}{6}$.

$$\frac{N(a+b)}{2} = \frac{16\left(\frac{1}{2}+3\right)}{2} = \frac{16\left(\frac{7}{2}\right)}{2} = \frac{56}{2} = 28$$

$$Na + \frac{N(N-1)d}{2} = 16\left(\frac{1}{2}\right) + \frac{16(15)\frac{1}{6}}{2} = 8 + \frac{40}{2} = 8 + 20 = 28$$

(42) Notes: Any term divided by the previous term equals $\frac{3}{2}$. For example, $\frac{8}{81} \div \frac{16}{243} = \frac{8}{81} \times \frac{243}{16} = \frac{3}{2}$ and $\frac{4}{27} \div \frac{8}{81} = \frac{4}{27} \times \frac{81}{8} = \frac{3}{2}$. (Recall that the way to divide by a fraction is to multiply by its reciprocal.) This is a geometric series with $N = 11$, $r = \frac{b_{n+1}}{b_n} = \frac{3}{2}$, $a = \frac{16}{243}$, and $b = \frac{243}{64}$.

$$\frac{a(1-r^N)}{1-r} = \frac{\frac{16}{243}\left[1-\left(\frac{3}{2}\right)^{11}\right]}{1-\frac{3}{2}} = \frac{\frac{16}{243}\left(1-\frac{177,147}{2048}\right)}{-1/2} = \frac{\frac{16}{243}\left(-\frac{175,099}{2048}\right)}{-1/2}$$

$$= \frac{175,099}{31,104} = \frac{175,099}{31,104} \div \frac{1}{2} = \frac{175,099}{31,104} \times \frac{2}{1} = \frac{175,099}{15,552} \approx 11.26$$

$$\frac{a-br}{1-r} = \frac{\frac{16}{243} - \frac{243}{64}\left(\frac{3}{2}\right)}{1-\frac{3}{2}} = \frac{\frac{16}{243} - \frac{729}{128}}{-1/2} = \frac{-175,099/31,104}{-1/2}$$

$$= \frac{175,099}{31,104} \times \frac{2}{1} = \frac{175,099}{15,552} \approx 11.26$$

(43) Notes: This is a geometric series. Before we identify the constants, we need to write $\frac{9}{2^k} = 9\left(\frac{1}{2}\right)^k$ in the standard form ar^{k-1}. There are two ways to do this. One way is to note that $\left(\frac{1}{2}\right)^k = \frac{1}{2}\left(\frac{1}{2}\right)^{k-1}$. The alternative is to shift the series to $n = k - 1$ using the method of Example 4 and Problems 23-26. We'll use the first method, rewriting $\frac{9}{2^k} = 9\left(\frac{1}{2}\right)^k$ as $9\left(\frac{1}{2}\right)\left(\frac{1}{2}\right)^{k-1} = \frac{9}{2}\left(\frac{1}{2}\right)^{k-1}$. Compare the general form ar^{k-1} with $\frac{9}{2}\left(\frac{1}{2}\right)^k$ to see that $a = \frac{9}{2}$ and $r = \frac{1}{2}$. Looking at the sum, $N = 20$. To double-check the constants, write the sum out longhand.

$$\sum_{k=1}^{20} \frac{9}{2^k} = \frac{9}{2} + \frac{9}{4} + \frac{9}{8} + \frac{9}{16} + \frac{9}{32} + \cdots + \frac{9}{1,048,576}$$

Now it may be easier to see that $a = \frac{9}{2}$ and $r = \frac{1}{2}$.

$$\frac{a(1-r^N)}{1-r} = \frac{\frac{9}{2}\left[1-\left(\frac{1}{2}\right)^{20}\right]}{1-\frac{1}{2}} = \frac{\frac{9}{2}\left(1-\frac{1}{1,048,576}\right)}{\frac{1}{2}} = \frac{\frac{9}{2}\left(\frac{1,048,575}{1,048,576}\right)}{\frac{1}{2}}$$

$$= \frac{9}{2}\left(\frac{1,048,575}{1,048,576}\right) \div \frac{1}{2} = \frac{9}{2}\left(\frac{1,048,575}{1,048,576}\right) \times \frac{2}{1} = 9\left(\frac{1,048,575}{1,048,576}\right) = \frac{9,437,175}{1,048,576}$$

Alternatively, first find the last term: $b = ar^{N-1} = \left(\frac{9}{2}\right)\left(\frac{1}{2}\right)^{20-1} = \left(\frac{9}{2}\right)\left(\frac{1}{2}\right)^{19} = \frac{9}{2}\left(\frac{1}{2^{19}}\right) = \frac{9}{2^{20}} = \frac{9}{1,048,576}$.

$$\frac{a-br}{1-r}=\frac{\frac{9}{2}-\frac{9}{1{,}048{,}576}\left(\frac{1}{2}\right)}{1-\frac{1}{2}}=\frac{\frac{9}{2}\left(1-\frac{1}{1{,}048{,}576}\right)}{\frac{1}{2}}=\frac{\frac{9}{2}\left(\frac{1{,}048{,}575}{1{,}048{,}576}\right)}{\frac{1}{2}}$$

$$=\frac{9}{2}\left(\frac{1{,}048{,}575}{1{,}048{,}576}\right)\div\frac{1}{2}=\frac{9}{2}\left(\frac{1{,}048{,}575}{1{,}048{,}576}\right)\times\frac{2}{1}=9\left(\frac{1{,}048{,}575}{1{,}048{,}576}\right)=\frac{9{,}437{,}175}{1{,}048{,}576}$$

(44) This isn't a purely arithmetic series because the difference between consecutive terms changes. This isn't a purely geometric series because the ratio between consecutive terms changes. For example, $\frac{32}{6}\approx 5.33$ whereas $\frac{3584}{768}\approx 4.67$. It turns out to be an arithmetic-geometric series, which has the form $[a+(n-1)d]r^{n-1}$. The first term is $a=6$. The second term is $[a+(2-1)d]r^{2-1}=(6+d)r=32$ and the third term is $[a+(3-1)d]r^{3-1}=(6+2d)r^2=160$. Looking at the two equations $(6+d)r=32$ and $(6+2d)r^2=160$ makes it easier to determine the constants d and r. (If you want to "solve" for them, you can. Isolate r in the first equation and plug it into the second equation to get a quadratic equation. It's a bit tedious and easy to make a mistake, but if you do the algebra carefully and correctly, it will lead you to the correct answers.) Looking at $(6+d)r=32$, what could d and r be to make the left-hand side equal 32? Since $(8)(4)=32$, it seems like $d=2$ and $r=4$ would be a good fit, but first we have to see if this combination works for the other terms. Observe that $d=2$ and $r=4$ also work in the second equation: $(6+2d)r^2=[6+2(2)]4^2=10(16)=160$. You can verify that this works for all of the terms. For example, $(6+3d)r^3=[6+3(2)]4^3=12(64)=768$. Plug $N=9$, $d=2$, $r=4$, $a=6$, and $b=1{,}441{,}792$ into the formula for an arithmetic-geometric series.

$$\frac{a+(d-a)r+br^2-(a+Nd)r^N}{(1-r)^2}=\frac{6+(2-6)(4)+1{,}441{,}792(4)^2-[6+9(2)]4^9}{(1-4)^2}$$

$$\frac{6-16+23{,}068{,}672-6{,}291{,}456}{(-3)^2}=\frac{16{,}777{,}206}{9}=1{,}864{,}134$$

Alternatively, use the other formula:

$$\frac{a(1-r^N)}{1-r}+\frac{rd[1-Nr^{N-1}+(N-1)r^N]}{(1-r)^2}=\frac{6(1-4^9)}{1-4}+\frac{4(2)[1-9(4)^8+8(4)^9]}{(1-4)^2}$$

$$=\frac{-1{,}572{,}858}{-3}+\frac{8(1{,}507{,}329)}{9}=524{,}286+1{,}339{,}848=1{,}864{,}134$$

(45) Notes: This is an arithmetic-geometric series. Compare $[5 + 2(m-1)]3^{m-1}$ with the general form $[a + (m-1)d]r^{m-1}$ to see that $d = 2, r = 3$, and $a = 5$. Look at the sum to see that $N = 12$.

$$\frac{a(1-r^N)}{1-r} + \frac{rd[1 - Nr^{N-1} + (N-1)r^N]}{(1-r)^2} = \frac{5(1-3^{12})}{1-3} + \frac{3(2)[1 - 12(3)^{11} + 11(3)^{12}]}{(1-3)^2}$$

$$= \frac{-2{,}657{,}200}{-2} + \frac{6(3{,}720{,}088)}{4} = 1{,}328{,}600 + 5{,}580{,}132 = 6{,}908{,}732$$

Alternatively, first find the last term.

$$b = [a + (N-1)d]r^{N-1} = [5 + 11(2)](3)^{12-1} = 27(177{,}147) = 4{,}782{,}969$$

$$\frac{a + (d-a)r + br^2 - (a+Nd)r^N}{(1-r)^2} = \frac{5 + (2-5)(3) + 4{,}782{,}969(3)^2 - [5 + 12(2)]3^{12}}{(1-3)^2}$$

$$= \frac{5 - 9 + 43{,}046{,}721 - 15{,}411{,}789}{4} = \frac{27{,}634{,}928}{4} = 6{,}908{,}732$$

(46) This is an arithmetic-geometric series. Before we identify the constants, we need to put this in the standard form $[a + (k-1)d]r^{k-1}$. Observe that $\frac{5k-3}{4^k} = (5k-3)\left(\frac{1}{4}\right)^k = (-3+5k)\left(\frac{1}{4}\right)^k = (-3+5k)\left(\frac{1}{4}\right)^{k-1}\left(\frac{1}{4}\right)^1 = \left(-\frac{3}{4} + \frac{5k}{4}\right)\left(\frac{1}{4}\right)^{k-1}$. It's still not quite in standard form because we have $5k$, but we need something proportional to $k-1$. The 'trick' is to add and subtract $\frac{5}{4}$. (We did something similar in the solution to Problem 33.) This gives us $\left(-\frac{3}{4} + \frac{5k}{4} - \frac{5}{4} + \frac{5}{4}\right)\left(\frac{1}{4}\right)^{k-1} = \left(\frac{2}{4} + \frac{5k}{4} - \frac{5}{4}\right)\left(\frac{1}{4}\right)^{k-1} = \left[\frac{1}{2} + \frac{5}{4}(k-1)\right]\left(\frac{1}{4}\right)^{k-1}$. Compare this to $[a + (k-1)d]r^{k-1}$ to identify $d = \frac{5}{4}, r = \frac{1}{4}$, and $a = \frac{1}{2}$. Look at the sum to see that $N = 9$.

$$\frac{a(1-r^N)}{1-r} + \frac{rd[1 - Nr^{N-1} + (N-1)r^N]}{(1-r)^2} = \frac{\frac{1}{2}\left(1 - \frac{1}{4^9}\right)}{1 - \frac{1}{4}} + \frac{\left(\frac{1}{4}\right)\left(\frac{5}{4}\right)\left[1 - 9\left(\frac{1}{4}\right)^8 + 8\left(\frac{1}{4}\right)^9\right]}{\left(1 - \frac{1}{4}\right)^2}$$

$$= \frac{\frac{1}{2}\left(1 - \frac{1}{262{,}144}\right)}{3/4} + \frac{\frac{5}{16}\left[1 - \frac{9}{65{,}536} + \frac{1}{32{,}768}\right]}{9/16}$$

$$= \frac{262{,}143}{524{,}288} \div \frac{3}{4} + \frac{5}{16}\left(\frac{65{,}529}{65{,}536}\right) \div \frac{9}{16} = \frac{262{,}143}{524{,}288} \times \frac{4}{3} + \frac{5}{16}\left(\frac{65{,}529}{65{,}536}\right) \times \frac{16}{9}$$

$$= \frac{87{,}381}{131{,}072} + \frac{36{,}405}{65{,}536} = \frac{87{,}381}{131{,}072} + \frac{72{,}810}{131{,}072} = \frac{160{,}191}{131{,}072} \approx 1.22216$$

Alternatively, first find the last term.

$$b = [a + (N-1)d]r^{N-1} = \left[\frac{1}{2} + 8\left(\frac{5}{4}\right)\right]\left(\frac{1}{4}\right)^{9-1} = \left(\frac{21}{2}\right)\left(\frac{1}{65{,}536}\right) = \frac{21}{131{,}072}$$

$$\frac{a + (d-a)r + br^2 - (a+Nd)r^N}{(1-r)^2} = \frac{\frac{1}{2} + \left(\frac{5}{4} - \frac{1}{2}\right)\left(\frac{1}{4}\right) + \frac{21}{131{,}072}\left(\frac{1}{4^2}\right) - \frac{\frac{1}{2} + 9\left(\frac{5}{4}\right)}{4^9}}{\left(1 - \frac{1}{4}\right)^2}$$

$$= \frac{\frac{1}{2} + \frac{3}{16} + \frac{21}{2{,}097{,}152} - \frac{47}{1{,}048{,}576}}{9/16} = \frac{1{,}441{,}719}{2{,}097{,}152} \div \frac{9}{16} = \frac{1{,}441{,}719}{2{,}097{,}152} \times \frac{16}{9} = \frac{160{,}191}{131{,}072}$$

(47) Notes: These are fourth powers with an overall minus sign: $-(1^4 + 2^4 + 3^4 + 4^4 + \cdots + 11^4)$. Use the formula for the sum of n^4. There are 11 terms.

$$= -\frac{11(11+1)[2(11)+1][3(11)^2 + 3(11) - 1]}{30} = -\frac{11(12)(23)(395)}{30} = -39{,}974$$

(48) Notes: Factor out the one-fifth.

$$\sum_{n=30}^{40} \frac{n^2}{5} = \frac{30^2}{5} + \frac{31^2}{5} + \cdots + \frac{40^2}{5} = \frac{1}{5}(30^2 + 31^2 + \cdots + 40^2)$$

Since this sum begins with 30, we first need to find the sum from 1 to 40, then find the sum from 1 to 29, and subtract.

$$\sum_{n=30}^{40} \frac{n^2}{5} = \frac{1}{5}\sum_{n=30}^{40} n^2 = \frac{1}{5}\left(\sum_{n=1}^{40} n^2 - \sum_{n=1}^{29} n^2\right) = \frac{1}{5}(S_1 - S_2) = S_3$$

Use the formula for the sum of n^2 to find S_1 and S_2 (defined above).

$$S_1 = \frac{40(40+1)[2(40)+1]}{6} = \frac{40(41)(81)}{6} = 22{,}140$$

$$S_2 = \frac{29(29+1)[2(29)+1]}{6} = \frac{29(30)(59)}{6} = 8555$$

$$S_3 = \frac{1}{5}(S_1 - S_2) = \frac{22{,}140 - 8555}{5} = \frac{13{,}585}{5} = 2717$$

(49) Notes: The 'trick' is to factor out the three: $3 + 24 + 81 + 192 + 375 + 648 + 1029 + 1536 = 3(1 + 8 + 27 + 64 + 125 + 216 + 343 + 512)$. After factoring out the 3, we have a sum of cubes. Use the formula for the sum of n^3. Multiply by 3 because we factored out a 3.

$$3\left[\frac{8^2(8+1)^2}{4}\right] = 3\left[\frac{64(81)}{4}\right] = 3(1296) = 3888$$

(50) Notes: Separate this into four different series.

$$\sum_{i=1}^{12}(i^3 - 2i^2 + i - 4) = \sum_{i=1}^{12} i^3 - 2\sum_{i=1}^{12} i^2 + \sum_{i=1}^{12} i - \sum_{i=1}^{12} 4$$

We'll do this one series at a time:

$$S_1 = \sum_{i=1}^{12} i^3 = \frac{12^2(12+1)^2}{4} = \frac{144(169)}{4} = 6084$$

$$S_2 = \sum_{i=1}^{12} i^2 = \frac{12(12+1)[2(12)+1]}{6} = \frac{12(13)(25)}{6} = 650$$

$$S_3 = \sum_{i=1}^{12} i = \frac{12(12+1)}{2} = 78$$

$$S_4 = \sum_{i=1}^{12} 4 = 4 + 4 + 4 + \cdots + 4 = 12(4) = 48$$

$$S_5 = S_1 - 2S_2 + S_3 - S_4 = 6084 - 2(650) + 78 - 48 = 6084 - 1300 + 30 = 4814$$

(51) Notes: Use the method of partial fractions. Note that the sum begins at 2.

$$\frac{3}{k^2 - k} = \frac{3}{k(k-1)} = \frac{x}{k} + \frac{y}{k-1}$$

Multiply both sides by $k(k-1)$.

$$3 = x(k-1) + ky = kx - x + ky = k(x+y) - x$$

This is true for all values of k provided that $x = -3$ and $y = -x = -(-3) = 3$.

$$\sum_{k=2}^{30} \frac{3}{k^2 - k} = \sum_{k=2}^{30}\left(-\frac{3}{k} + \frac{3}{k-1}\right)$$

$$= -\frac{3}{2} + \frac{3}{1} - \frac{3}{3} + \frac{3}{2} - \frac{3}{4} + \frac{3}{3} - \frac{3}{5} + \frac{3}{4} + \cdots - \frac{3}{29} + \frac{3}{28} - \frac{3}{30} + \frac{3}{29}$$

Everything cancels out except for:

$$\frac{3}{1} - \frac{3}{30} = \frac{3}{1} - \frac{1}{10} = \frac{30}{10} - \frac{1}{10} = \frac{29}{10} = 2.9$$

(52) Notes: Use the method of partial fractions.

$$\frac{2}{4n^2 - 1} = \frac{2}{(2n+1)(2n-1)} = \frac{x}{2n+1} + \frac{y}{2n-1}$$

Multiply both sides by $(2n + 1)(2n - 1)$.

$$2 = x(2n - 1) + y(2n + 1) = 2nx - x + 2ny + y = 2n(x + y) - x + y$$

This is true for all values of n provided that $x + y = 0$ and $-x + y = 2$. Solve this system to find that $x = -1$ and $y = -x = -(-1) = 1$.

$$\sum_{n=1}^{18} \frac{2}{4n^2 - 1} = \sum_{n=1}^{18} \left(-\frac{1}{2n+1} + \frac{1}{2n-1}\right)$$

$$= -\frac{1}{3} + \frac{1}{1} - \frac{1}{5} + \frac{1}{3} - \frac{1}{7} + \frac{1}{5} - \frac{1}{9} + \frac{1}{7} + \cdots - \frac{1}{35} + \frac{1}{33} - \frac{1}{37} + \frac{1}{35}$$

Everything cancels out except for:

$$\frac{1}{1} - \frac{1}{37} = \frac{37}{37} - \frac{1}{37} = \frac{36}{37}$$

Chapter 4 Answers

(1) This series **diverges** because the corresponding sequence $\left\{\frac{k^2}{9}\right\}$ diverges.

(2) The simple divergence test is **inconclusive** because the corresponding sequence $\left\{\frac{9}{(j+1)^2}\right\}$ converges to zero.

(3) This series **diverges** because the corresponding sequence $\left\{\cos\left(\frac{n\pi}{4}\right)\right\}$ diverges. The series is $1+\frac{\sqrt{2}}{2}+0-\frac{\sqrt{2}}{2}-1-\frac{\sqrt{2}}{2}+0+\frac{\sqrt{2}}{2}+1+\frac{\sqrt{2}}{2}+0+\cdots$ (where the angles were given in radians) involves a sequence that repeats forever without converging to any one value.

(4) This series **diverges** because the corresponding sequence $\left\{\cos\left(\frac{4}{m}\right)\right\}$ doesn't converge to zero. The sequence $\left\{\cos\left(\frac{4}{m}\right)\right\}$ converges to one (which isn't zero) because the angle $\frac{4}{m}$ converges to zero and $\cos 0 = 1$.

(5) This series **diverges** because the corresponding sequence $\{97, -87, 73, -61, 49, \ldots\}$ diverges. Don't be fooled into thinking that this sequence converges to zero; it doesn't. To see this, try adding more terms. The sequence is $a_{n+1} = (-1)^{n+1}(|a_n| - 12)$; that is, take the absolute value of the previous term, subtract 12, and alternate signs.

$$97 - 85 + 73 - 61 + 49 - 37 + 25 - 13 + 1 - 11 + 23 - 35 + 47 - 59 + 71 - 83$$
$$-95 + 107 - 119 + 131 - 143 + 155 - 167 + 179 - 191 + 203 - 215 + \cdots$$

At the tail of the corresponding sequence, terms are alternately growing to ∞ and $-\infty$.

(6) The simple divergence test is **inconclusive** because the corresponding sequence converges to zero. The sequence is $a_{n+1} = \frac{a_n}{n+1}$.

(7) This series **diverges** because the corresponding sequence $\left\{\frac{5i-2}{4i+3}\right\}$ doesn't converge to zero. The sequence $\left\{\frac{5i-2}{4i+3}\right\}$ converges to $\frac{5}{4}$ (which isn't zero). To see this, divide by i in the numerator and denominator to get $\frac{5-\frac{2}{i}}{4+\frac{3}{i}}$, which converges to $\frac{5-0}{4+0} = \frac{5}{4}$.

(8) The simple divergence test is **inconclusive** because the corresponding sequence $\left\{\frac{3\ell-6}{7-\ell^2}\right\}$ converges to zero. To see this, divide by ℓ^2 in the numerator and denominator

to get $\frac{\frac{3}{\ell}-\frac{6}{\ell^2}}{\frac{7}{\ell^2}-1}$, which converges to $\frac{0-0}{0-1} = \frac{0}{-1} = 0$.

(9) This series **diverges** because the corresponding sequence $\left\{\frac{5n^3}{8n^2-5n+2}\right\}$ diverges. To see this, divide by n^3 in the numerator and denominator to get $\frac{5}{\frac{8}{n}-\frac{5}{n^2}+\frac{2}{n^3}}$, which diverges since $\frac{5}{0-0+0} = \frac{5}{0} = \infty$.

(10) This series **diverges** because the corresponding sequence diverges. The sequence is $\{0.1n\} = \left\{\frac{n}{10}\right\}$.

(11) The simple divergence test is **inconclusive** because the corresponding sequence converges to zero. The sequence is $\left\{\frac{10}{(-2)^{n-1}}\right\} = \left\{10\left(-\frac{1}{2}\right)^{n-1}\right\}$.

(12) This series **diverges** because the corresponding sequence $\left\{1-\frac{1}{i}\right\}$ doesn't converge to zero. The sequence $\left\{1-\frac{1}{i}\right\}$ converges to 1 (which isn't zero).

(13) This series **diverges** because the corresponding sequence $\{m^2 - 4m + 11\}$ diverges.

(14) This series **diverges** because the corresponding sequence $\left\{\sqrt{\frac{k+1}{k-1}}\right\}$ doesn't converge to zero. The sequence $\left\{\sqrt{\frac{k+1}{k-1}}\right\}$ converges to 1 (which isn't zero). Divide by k in the numerator and denominator to see that $\frac{k+1}{k-1} = \frac{1+\frac{1}{k}}{1-\frac{1}{k}}$ converges to 1, such that $\sqrt{\frac{k+1}{k-1}}$ converges to $\sqrt{1} = 1$.

(15) This series **diverges** because the corresponding sequence $\left\{(-1)^j \frac{2j+1}{4j-3}\right\}$ diverges. Divide by j in the numerator and denominator to see that $\frac{2j+1}{4j-3} = \frac{2+\frac{1}{j}}{4-\frac{3}{j}}$ converges to $\frac{2}{4} = \frac{1}{2}$. The tail of this sequence alternates between approximately $\frac{1}{2}$ and $-\frac{1}{2}$, without ever converging to any one value.

(16) The simple divergence test is **inconclusive** because the corresponding sequence $\left\{\frac{\ell^2}{\ell!}\right\}$ converges to zero. (We saw sequences of this kind in Chapter 2.)

(17) This series **diverges**. Compare ar^{n-1} with $\frac{2^{n-1}}{5}$ to see that this is a geometric series with $a = \frac{1}{5}$ and $r = 2$. Since $|r| \geq 1$ (and $a \neq 0$), this series diverges.

(18) This series **converges**. Compare ar^{i-1} with $2\left(\frac{1}{5}\right)^{i-1}$ to see that this is a geometric series with $a = 2$ and $r = \frac{1}{5}$. Since $|r| < 1$, this series converges. Before using the formula, you need to rewrite the sum starting at $i = 1$. Do this by subtracting and adding the $i = 1$ term. Once we write the sum starting at $i = 1$, we may use the formula $L = \frac{a}{1-r}$.

$$\sum_{i=2}^{\infty} 2\left(\frac{1}{5}\right)^{i-1} = -2\left(\frac{1}{5}\right)^{1-1} + 2\left(\frac{1}{5}\right)^{1-1} + \sum_{i=2}^{\infty} 2\left(\frac{1}{5}\right)^{i-1} = -2\left(\frac{1}{5}\right)^{1-1} + \sum_{i=1}^{\infty} 2\left(\frac{1}{5}\right)^{i-1}$$

$$= -2\left(\frac{1}{5}\right)^{0} + \frac{a}{1-r} = -2(1) + \frac{2}{1-\frac{1}{5}} = -2 + \frac{2}{4/5} = -2 + 2 \div \frac{4}{5}$$

$$= -2 + 2 \times \frac{5}{4} = -2 + \frac{5}{2} = -\frac{4}{2} + \frac{5}{2} = \frac{1}{2}$$

(19) Use the simple divergence test. This series **diverges** because the corresponding sequence $\left\{\frac{\ell}{1{,}000{,}000}\right\}$ diverges. Here, ℓ grows to infinity, even after dividing by a million.

(20) This series **converges**. This is a geometric series, but it isn't in standard form. Shift the index using $k = n - 1$. Note that $k + 1 = n$. When $k = -1$, $n = -1 + 1 = 0$.

$$\sum_{k=-1}^{\infty} \frac{12}{4^k} = \sum_{n=0}^{\infty} \frac{12}{4^{n-1}} = \sum_{n=0}^{\infty} 12\left(\frac{1}{4}\right)^{n-1}$$

Compare ar^{n-1} with $12\left(\frac{1}{4}\right)^{n-1}$ to see that $a = 12$ and $r = \frac{1}{4}$. Since $|r| < 1$, this series converges. Before using the formula, the sum needs to start at $n = 1$. The given sum has one extra term. Just write out the $n = 0$ term first.

$$\sum_{n=0}^{\infty} 12\left(\frac{1}{4}\right)^{n-1} = 12\left(\frac{1}{4}\right)^{0-1} + \sum_{n=0}^{\infty} 12\left(\frac{1}{4}\right)^{n-1} = 12\left(\frac{1}{4}\right)^{-1} + \frac{a}{1-r}$$

$$= 12\left(\frac{4}{1}\right) + \frac{12}{1-\frac{1}{4}} = \frac{48}{1} + \frac{12}{3/4} = 48 + 12 \div \frac{3}{4} = 48 + 12 \times \frac{4}{3}$$

$$= 48 + \frac{48}{3} = 48 + 16 = 64$$

(21) This series **converges**. This is a geometric series, but it isn't in standard form. Shift the sum using $j + 2 = n - 1$. Note that $j + 3 = n$. When $j = 1, n = 1 + 3 = 4$. We can check this as follows. Each series below has the same first term; the left sum begins with $\frac{9}{3^{1+2}} = \frac{9}{3^3}$ and the right sum begins with $\frac{9}{3^{4-1}} = \frac{9}{3^3}$. When shifting an index, it's good to check that the first terms are identical before and after.

$$\sum_{j=1}^{\infty} \frac{9}{3^{j+2}} = \sum_{n=4}^{\infty} \frac{9}{3^{n-1}} = \sum_{n=4}^{\infty} 9\left(\frac{1}{3}\right)^{n-1}$$

Compare ar^{n-1} with $9\left(\frac{1}{3}\right)^{n-1}$ to see that $a = 9$ and $r = \frac{1}{3}$. Since $|r| < 1$, this series converges. Before using the formula, the sum needs to start at $n = 1$. Do this by subtracting and adding the first three terms.

$$\sum_{n=4}^{\infty} 9\left(\frac{1}{3}\right)^{n-1} = -9\left(\frac{1}{3}\right)^{1-1} - 9\left(\frac{1}{3}\right)^{2-1} - 9\left(\frac{1}{3}\right)^{3-1} + 9\left(\frac{1}{3}\right)^{1-1} + 9\left(\frac{1}{3}\right)^{2-1} + 9\left(\frac{1}{3}\right)^{3-1} + \sum_{n=4}^{\infty} 9\left(\frac{1}{3}\right)^{n-1}$$

$$= -9\left(\frac{1}{3}\right)^{0} - 9\left(\frac{1}{3}\right)^{1} - 9\left(\frac{1}{3}\right)^{2} + \sum_{n=1}^{\infty} 9\left(\frac{1}{3}\right)^{n-1} = -9(1) - 9\left(\frac{1}{3}\right) - 9\left(\frac{1}{9}\right) + \frac{a}{1-r}$$

$$= -9 - \frac{9}{3} - \frac{9}{9} + \frac{9}{1-\frac{1}{3}} = -9 - 3 - 1 + \frac{9}{2/3} = -13 + 9 \div \frac{2}{3}$$

$$= -13 + 9 \times \frac{3}{2} = -13 + \frac{27}{2} = -\frac{26}{2} + \frac{27}{2} = \frac{1}{2}$$

Alternate solution: Use $3^{j+2} = 3^2 3^j$ and $3^j = 3^1 3^{j-1}$ as follows.

$$\sum_{j=1}^{\infty} \frac{9}{3^{j+2}} = \sum_{j=1}^{\infty} \frac{9}{3^2 3^j} = \sum_{j=1}^{\infty} \frac{1}{3^j} = \sum_{j=1}^{\infty} \frac{1}{3^1 3^{j-1}} = \sum_{j=1}^{\infty} \frac{1}{3}\left(\frac{1}{3}\right)^{j-1}$$

$$\frac{a}{1-r} = \frac{1/3}{1-1/3} = \frac{1/3}{2/3} = \frac{1}{3} \div \frac{2}{3} = \frac{1}{3} \times \frac{3}{2} = \frac{1}{2}$$

(22) This series **converges**. Use the method of partial fractions.

$$\frac{3}{m(m+3)} = \frac{x}{m} + \frac{y}{m+3}$$

Multiply by $m(m+3)$ on both sides.

$$3 = x(m+3) + my = mx + 3x + my = 3x + m(x+y)$$

In order for the above equation to be true for any value of m, we need $3 = 3x$ (for which $x = 1$) and $(x+y) = 0$ (for which $y = -x = -1$). Use $x = 1$ and $y = -1$ in the partial fraction formula.

$$\sum_{m=2}^{\infty} \frac{3}{m(m+3)} = \sum_{m=2}^{\infty} \left(\frac{1}{m} - \frac{1}{m+3}\right)$$

The partial sum S_m for this series is:

$$S_m = \sum_{k=2}^{m} \left(\frac{1}{k} - \frac{1}{k+3}\right)$$

$$= \frac{1}{2} - \frac{1}{5} + \frac{1}{3} - \frac{1}{6} + \frac{1}{4} - \frac{1}{7} + \frac{1}{5} - \frac{1}{8} + \frac{1}{6} - \frac{1}{9} + \frac{1}{7} - \frac{1}{10} + \cdots + \frac{1}{m} - \frac{1}{m+3}$$

All of the terms cancel except for 3 terms in the beginning and 3 terms at the end; since m goes to infinity, the 3 terms at the end approach zero (that is, $\frac{1}{m+3}$ approaches zero), so only the 3 terms at the beginning matter.

$$\sum_{m=2}^{\infty} \frac{3}{m(m+3)} = \frac{1}{2} + \frac{1}{3} + \frac{1}{4} = \frac{6}{12} + \frac{4}{12} + \frac{3}{12} = \frac{13}{12}$$

(23) Use the simple divergence test. This series **diverges** because the corresponding sequence $\left\{\frac{i+3}{i}\right\}$ doesn't converge to zero. The sequence $\left\{\frac{i+3}{i}\right\}$ converges to 1 (which isn't zero). To see this, divide by i in the numerator and denominator to get $\frac{1+\frac{3}{i}}{1}$, which converges to $\frac{1+0}{1} = \frac{1}{1} = 1$.

(24) (Unlike Problem 23, here the simple divergence test would be inconclusive.) The 'trick' is to note that $n^2 - 1$ can be factored as $(n+1)(n-1) = n^2 - 1$. This series **converges**. Use the method of partial fractions.

$$\frac{4}{n^2 - 1} = \frac{x}{n+1} + \frac{y}{n-1}$$

Multiply both sides by $(n+1)(n-1)$, which equals $n^2 - 1$.

$$4 = x(n-1) + y(n+1) = nx - x + ny + y = -x + y + n(x+y)$$

In order for the above equation to be true for any value of n, we need both of the equations below to be true.

$$-x+y=4 \quad , \quad x+y=0$$

The second equation gives $y=-x$. Plug this into the first equation to get $-x-x=4$, which simplifies to $-2x=4$, for which $x=\frac{4}{-2}=-2$. Then $y=-(-2)=2$. Use $x=-2$ and $y=2$ in the partial fraction formula.

$$\sum_{n=2}^{\infty}\frac{4}{n^2-1}=\sum_{n=2}^{\infty}\left(-\frac{2}{n+1}+\frac{2}{n-1}\right)$$

The partial sum S_n for this series is:

$$S_n=\sum_{k=2}^{n}\left(-\frac{2}{k+1}+\frac{2}{k-1}\right)$$

$$=-\frac{2}{3}+\frac{2}{1}-\frac{2}{4}+\frac{2}{2}-\frac{2}{5}+\frac{2}{3}-\frac{2}{6}+\frac{2}{4}-\frac{2}{7}+\frac{2}{5}-\frac{2}{8}+\frac{2}{6}+\cdots-\frac{2}{n+1}+\frac{2}{n-1}$$

All of the terms cancel except for 2 terms in the beginning and 2 terms at the end; since n goes to infinity, the 2 terms at the end approach zero (since $\frac{2}{n+1}$ approaches zero), so only the 2 terms at the beginning matter.

$$\sum_{n=2}^{\infty}\frac{4}{n^2-1}=\frac{2}{1}+\frac{2}{2}=2+1=3$$

(25) This series **diverges**. Observe that this is a geometric series:

$$\sum_{n=1}^{\infty}5(-4)^{n-1}=5-20+80-320+1280-5120+20{,}480+\cdots$$

Compare ar^{n-1} with $5(-4)^{n-1}$ to see that this is a geometric series with $a=5$ and $r=-4$. Since $|r|\geq 1$ (and $a\neq 0$), this series diverges.

(26) This series **converges**. Observe that this is a geometric series. We may express it either in terms of fractions or decimals; the two series below are equivalent.

$$\sum_{n=1}^{\infty}\frac{1}{5}\left(\frac{1}{5}\right)^{n-1}=\frac{1}{5}+\frac{1}{25}+\frac{1}{125}+\frac{1}{625}+\frac{1}{3125}+\frac{1}{15{,}625}+\frac{1}{78{,}125}+\cdots$$

$$\sum_{n=1}^{\infty}0.2(0.2)^{n-1}=0.2+0.04+0.008+0.0016+0.00032+0.000064+0.0000128+\cdots$$

Compare ar^{i-1} with $\frac{1}{5}\left(\frac{1}{5}\right)^{n-1} = 0.2(0.2)^{n-1}$ to see that this is a geometric series with $a = \frac{1}{5} = 0.2$ and $r = \frac{1}{5} = 0.2$. Since $|r| < 1$, this series converges.

$$\sum_{n=1}^{\infty} \frac{1}{5}\left(\frac{1}{5}\right)^{n-1} = \frac{a}{1-r} = \frac{0.2}{1-0.2} = \frac{0.2}{0.8} = \frac{1}{4} = 0.25$$

(27) Use the simple divergence test. This series **diverges** because the corresponding sequence $\left\{\frac{e^k}{k^2}\right\}$ diverges. You can see this by applying l'Hôpital's rule (Chapter 2). We applied l'Hôpital's rule twice below (since it was still an indeterminate form after only taking the first derivative). The main idea is that the exponential grows much more rapidly than the polynomial.

$$\lim_{x\to\infty} \frac{e^x}{x^2} = \frac{\frac{d}{dx}e^x\Big|_{x\to\infty}}{\frac{d}{dx}x^2\Big|_{x\to\infty}} = \frac{\lim\limits_{x\to\infty} e^x}{\lim\limits_{x\to\infty} 2x} = \frac{\frac{d}{dx}e^x\Big|_{x\to\infty}}{\frac{d}{dx}2x\Big|_{x\to\infty}} = \frac{\lim\limits_{x\to\infty} e^x}{\lim\limits_{x\to\infty} 2} = \frac{\infty}{2} = \infty$$

(28) This series **converges**. The partial sum S_m for this series is:

$$S_m = \sum_{k=2}^{m} \left[\frac{1}{k^2} - \frac{1}{(k+4)^2}\right]$$

$$= \frac{1}{2^2} - \frac{1}{6^2} + \frac{1}{3^2} - \frac{1}{7^2} + \frac{1}{4^2} - \frac{1}{8^2} + \frac{1}{5^2} - \frac{1}{9^2} + \frac{1}{6^2} - \frac{1}{10^2} + \frac{1}{7^2} - \frac{1}{11^2} + \cdots + \frac{1}{m^2} - \frac{1}{(m+4)^2}$$

All of the terms cancel except for the first 4 terms of $\frac{1}{m^2}$ and the last 4 terms of $\frac{1}{(m+4)^2}$; since m goes to infinity, the last 4 terms of $\frac{1}{(m+4)^2}$ each approach zero.

$$\sum_{m=2}^{\infty} \left[\frac{1}{m^2} - \frac{1}{(m+4)^2}\right] = \frac{1}{2^2} + \frac{1}{3^2} + \frac{1}{4^2} + \frac{1}{5^2} = \frac{1}{4} + \frac{1}{9} + \frac{1}{16} + \frac{1}{25}$$

$$= \frac{900 + 400 + 225 + 144}{3600} = \frac{1669}{3600}$$

(29) This series **converges**. The 'trick' is to realize that the $\frac{1}{8}$'s add up to $\frac{2}{8} = \frac{1}{4}$, the $\frac{1}{32}$'s add up to $\frac{4}{32} = \frac{1}{8}$, the $\frac{1}{64}$'s add up to $\frac{8}{128} = \frac{1}{16}$, there will be sixteen $\frac{1}{512}$'s that add up to $\frac{16}{512} = \frac{1}{32}$, there will be thirty-two $\frac{1}{2048}$'s that add up to $\frac{32}{2048} = \frac{1}{64}$, etc. The given series is equivalent to the series below.

$$\frac{1}{2}+\frac{1}{4}+\frac{1}{8}+\frac{1}{16}+\frac{1}{32}+\frac{1}{64}+\cdots=\sum_{n=1}^{\infty}\frac{1}{2}\left(\frac{1}{2^{n-1}}\right)=\sum_{n=1}^{\infty}\frac{1}{2}\left(\frac{1}{2}\right)^{n-1}$$

This is a geometric series. Compare ar^{n-1} with $\left(\frac{1}{2}\right)^{n-1}=1\left(\frac{1}{2}\right)^{n-1}$ to see that $a=\frac{1}{2}$ and $r=\frac{1}{2}$. Since $|r|<1$, this series converges.

$$\sum_{n=1}^{\infty}\left(\frac{1}{2}\right)^{n-1}=\frac{a}{1-r}=\frac{1/2}{1-\frac{1}{2}}=\frac{1/2}{1/2}=\frac{0.5}{0.5}=1$$

(30) This series **diverges**. The 'trick' is to observe that the thirds add up to $\frac{2}{3}$, the fourths add up to $\frac{3}{4}$, the fifths add up to $\frac{4}{5}$, the sixths add up to $\frac{5}{6}$, etc. The given series is equivalent to the series below.

$$\frac{1}{2}+\frac{2}{3}+\frac{3}{4}+\frac{4}{5}+\frac{5}{6}+\cdots=\sum_{n=1}^{\infty}\frac{n}{n+1}$$

Use the simple divergence test. This series diverges because the corresponding sequence $\left\{\frac{n}{n+1}\right\}$ doesn't converge to zero. The sequence $\left\{\frac{n}{n+1}\right\}$ converges to $\frac{n}{n+1}=\frac{1}{1+\frac{1}{n}}=\frac{1}{1+0}=\frac{1}{1}=1$ (which isn't zero).

(31) Two geometric series are combined together. Apply the property that two converging series added together make a **converging** series: $\sum a_n=L_a, \sum b_n=L_b \rightarrow \sum(a_n+b_n)=L_a+L_b$. Note that $\frac{1}{1.5}=\frac{1}{1.5}\frac{2}{2}=\frac{2}{3}$ and $\frac{1}{1.25}=\frac{1}{1.25}\frac{4}{4}=\frac{4}{5}$.

$$\sum_{\ell=1}^{\infty}\left(\frac{4}{1.5^{\ell-1}}-\frac{2}{1.25^{\ell-1}}\right)=\sum_{\ell=1}^{\infty}\frac{4}{1.5^{\ell-1}}-\sum_{\ell=1}^{\infty}\frac{2}{1.25^{\ell-1}}$$

$$=\sum_{\ell=1}^{\infty}4\left(\frac{1}{1.5}\right)^{\ell-1}-\sum_{\ell=1}^{\infty}2\left(\frac{1}{1.25}\right)^{\ell-1}=\sum_{\ell=1}^{\infty}4\left(\frac{2}{3}\right)^{\ell-1}-\sum_{\ell=1}^{\infty}2\left(\frac{4}{5}\right)^{\ell-1}$$

The first series has $a_1=4$ and $r_1=\frac{2}{3}$ while the second series has $a_2=2$ and $r_2=\frac{4}{5}$.

$$\sum_{\ell=1}^{\infty}4\left(\frac{2}{3}\right)^{\ell-1}-\sum_{\ell=1}^{\infty}2\left(\frac{4}{5}\right)^{\ell-1}=\frac{a_1}{1-r_1}-\frac{a_2}{1-r_2}=\frac{4}{1-\frac{2}{3}}-\frac{2}{1-\frac{4}{5}}=\frac{4}{1/3}-\frac{2}{1/5}$$

$$4\div\frac{1}{3}-2\div\frac{1}{5}=4\times\frac{3}{1}-2\times\frac{5}{1}=12-10=2$$

(32) This series **diverges**. This is similar to the solution to Problem 31, except that the second series diverges because $r_2 = 1.1 \geq 1$. The sum (or difference) of a converging series and a diverging series is a diverging series.

(33) This series **converges**. By itself, each series would diverge, but combined together in this way they converge because almost all of the terms cancel out. Write these out longhand, noting the different initial values.

$$\sum_{k=5}^{\infty} \frac{1}{2k+2} - \sum_{m=3}^{\infty} \frac{1}{2m} = \frac{1}{12} + \frac{1}{14} + \frac{1}{16} + \frac{1}{18} + \frac{1}{20} + \cdots - \frac{1}{6} - \frac{1}{8} - \frac{1}{10} - \frac{1}{12} - \frac{1}{14} - \frac{1}{16} - \cdots$$

All of the terms cancel except for the first 3 terms of $-\frac{1}{2m}$ (and the last 3 terms of $\frac{1}{2k+2}$ in the partial sum, but since $\frac{1}{2k+2}$ approaches zero as k becomes infinite, only the first 3 terms matter).

$$\sum_{k=5}^{\infty} \frac{1}{2k+2} - \sum_{m=3}^{\infty} \frac{1}{2m} = -\frac{1}{6} - \frac{1}{8} - \frac{1}{10} = -\frac{20}{120} - \frac{15}{120} - \frac{12}{120} = -\frac{47}{120}$$

(34) Consider what happens if you shift the index to $n = -i$. The sum will contain all of the same terms, but n will vary from 0 to $+\infty$. Since $i = -n$, the exponent will change to $-n - 4$. Note that $\frac{1}{2^{-n-4}} = \frac{1}{2^{-(n+4)}} = 2^{n+4}$ according to the rules $x^{-p} = \frac{1}{x^p}$ and $\frac{1}{x^{-q}} = x^q$. This is a geometric series, but it isn't in standard form. If needed, you can shift the index to $n + 4 = k - 1$, but since the base will still be $r = 2$, this series will **diverge** since $r \geq 1$.

$$\sum_{i=-\infty}^{0} \frac{1}{2^{i-4}} = \sum_{n=0}^{\infty} 2^{n+4}$$

Another way to approach this is to simply consider what happens to $\frac{1}{2^{i-4}}$ in the limit that i approaches $-\infty$. When $i = -10$, we get $\frac{1}{2^{-10-4}} = \frac{1}{2^{-14}} = 2^{14} = 16{,}384$. The more negative i becomes, the larger $\frac{1}{2^{i-4}}$ grows.

(35) We may split the given series into separate sums as follows:

$$\sum_{m=-\infty}^{\infty} \frac{2}{3^{|m|}} = \sum_{m=-\infty}^{-1} \frac{2}{3^{|m|}} + \frac{2}{3^0} + \sum_{m=1}^{\infty} \frac{2}{3^{|m|}}$$

With the absolute values around the index, $|m|$ will be the same regardless of whether m is negative or positive. This means that the first sum and the last sum on the previous line are identical; they include all of the same terms added together.

$$\sum_{m=-\infty}^{\infty} \frac{2}{3^{|m|}} = \sum_{m=-\infty}^{-1} \frac{2}{3^{|m|}} + \frac{2}{3^0} + \sum_{m=1}^{\infty} \frac{2}{3^{|m|}} = \frac{2}{1} + 2\sum_{m=1}^{\infty} \frac{2}{3^m} = 2 + 2\sum_{m=1}^{\infty} \frac{2}{3^m}$$

(We removed the absolute values now that m is strictly positive.) This series <u>**converges**</u>. The remaining sum is a geometric series, but it isn't in standard form. You could shift the index using $m = k - 1$, or note that $3^m = 3^1 3^{m-1}$.

$$\sum_{m=-\infty}^{\infty} \frac{2}{3^{|m|}} = 2 + 2\sum_{m=1}^{\infty} \frac{2}{3^m} = 2 + 2\sum_{m=1}^{\infty} \frac{2}{3^1 3^{m-1}} = 2 + 2\sum_{m=1}^{\infty} \frac{2}{3}\left(\frac{1}{3}\right)^{m-1}$$

Compare ar^{n-1} with $\frac{2}{3}\left(\frac{1}{3}\right)^{m-1}$ to see that $a = \frac{2}{3}$ and $r = \frac{1}{3}$. Since $|r| < 1$, this series converges.

$$\sum_{m=-\infty}^{\infty} \frac{2}{3^{|m|}} = 2 + 2\sum_{m=1}^{\infty} \frac{2}{3}\left(\frac{1}{3}\right)^{m-1} = 2 + 2\frac{a}{1 - r} = 2 + 2\frac{2/3}{1 - 1/3}$$

$$= 2 + 2\frac{2/3}{2/3} = 2 + 2(1) = 2 + 2 = 4$$

Check the answer: Start by checking this geometric series:

$$\sum_{m=1}^{\infty} \frac{2}{3^m} = \frac{2}{3^1} + \frac{2}{3^2} + \frac{2}{3^3} + \frac{2}{3^4} + \frac{2}{3^5} + \frac{2}{3^6} + \cdots = \frac{2}{3} + \frac{2}{9} + \frac{2}{27} + \frac{2}{81} + \frac{2}{243} + \frac{2}{729} + \cdots$$

$$\approx 0.6667 + 0.2222 + 0.0741 + 0.0247 + 0.0082 + 0.0027 = 0.9986$$

This shows that this geometric series converges to 1, such that the complete series converges to

$$\sum_{m=-\infty}^{\infty} \frac{2}{3^{|m|}} = 2 + 2\sum_{m=1}^{\infty} \frac{2}{3^m} = 2 + 2(1) = 2 + 2 = 4$$

Chapter 5 Answers

(1) Let $a_n = \frac{1}{n^2+3}$ and $c_n = \frac{1}{n^2}$. Observe that $a_n < c_n$ for all n. Example: $a_2 = \frac{1}{2^2+3} = \frac{1}{4+3} = \frac{1}{7} \approx 0.143$ is less than $c_2 = \frac{1}{2^2} = \frac{1}{4} = 0.25$. Since $\sum c_n = \sum \frac{1}{n^2}$ converges, it follows that $\sum a_n = \sum \frac{1}{n^2+3}$ **converges**.

(2) Let $a_i = \frac{4}{i-\sqrt{i}}$ and $d_i = \frac{4}{i}$. Observe that $a_i > d_i$ for all i. Example: $a_4 = \frac{4}{4-\sqrt{4}} = \frac{4}{4-2} = \frac{4}{2} = 2$ is greater than $d_2 = \frac{4}{4} = 1$. Since $\sum d_i = \sum \frac{4}{i}$ diverges (since it is a multiple of $\sum \frac{1}{i}$), it follows that $\sum a_i = \sum \frac{4}{i-\sqrt{i}}$ **diverges**.

(3) Let $a_m = \frac{5m+1}{m^3-2m}$ and $c_m = \frac{1}{m^2}$. There is a small issue to deal with; $a_1 = \frac{5(1)+1}{1^3-2(1)} = \frac{6}{-1} = -6$ is a negative term. To deal with this, we can write the given sum as

$$\sum_{m=1}^{\infty} \frac{5m+1}{m^3-2m} = -6 + \sum_{m=2}^{\infty} \frac{5m+1}{m^3-2m}$$

where we have separated the first term from the rest of the series. The series on the right doesn't contain any negative terms. If the series on the right converges, subtracting 6 won't change that, so either both sums above converge or both sums above diverge.

$$\lim_{m\to\infty} \frac{a_m}{c_m} = \lim_{m\to\infty} \frac{\frac{5m+1}{m^3-2m}}{\frac{1}{m^2}} = \lim_{m\to\infty} \left(\frac{5m+1}{m^3-2m} \div \frac{1}{m^2}\right) = \lim_{m\to\infty} \left(\frac{5m+1}{m^3-2m} \times \frac{m^2}{1}\right)$$

$$= \lim_{m\to\infty} \frac{5m^3+m^2}{m^3-2m} = \lim_{m\to\infty} \frac{5+\frac{1}{m}}{1-\frac{2}{m^2}} = \frac{5+0}{1-0} = 5$$

Since $\sum c_m = \sum \frac{1}{m^2}$ converges and $\lim_{m\to\infty} \frac{a_m}{c_m} = 5$ is finite, it follows that $\sum a_m = \sum \frac{5m+1}{m^3-2m}$ **converges**. (This is similar to Example 2.)

(4) Let $a_j = \frac{9}{6+\sqrt{j}}$ and $d_j = \frac{1}{\sqrt{j}}$. Use the limit comparison test (since a_j isn't greater than d_j).

$$\lim_{j\to\infty}\frac{a_j}{d_j}=\lim_{j\to\infty}\frac{\frac{9}{6+\sqrt{j}}}{\frac{1}{\sqrt{j}}}=\lim_{j\to\infty}\left(\frac{9}{6+\sqrt{j}}\div\frac{1}{\sqrt{j}}\right)=\lim_{j\to\infty}\left(\frac{9}{6+\sqrt{j}}\times\frac{\sqrt{j}}{1}\right)$$

$$=\lim_{j\to\infty}\left(\frac{9\sqrt{j}}{6+\sqrt{j}}\right)=\lim_{j\to\infty}\left(\frac{9}{\frac{6}{\sqrt{j}}+1}\right)=\frac{9}{0+1}=\frac{9}{1}=9$$

Since $\sum d_j=\sum\frac{1}{\sqrt{j}}$ diverges (since $\frac{1}{\sqrt{j}}=\frac{1}{j^{1/2}}$ and $\sum\frac{1}{j^p}$ diverges if $p\le 1$) and $\lim_{j\to\infty}\frac{a_j}{d_j}=9$ is positive, it follows that $\sum a_j=\sum\frac{9}{6+\sqrt{j}}$ **diverges**. (This is similar to Example 4.)

(5) Separate this into two sums as follows:

$$\sum_{\ell=1}^{\infty}\frac{3-4^{\ell+1}}{5^{\ell-1}}=\sum_{\ell=1}^{\infty}\frac{3}{5^{\ell-1}}-\sum_{\ell=1}^{\infty}\frac{4^{\ell+1}}{5^{\ell-1}}$$

Note that $4^{\ell+1}=4^\ell 4^1=4^{\ell-1}4^1 4^1=(16)4^{\ell-1}$. With this substitution, the sums become:

$$\sum_{\ell=1}^{\infty}\frac{3}{5^{\ell-1}}-\sum_{\ell=1}^{\infty}\frac{(16)4^{\ell-1}}{5^{\ell-1}}=\sum_{\ell=1}^{\infty}\frac{3}{5^{\ell-1}}-\sum_{\ell=1}^{\infty}16\left(\frac{4}{5}\right)^{\ell-1}$$

These are geometric series of the form $\sum ar^{\ell-1}$. The first sum, $\sum\frac{3}{5^{\ell-1}}$, has $a=3$ and $r=\frac{1}{5}$ while the second sum, $\sum 16\left(\frac{4}{5}\right)^{\ell-1}$, has $a=16$ and $r=\frac{4}{5}$. In each case, $|r|<1$, which means that each series converges. As we learned in Chapter 4, the sum or difference of two converging series also **converges**.

(6) Let $a_k=\frac{2}{k-\ln k}$ and $d_k=\frac{2}{k}$. Observe that $a_k>d_k$ for all k. Example: $a_3=\frac{3}{3-\ln 3}\approx\frac{3}{3-1.0986}\approx\frac{3}{1.9014}\approx 1.58$ is greater than $d_3=\frac{3}{2}=1.5$. Since $\sum d_k=\sum\frac{2}{k}$ diverges (since it is a multiple of $\sum\frac{1}{k}$), it follows that $\sum a_k=\sum\frac{2}{k-\ln k}$ **diverges**.

(7) Let $a_i=\frac{1}{i^2}\sin\left(\frac{1}{i}\right)$ and $c_i=\frac{1}{i^2}$. (Here, $\frac{1}{i}$ is in radians, but the series would still converge if it were in degrees.)

$$\lim_{i\to\infty}\frac{a_i}{c_i}=\lim_{i\to\infty}\frac{\frac{1}{i^2}\sin\left(\frac{1}{i}\right)}{\frac{1}{i^2}}=\lim_{i\to\infty}\left[\frac{1}{i^2}\sin\left(\frac{1}{i}\right)\div\frac{1}{i^2}\right]=\lim_{i\to\infty}\left[\frac{1}{i^2}\sin\left(\frac{1}{i}\right)\times\frac{i^2}{1}\right]=\lim_{i\to\infty}\sin\left(\frac{1}{i}\right)=0$$

As i grows to infinity, $\frac{1}{i}$ approaches zero and $\sin 0 = 0$. Since $\sum c_i = \sum \frac{1}{i^2}$ converges and $\lim\limits_{i\to\infty} \frac{a_i}{c_i} = 0$ is finite, it follows that $\sum a_i = \sum \frac{1}{i^2} \sin\left(\frac{1}{i}\right)$ **converges**. (This is similar to Example 2.) In the case of convergence, we want the limit to be finite (that is, not infinity), whereas in the case of divergence, we want the limit to be positive (that is, greater than zero). Since we tested convergence, zero is perfectly fine.

(8) Let $a_m = \frac{\cos^2 m}{m^2}$ and $c_m = \frac{1}{m^2}$. Since the cosine function is **squared**, we don't need to worry about negative values. (Although cosine may be negative, the square of cosine is strictly nonnegative.) Since $0 \le \cos^2 m \le 1$, $a_m \le c_m$ for all m. Since $\sum c_m = \sum \frac{1}{m^2}$ converges, it follows that $\sum a_m = \sum \frac{\cos^2 m}{m^2}$ **converges**.

(9) The sequence $\left\{\frac{2^n-5}{5(2^n-3)}\right\}$ doesn't converge to zero, as shown below.

$$\lim_{n\to\infty} \frac{2^n - 5}{5(2^n - 3)} = \lim_{n\to\infty} \frac{1 - \frac{5}{2^n}}{5\left(1 - \frac{3}{2^n}\right)} = \frac{1-0}{5(1-0)} = \frac{1}{5(1)} = \frac{1}{5}$$

According to the simple divergence test (Chapter 4), this series **diverges** because the corresponding sequence doesn't converge to zero (it converges to $\frac{1}{5}$).

(10) Let $a_k = \frac{k^3-4k+5}{3k^5+7k^3-1}$ and $c_k = \frac{1}{k^2}$.

$$\lim_{k\to\infty} \frac{a_k}{c_k} = \lim_{k\to\infty} \frac{\frac{k^3 - 4k + 5}{3k^5 + 7k^3 - 1}}{\frac{1}{k^2}} = \lim_{k\to\infty} \left(\frac{k^3 - 4k + 5}{3k^5 + 7k^3 - 1} \div \frac{1}{k^2}\right) = \lim_{k\to\infty} \left(\frac{k^3 - 4k + 5}{3k^5 + 7k^3 - 1} \times \frac{k^2}{1}\right)$$

$$\lim_{k\to\infty} \frac{k^5 - 4k^3 + 5k^2}{3k^5 + 7k^3 - 1} = \lim_{k\to\infty} \frac{1 - \frac{4}{k^2} + \frac{5}{k^3}}{3 + \frac{7}{k^2} - \frac{1}{k^5}} = \frac{1 - 0 + 0}{3 + 0 - 0} = \frac{1}{3}$$

Since $\sum c_k = \sum \frac{1}{k^2}$ converges and $\lim\limits_{k\to\infty} \frac{a_k}{c_k} = \frac{1}{3}$ is finite, it follows that $\sum a_k = \sum \frac{k^3-4k+5}{3k^5+7k^3-1}$ **converges**. (This is similar to Example 2.)

(11) Let $a_j = \frac{\ln j}{j^2}$ and $c_j = \frac{1}{j^2}$. Since the numerator ($\ln j$) and denominator (j^2) both grow to infinity as $j \to \infty$, we need to apply l'Hôpital's rule for this indeterminate form. (Recall Chapter 2.)

$$\lim_{j\to\infty}\frac{a_j}{c_j}=\lim_{j\to\infty}\frac{\ln j}{j^2}=\frac{\frac{d}{dx}\ln x\Big|_{x\to\infty}}{\frac{d}{dx}x^2\Big|_{x\to\infty}}=\frac{\lim\limits_{x\to\infty}\frac{1}{x}}{\lim\limits_{x\to\infty}2x}=\frac{0}{\infty}=0$$

Since $\sum c_j=\sum\frac{1}{j^2}$ converges and $\lim\limits_{j\to\infty}\frac{a_j}{c_j}=0$ is finite, it follows that $\sum a_j=\sum\frac{\ln j}{j^2}$ **converges**. (This is similar to Example 2.) In the case of convergence, we want the limit to be finite (that is, not infinity), whereas in the case of divergence, we want the limit to be positive (that is, greater than zero). Since we tested convergence, zero is perfectly fine.

(12) Let $a_\ell=\frac{\ell^5}{e^\ell}$ and $c_\ell=\frac{1}{\ell^2}$.

$$\lim_{\ell\to\infty}\frac{a_\ell}{c_\ell}=\lim_{\ell\to\infty}\frac{\ell^5/e^\ell}{1/\ell^2}=\lim_{\ell\to\infty}\left(\frac{\ell^5}{e^\ell}\div\frac{1}{\ell^2}\right)=\lim_{\ell\to\infty}\left(\frac{\ell^5}{e^\ell}\times\frac{\ell^2}{1}\right)=\lim_{\ell\to\infty}\frac{\ell^7}{e^\ell}$$

Since the numerator (ℓ^7) and denominator (e^ℓ) both grow to infinity as $\ell\to\infty$, we need to apply l'Hôpital's rule for this indeterminate form. (Recall Chapter 2.) It will take 7 consecutive derivatives before the numerator is reduced to a constant; the denominator remains unchanged since $\frac{d}{dx}e^x=e^x$. In the numerator, the first derivative gives $7x^6$, the second derivative gives $42x^5$, the third derivative gives $210x^4$, and so on until the sixth derivative gives $5040x$, after which the seventh derivative yields the constant 5040 (which is equal to 7 factorial). The main idea here is that the ratio $\frac{\ell^7}{e^\ell}$ approaches zero as $\ell\to\infty$ because the exponential function grows more rapidly than the polynomial ℓ^7 when ℓ is large.

$$\lim_{\ell\to\infty}\frac{a_\ell}{c_\ell}=\frac{\frac{d^7}{dx^7}(x^7)\Big|_{x\to\infty}}{\frac{d^7}{dx^7}e^x\Big|_{x\to\infty}}=\frac{\lim\limits_{x\to\infty}5040}{\lim\limits_{x\to\infty}e^x}=\frac{5040}{\infty}=0$$

Since $\sum c_\ell=\sum\frac{1}{\ell^2}$ converges and $\lim\limits_{\ell\to\infty}\frac{a_\ell}{c_\ell}=0$ is finite, it follows that $\sum a_\ell=\sum\frac{\ell^5}{e^\ell}$ **converges**. (This is similar to Example 2.) In the case of convergence, we want the limit to be finite (that is, not infinity), whereas in the case of divergence, we want the limit to be positive (that is, greater than zero). Since we tested convergence, zero is perfectly fine.

(13) Let $a_n = \sqrt{\frac{n^2+5n+2}{n^4-4n^2+3}}$ and $d_n = \frac{1}{n}$.

$$\lim_{n\to\infty}\frac{a_n}{d_n} = \lim_{n\to\infty}\frac{\sqrt{\frac{n^2+5n+2}{n^4-4n^2+3}}}{\frac{1}{n}} = \lim_{n\to\infty}\left(\sqrt{\frac{n^2+5n+2}{n^4-4n^2+3}} \div \frac{1}{n}\right) = \lim_{n\to\infty}\left(\sqrt{\frac{n^2+5n+2}{n^4-4n^2+3}} \times \frac{n}{1}\right)$$

$$= \lim_{n\to\infty}\left(\sqrt{\frac{n^2+5n+2}{n^4-4n^2+3}} \times \sqrt{n^2}\right) = \lim_{n\to\infty}\sqrt{\frac{(n^2+5n+2)n^2}{n^4-4n^2+3}} = \lim_{n\to\infty}\sqrt{\frac{n^4+5n^3+2n^2}{n^4-4n^2+3}}$$

$$= \lim_{n\to\infty}\sqrt{\frac{1+\frac{5}{n}+\frac{2}{n^2}}{1-\frac{4}{n^2}+\frac{3}{n^4}}} = \sqrt{\frac{1+0+0}{1-0+0}} = \sqrt{\frac{1}{1}} = \sqrt{1} = 1$$

Since $\sum d_n = \sum \frac{1}{n}$ diverges and $\lim\limits_{n\to\infty}\frac{a_n}{d_n} = 1$ is positive, it follows that $\sum a_n = \sum \sqrt{\frac{n^2+5n+2}{n^4-4n^2+3}}$ **<u>diverges</u>**. (This is similar to Example 4.) Note: On the bottom line, we divided the numerator and denominator each by n^4 (a strategy that we applied to evaluate limits back in Chapter 2).

(14) Let $a_i = \left(\frac{i-1}{2i+3}\right)\left(\frac{1}{4^i}\right)$ and $c_i = \frac{1}{4^i}$.

$$\lim_{i\to\infty}\frac{a_i}{c_i} = \lim_{i\to\infty}\frac{\left(\frac{i-1}{2i+3}\right)\left(\frac{1}{4^i}\right)}{\frac{1}{4^i}} = \lim_{i\to\infty}\left[\left(\frac{i-1}{2i+3}\right)\left(\frac{1}{4^i}\right) \div \left(\frac{1}{4^i}\right)\right] = \lim_{i\to\infty}\left[\left(\frac{i-1}{2i+3}\right)\left(\frac{1}{4^i}\right) \times \frac{4^i}{1}\right]$$

$$= \lim_{i\to\infty}\frac{i-1}{2i+3} = \lim_{i\to\infty}\frac{1-\frac{1}{i}}{2+\frac{3}{i}} = \frac{1-0}{2+0} = \frac{1}{2}$$

Since $\sum c_i = \sum \frac{1}{4^i}$ converges and $\lim\limits_{i\to\infty}\frac{a_i}{c_i} = \frac{1}{2}$ is finite, it follows that $\sum a_i = \sum\left(\frac{i-1}{2i+3}\right)\left(\frac{1}{4^i}\right)$ **<u>converges</u>**. (This is similar to Example 2.)

(15) Let $a_k = \frac{4}{(k-1)\sqrt{k+1}}$ and $c_k = \frac{1}{k^{3/2}} = \frac{1}{k\sqrt{k}}$.

$$\lim_{k\to\infty}\frac{a_k}{c_k} = \lim_{k\to\infty}\frac{\frac{4}{(k-1)\sqrt{k+1}}}{\frac{1}{k\sqrt{k}}} = \lim_{k\to\infty}\left(\frac{4}{(k-1)\sqrt{k+1}} \div \frac{1}{k\sqrt{k}}\right)$$

$$= \lim_{k\to\infty}\left(\frac{4}{(k-1)\sqrt{k+1}} \times \frac{k\sqrt{k}}{1}\right) = \lim_{k\to\infty}\frac{4k\sqrt{k}}{(k-1)\sqrt{k+1}} = \lim_{k\to\infty}\left(\frac{4k}{k-1}\sqrt{\frac{k}{k+1}}\right)$$

$$= \lim_{k\to\infty}\left(\frac{4}{1-\frac{1}{k}}\sqrt{\frac{1}{1+\frac{1}{k}}}\right) = \frac{4}{1-0}\sqrt{\frac{1}{1+0}} = \frac{4}{1}\sqrt{\frac{1}{1}} = 4\sqrt{1} = 4(1) = 4$$

Since $\sum c_k = \sum \frac{1}{k^{3/2}} = \sum \frac{1}{k\sqrt{k}}$ converges (where $k^{3/2} = k^1 k^{1/2}$ and $\sqrt{k} = k^{1/2}$) and $\lim_{k\to\infty}\frac{a_k}{c_k} = 4$ is finite, it follows that $\sum a_k = \sum \frac{4}{(k-1)\sqrt{k+1}}$ **converges**. (This is similar to Example 2.) Note: On the bottom line, we divided the numerator and denominator each by k for both fractions, including the fraction inside of the radical (this is a strategy that we applied to evaluate limits back in Chapter 2).

(16) Let $a_m = \frac{3m-1}{6^m(m+2)}$ and $c_m = \frac{1}{6^m}$.

$$\lim_{m\to\infty}\frac{a_m}{c_m} = \lim_{m\to\infty}\frac{\frac{3m-1}{6^m(m+2)}}{\frac{1}{6^m}} = \lim_{m\to\infty}\left[\frac{3m-1}{6^m(m+2)} \div \frac{1}{6^m}\right] = \lim_{m\to\infty}\left[\frac{3m-1}{6^m(m+2)} \times \frac{6^m}{1}\right]$$

$$= \lim_{m\to\infty}\frac{3m-1}{m+2} = \frac{3-\frac{1}{m}}{1+\frac{2}{m}} = \frac{3-0}{1+0} = \frac{3}{1} = 3$$

Since $\sum c_m = \sum \frac{1}{6^m}$ converges and $\lim_{m\to\infty}\frac{a_m}{c_m} = 3$ is finite, it follows that $\sum a_m = \sum \frac{3m-1}{6^m(m+2)}$ **converges**. (This is similar to Example 2.)

(17) Let $a_j = \frac{1}{j!}$ and $c_j = \frac{1}{2^j}$. Observe that $a_j < c_j$ once we reach $j > 3$. Examples:

- $a_4 = \frac{1}{4!} = \frac{1}{24} \approx 0.0417$ is less than $c_4 = \frac{1}{2^4} = \frac{1}{16} = 0.0625$.
- $a_5 = \frac{1}{5!} = \frac{1}{120} \approx 0.0083$ is less than $c_5 = \frac{1}{2^5} = \frac{1}{32} = 0.03125$.
- $a_6 = \frac{1}{6!} = \frac{1}{720} \approx 0.0014$ is less than $c_6 = \frac{1}{2^6} = \frac{1}{64} = 0.015625$.

Since $\sum c_j = \sum \frac{1}{2^j}$ converges, it follows that $\sum a_j = \sum \frac{1}{j!}$ **converges**.

(18) Let $a_\ell = \frac{3\ell+8}{(\ell-1)^2}$ and $d_\ell = \frac{1}{\ell}$.

$$\lim_{\ell\to\infty}\frac{a_\ell}{d_\ell}=\lim_{\ell\to\infty}\frac{\frac{3\ell+8}{(\ell-1)^2}}{\frac{1}{\ell}}=\lim_{\ell\to\infty}\left[\frac{3\ell+8}{(\ell-1)^2}\div\frac{1}{\ell}\right]=\lim_{\ell\to\infty}\left[\frac{3\ell+8}{(\ell-1)^2}\times\frac{\ell}{1}\right]=\lim_{\ell\to\infty}\frac{3\ell^2+8\ell}{(\ell-1)^2}$$

$$=\lim_{\ell\to\infty}\frac{3\ell^2+8\ell}{\ell^2-2\ell+1}=\lim_{\ell\to\infty}\frac{3+\frac{8}{\ell}}{1-\frac{2}{\ell}+\frac{1}{\ell^2}}=\frac{3+0}{1-0+1}=\frac{3}{1}=3$$

Since $\sum d_\ell=\sum\frac{1}{\ell}$ diverges and $\lim_{\ell\to\infty}\frac{a_\ell}{d_\ell}=3$ is positive, it follows that $\sum a_n=\sum\frac{3\ell+8}{(\ell-1)^2}$ **diverges**. (This is similar to Example 4.)

(19) If you note that $4\sqrt{4}=4(2)=8$, $8\sqrt{8}=8\sqrt{(4)(2)}=8\sqrt{4}\sqrt{2}=8(2)\sqrt{2}=16\sqrt{2}$, and $9\sqrt{9}=9(3)=27$, the given series may be rewritten as

$$\frac{1}{2\sqrt{2}}+\frac{2}{3\sqrt{3}}+\frac{3}{4\sqrt{4}}+\frac{4}{5\sqrt{5}}+\frac{5}{6\sqrt{6}}+\frac{6}{7\sqrt{7}}+\frac{7}{8\sqrt{8}}+\frac{8}{9\sqrt{9}}+\cdots$$

Rewrite the above series using summation notation.

$$\sum_{n=2}^{\infty}\frac{n-1}{n\sqrt{n}}$$

Let $a_n=\frac{n-1}{n\sqrt{n}}$ and $d_n=\frac{1}{n^{1/2}}=\frac{1}{\sqrt{n}}$. Use the limit comparison test (since a_n isn't greater than d_n).

$$\lim_{n\to\infty}\frac{a_n}{d_n}=\lim_{n\to\infty}\frac{\frac{n-1}{n\sqrt{n}}}{\frac{1}{\sqrt{n}}}=\lim_{n\to\infty}\left(\frac{n-1}{n\sqrt{n}}\div\frac{1}{\sqrt{n}}\right)=\lim_{n\to\infty}\left(\frac{n-1}{n\sqrt{n}}\times\frac{\sqrt{n}}{1}\right)$$

$$=\lim_{n\to\infty}\frac{n-1}{n}=\frac{1-\frac{1}{n}}{1}=\frac{1-0}{1}=\frac{1}{1}=1$$

Since $\sum d_n=\sum\frac{1}{n^{1/2}}=\sum\frac{1}{\sqrt{n}}$ diverges (since $\sum\frac{1}{n^p}$ diverges if $p\le 1$) and $\lim_{n\to\infty}\frac{a_n}{d_n}=1$ is positive, it follows that $\sum a_n=\sum\frac{n-1}{n\sqrt{n}}$ **diverges**. (This is similar to Example 4.)

(20) We can express the given series in summation notation as follows.

$$\sum_{n=1}^{\infty}\frac{1}{2^n+1}=\frac{1}{3}+\frac{1}{5}+\frac{1}{9}+\frac{1}{17}+\frac{1}{33}+\frac{1}{65}+\cdots$$

Let $a_n=\frac{1}{2^n+1}$ and $c_n=\frac{1}{2^n}$. Observe that $a_n<c_n$ for all n. Example: $a_3=\frac{1}{2^3+1}=$

$\frac{1}{8+1} = \frac{1}{9} \approx 0.111$ is less than $c_3 = \frac{1}{2^3} = \frac{1}{8} = 0.125$. Since $\sum c_n = \sum \frac{1}{2^n}$ converges, it follows that $\sum a_n = \sum \frac{1}{2^n+1}$ **converges**.

(21) Note that the numerators begin 13, 15, 17, 19, ... How do you explain the numerators that are 7, 5, and 3? Those fractions were reduced. Note, for example, that $\frac{7}{40} = \frac{21}{120}$. We can rewrite the given series as

$$\frac{13}{1} + \frac{15}{2} + \frac{17}{6} + \frac{19}{24} + \frac{21}{120} + \frac{23}{720} + \frac{25}{5040} + \frac{27}{40{,}320} + \frac{29}{362{,}880} + \cdots$$

Now the numerators are odd numbers. We can make the numerators with the formula $2n + 11$. The denominators above are factorials. For example, $6! = 6(5)(4)(3)(2)(1) = 720$. Using summation notation, the above series is:

$$\sum_{n=1}^{\infty} \frac{2n + 11}{n!}$$

Let $a_n = \frac{2n+11}{n!}$ and $c_n = \frac{1}{2^n}$. Observe that $a_n < c_n$ once we reach $n > 6$. Examples:

- $a_7 = \frac{2(7)+11}{7!} = \frac{25}{5040} \approx 0.00496$ is less than $c_7 = \frac{1}{2^7} = \frac{1}{128} = 0.0078125$.
- $a_8 = \frac{2(8)+11}{8!} = \frac{27}{40{,}320} \approx 0.00067$ is less than $c_8 = \frac{1}{2^8} = \frac{1}{256} = 0.00390625$.
- $a_9 = \frac{2(9)+11}{9!} = \frac{29}{362{,}880} \approx 0.00008$ is less than $c_9 = \frac{1}{2^6} = \frac{1}{512} = 0.001953125$.

Since $\sum c_n = \sum \frac{1}{2^n}$ converges, it follows that $\sum a_j = \sum \frac{2n+11}{n!}$ **converges**.

Chapter 6 Answers

(1) Let $f(x) = \frac{1}{x^2+1}$ and $a_n = f(n) = \frac{1}{n^2+1}$.

$$\int_{x=1}^{\infty} f(x)\,dx = \int_{x=1}^{\infty} \frac{dx}{x^2+1}$$

One way to perform this integral is to make the trigonometric substitution $x = \tan\theta$. Why? Divide by $\cos^2\theta$ on both sides of $\sin^2\theta + \cos^2\theta = 1$ to see that $\tan^2\theta + 1 = \sec^2\theta$. By setting $x = \tan\theta$, the new denominator will have a single term instead of two terms, which will make the integral simpler. Implicitly differentiate both sides of $x = \tan\theta$ to get $dx = \sec^2\theta\,d\theta$. Make the substitutions $dx = \sec^2\theta\,d\theta$ and $x = \tan\theta$ in the above integral. Make the substitution $\tan^2\theta + 1 = \sec^2\theta$ in the new denominator. Note that $\theta = \tan^{-1}x$. The new lower limit will be $\tan^{-1}1 = \frac{\pi}{4}$ radians (equivalent to 45°) and the new upper limit will be $\lim_{x\to\infty}\tan^{-1}x$. What does this equal? We can figure it out by remembering that $\tan\theta = \frac{\sin\theta}{\cos\theta}$. As θ grows to $\frac{\pi}{2}$ radians (equivalent to 90°), $\sin\theta$ gets larger (approaching one) while $\cos\theta$ gets smaller (approaching zero), such that $\tan\theta$ grows infinite. For example, when $\theta = 0.99999 \times \frac{\pi}{2}$ radians (or 89.9991°), $\tan\theta$ equals 63,662. Since $x = \tan\theta$ grows to ∞ as θ approaches $\frac{\pi}{2}$ radians, this shows that $\lim_{x\to\infty}\tan^{-1}x$ (which is the upper limit of θ) equals $\frac{\pi}{2}$ radians. You're probably used to thinking that $\tan\frac{\pi}{2}$ is undefined, but it's more precise to say that $\tan\theta$ becomes infinite as θ approaches $\frac{\pi}{2}$ radians.

$$\int_{x=1}^{\infty} \frac{dx}{x^2+1} = \int_{\theta=\tan^{-1}1}^{\pi/2} \frac{\sec^2\theta\,d\theta}{\tan^2\theta+1} = \int_{\theta=\pi/4}^{\pi/2} \frac{\sec^2\theta\,d\theta}{\sec^2\theta} = \int_{\theta=\pi/4}^{\pi/2} d\theta = [\theta]_{x=\pi/4}^{\pi/2} = \frac{\pi}{2} - \frac{\pi}{4} = \frac{\pi}{4}$$

Since the integral is finite, the corresponding series **converges**. Note: Although you could solve this problem with a comparison test, the instructions ask you to use the integral test.

(2) Let $f(x) = \frac{x}{x^2+3}$ and $a_k = f(k) = \frac{k}{k^2+3}$.

$$\int_{x=2}^{\infty} f(x)\,dx = \int_{x=2}^{\infty} \frac{x}{x^2+3}\,dx$$

The 'easy' way to do this integral is to think about taking a derivative of the denominator. If you take a derivative of $x^2 + 3$ with respect to x, you get $3x$, which isn't too different from the numerator. This suggests trying the substitution $u = x^2 + 3$. Implicitly differentiate both sides to see that $du = 2x\,dx$. Replace $x\,dx$ with $\frac{du}{2}$ and replace $x^2 + 3$ with u in the above integral. Since $u = x^2 + 3$, the new lower limit will be $2^2 + 3 = 4 + 3 = 7$.

$$\int_{x=2}^{\infty} \frac{x}{x^2+3}\,dx = \int_{x=2}^{\infty} \frac{du}{2u} = \frac{1}{2}[\ln|u|]_{u=7}^{\infty} = \frac{\ln\infty - \ln 7}{2} = \infty$$

Since the integral diverges, the corresponding series **diverges**.

(3) Let $f(x) = \frac{1}{\sqrt{x+1}}$ and $a_m = f(m) = \frac{1}{\sqrt{m+1}}$. Recall that $x^{-1/2} = \frac{1}{\sqrt{x}}$.

$$\int_{x=0}^{\infty} f(x)\,dx = \int_{x=0}^{\infty} \frac{dx}{\sqrt{x+1}} = \int_{x=0}^{\infty} (x+1)^{-1/2}\,dx = \left[\frac{(x+1)^{-1/2+1}}{-1/2+1}\right]_{x=0}^{\infty} = \left[\frac{(x+1)^{1/2}}{1/2}\right]_{x=0}^{\infty}$$

$$= \left[\frac{\sqrt{x+1}}{1/2}\right]_{x=0}^{\infty} = \left[\sqrt{x+1} \div \frac{1}{2}\right]_{x=0}^{\infty} = \left[\sqrt{x+1} \times \frac{2}{1}\right]_{x=0}^{\infty} = \left[2\sqrt{x+1}\right]_{x=0}^{\infty} = 2\sqrt{\infty} - 2\sqrt{1} = \infty$$

Alternate solution: Let $k = m + 1$ (recall Chapter 3). This sum begins at $k = 0 + 1 = 1$.

$$\sum_{m=0}^{\infty} \frac{1}{\sqrt{m+1}} = \sum_{k=1}^{\infty} \frac{1}{\sqrt{k}}$$

Let $f(x) = \frac{1}{\sqrt{x}}$ and $a_k = f(k) = \frac{1}{\sqrt{k}}$. Since $k = m + 1$, this integral begins at 1 (not 0).

$$\int_{x=1}^{\infty} f(x)\,dx = \int_{x=1}^{\infty} \frac{dx}{\sqrt{x}} = \int_{x=1}^{\infty} x^{-1/2}\,dx = \left[\frac{x^{-1/2+1}}{-1/2+1}\right]_{x=1}^{\infty} = \left[\frac{x^{1/2}}{1/2}\right]_{x=1}^{\infty} = \left[\frac{\sqrt{x}}{1/2}\right]_{x=1}^{\infty}$$

$$= \left[\sqrt{x} \div \frac{1}{2}\right]_{x=1}^{\infty} = \left[\sqrt{x} \times \frac{2}{1}\right]_{x=1}^{\infty} = \left[2\sqrt{x}\right]_{x=1}^{\infty} = 2\sqrt{\infty} - 2\sqrt{1} = \infty$$

Since the integral diverges, the corresponding series **diverges**. Note: You should recognize that this is a $\sum \frac{1}{x^p}$ series with $p = \frac{1}{2}$; it diverges because $p \le 1$.

(4) Let $f(x) = \frac{6}{x\sqrt{x}}$ and $a_j = f(j) = \frac{6}{j\sqrt{j}}$. Since $\sqrt{x} = x^{1/2}$, it follows that $\frac{1}{x\sqrt{x}} = \frac{1}{x(x^{1/2})} = \frac{1}{x^{1+1/2}} = \frac{1}{x^{3/2}} = x^{-3/2}$.

$$\int_{x=4}^{\infty} f(x)\,dx = \int_{x=4}^{\infty} \frac{6}{x\sqrt{x}}\,dx = 6\int_{x=4}^{\infty} x^{-3/2}\,dx = 6\left[\frac{x^{-3/2+1}}{-\frac{3}{2}+1}\right]_{x=4}^{\infty} = 6\left[\frac{x^{-1/2}}{-1/2}\right]_{x=4}^{\infty}$$

$$= 6\left[-x^{-1/2} \div \frac{1}{2}\right]_{x=4}^{\infty} = -6\left[x^{-1/2} \times \frac{2}{1}\right]_{x=4}^{\infty} = -12\left[x^{-1/2}\right]_{x=4}^{\infty} = -12\left[\frac{1}{x^{1/2}}\right]_{x=4}^{\infty}$$

$$= -12\left[\frac{1}{\sqrt{x}}\right]_{x=4}^{\infty} = -12\left(\frac{1}{\sqrt{\infty}} - \frac{1}{\sqrt{4}}\right) = -12\left(0 - \frac{1}{2}\right) = -12\left(-\frac{1}{2}\right) = 6$$

Since the integral is finite, the corresponding series **converges**.

(5) Let $f(x) = \frac{8}{x(x+4)}$ and $a_i = f(i) = \frac{8}{i(i+4)}$.

$$\int_{x=4}^{\infty} f(x)\,dx = \int_{x=4}^{\infty} \frac{8}{x(x+4)}\,dx$$

Use the method of partial fractions. (Recall Chapter 4.)

$$\frac{8}{x(x+4)} = \frac{p}{x} + \frac{q}{x+4}$$

Multiply both sides by $x(x+4)$.

$$8 = p(x+4) + qx = px + 4p + xq = (p+q)x + 4p$$

Both sides of the above equation will be equal for any value of x if $p + q = 0$ and $8 = 4p$, which means that $p = \frac{8}{4} = 2$ and $q = -p = -2$.

$$\int_{x=4}^{\infty} \frac{8}{x(x+4)}\,dx = \int_{x=4}^{\infty} \frac{2}{x}\,dx - \int_{x=4}^{\infty} \frac{2}{x+4}\,dx = 2[\ln|x|]_{x=4}^{\infty} - 2[\ln|x+4|]_{x=4}^{\infty}$$

Recall the identity $\ln\left(\frac{t}{u}\right) = \ln t - \ln u$.

$$\int_{x=4}^{\infty} \frac{8}{x(x+4)}\,dx = 2\left[\ln\left(\frac{x}{x+4}\right)\right]_{x=4}^{\infty} = 2\lim_{x\to\infty}\left[\ln\left(\frac{x}{x+4}\right)\right] - 2\ln\left(\frac{4}{4+4}\right)$$

To evaluate the limit, divide the numerator and denominator each by x.

$$\int_{x=4}^{\infty} \frac{8}{x(x+4)}\,dx = 2\lim_{x\to\infty}\left[\ln\left(\frac{1}{1+\frac{4}{x}}\right)\right] - 2\ln\left(\frac{4}{8}\right) = 2\ln\left(\frac{1}{1+0}\right) - 2\ln\left(\frac{1}{2}\right)$$

$$= 2\ln\left(\frac{1}{1}\right) - 2\ln\left(\frac{1}{2}\right) = 2\ln(1) - 2\ln\left(\frac{1}{2}\right) = 2(0) - 2\ln\left(\frac{1}{2}\right) = -2\ln\left(\frac{1}{2}\right)$$

Since the integral is finite, the corresponding series **converges**. Note: This integral is positive: $-2\ln\left(\frac{1}{2}\right) = -2\ln 0.5 \approx -2(-0.693) \approx 2(0.693) \approx 1.386$.

(6) Let $f(x) = \frac{3x}{(x+2)(x+5)}$ and $a_\ell = f(\ell) = \frac{3\ell}{(\ell+2)(\ell+5)}$.

$$\int_{x=3}^{\infty} f(x)\,dx = \int_{x=3}^{\infty} \frac{3x}{(x+2)(x+5)}dx$$

Use the method of partial fractions. (Recall Chapter 4.)

$$\frac{3x}{(x+2)(x+5)} = \frac{p}{x+2} + \frac{q}{x+5}$$

Multiply both sides by $(x+2)(x+5)$.

$$3x = p(x+5) + q(x+2) = px + 5p + qx + 2q = (p+q)x + 5p + 2q$$

Note how this is different from the previous exercise and also different from Example 3. Here, we have a variable on the left-hand side. In order to make both sides of the equation equal for any value of x, we need $p + q = 3$ and $5p + 2q = 0$. The first equation gives $q = 3 - p$. Plug this into the second equation: $5p + 2(3 - p) = 0$. This simplifies to $5p + 6 - 2p = 0$ or $3p = -6$. This gives $p = -\frac{6}{3} = -2$ and $q = 3 - p = 3 - (-2) = 3 + 2 = 5$.

$$\int_{x=3}^{\infty} \frac{3x}{(x+2)(x+5)}dx = \int_{x=3}^{\infty} \frac{-2}{x+2}dx + \int_{x=3}^{\infty} \frac{5}{x+5}dx$$

$$= -2[\ln|x+2|]_{x=3}^{\infty} + 5[\ln|x+5|]_{x=3}^{\infty}$$

Unlike the previous exercise and Example 3, the first term has a coefficient of -2 while the second term has a coefficient of 5. Don't use the subtraction of logarithms identity until the coefficients match. Recall the identity $a \ln b = \ln b^a$.

$$\int_{x=3}^{\infty} \frac{3x}{(x+2)(x+5)}dx = -[\ln(x+2)^2]_{x=3}^{\infty} + [\ln(x+5)^5]_{x=3}^{\infty}$$

Now that the coefficients are -1 and 1, use the identity $\ln\left(\frac{t}{u}\right) = \ln t - \ln u$.

$$\int_{x=3}^{\infty} \frac{3x}{(x+2)(x+5)}dx = 2\left[\ln\frac{(x+5)^5}{(x+2)^2}\right]_{x=4}^{\infty} = \infty$$

How did we get ∞? Recall the technique from Chapter 2 for evaluating a limit with a ratio of polynomials. The numerator inside of the logarithm is $(x+5)^5$, which begins with x^5 if you multiply it out, whereas the denominator inside of the logarithm is $(x+2)^2 = x^2 + 4x + 4$. As discussed following Example 10 in Chapter 2, if the degree of the numerator is higher than the degree of the denominator, the limit diverges. Since the integral diverges, the corresponding series **diverges**.

(7) Let $f(x) = e^{cx}$ and $a_k = f(k) = e^{ck}$.

$$\int_{x=1}^{\infty} f(x)\,dx = \int_{x=1}^{\infty} e^{cx}\,dx$$

If $c \neq 0$, we get

$$\int_{x=1}^{\infty} e^{cx}\,dx = \left[\frac{e^{cx}}{c}\right]_{x=1}^{\infty} = \lim_{x\to\infty}\left(\frac{e^{cx}}{c}\right) - \frac{e^{c}}{c}$$

If $c > 0$, the integral diverges. If $c < 0$, the integral converges. For example, if $c = -1, \frac{e^{cx}}{c} = \frac{e^{-x}}{-1} = -e^{-x} = -\frac{1}{e^x}$ approaches zero as $x \to \infty$.

If $c = 0$, the integral diverges:

$$\int_{x=1}^{\infty} e^{cx}\,dx = \int_{x=1}^{\infty} e^{0}\,dx = \int_{x=1}^{\infty} 1\,dx = [x]_{x=1}^{\infty} = \infty - 1 = \infty$$

Combining these results, the series **diverges** if $c \geq 0$ and **converges** if $c < 0$.

Note: The case $c < 0$ is a geometric series. For example, if $c = -1$, we get $\sum \frac{1}{e^k}$, which converges because $|r| = \frac{1}{e} \approx \frac{1}{2.718} \approx 0.368 < 1$.

(8) Let $f(x) = \frac{1}{\sqrt{e^x}}$ and $a_n = f(n) = \frac{1}{\sqrt{e^n}}$. This integral is easy if you realize that $\frac{1}{\sqrt{e^x}} = \frac{1}{(e^x)^{1/2}} = \frac{1}{e^{x/2}} = e^{-x/2}$.

$$\int_{x=1}^{\infty} f(x)\,dx = \int_{x=1}^{\infty} \frac{dx}{\sqrt{e^x}} = \int_{x=1}^{\infty} e^{-x/2}\,dx = \left[\frac{e^{-x/2}}{-1/2}\right]_{x=1}^{\infty} = -2\left[e^{-x/2}\right]_{x=1}^{\infty}$$

$$= -2\lim_{x\to\infty}\left(e^{-x/2}\right) - 2\left(e^{-1/2}\right) = -2\lim_{x\to\infty}\left(\frac{1}{e^{x/2}}\right) - 2\left(-e^{-1/2}\right)$$

$$= 0 + 2e^{-1/2} = 2e^{-1/2} = \frac{2}{e^{1/2}} = \frac{2}{\sqrt{e}}$$

Since the integral is finite, the corresponding series **converges**.

(9) Let $f(x) = \frac{x}{e^{x^2}}$ and $a_m = f(m) = \frac{m}{e^{m^2}}$.

$$\int_{x=1}^{\infty} f(x)\, dx = \int_{x=1}^{\infty} \frac{x}{e^{x^2}}\, dx = \int_{x=1}^{\infty} x e^{-x^2}\, dx$$

Make the substitution $u = e^{-x^2}$. Take a derivative: $\frac{du}{dx} = -2xe^{-x^2}$. We used the **chain rule** here; let $g = x^2$ to see that $\frac{du}{dg}\frac{dg}{dx} = \frac{d}{dg}e^{-g}\frac{d}{dx}x^2 = e^{-g}(2x) = 2xe^{-x^2}$. If you're not familiar with the chain rule, then you probably know the rule as follows: After you take a derivative of the exponential, then take a derivative of its argument (that's where the $2x$ comes from). This rule is due to the chain rule. Anyway, this shows that $du = -2xe^{-x^2}dx$, which we can write as $-\frac{du}{2} = xe^{-x^2}dx$, which means that the integrand simply equals $-\frac{du}{2}$. The lower limit will be $e^{-1^2} = e^{-1} = \frac{1}{e}$. The upper limit will be zero since $u = e^{-x^2}$ approaches zero as $x \to \infty$.

$$\int_{x=1}^{\infty} x e^{-x^2}\, dx = -\int_{x=\frac{1}{e}}^{0} \frac{du}{2} = -\frac{0}{2} - \left(-\frac{1}{2e}\right) = \frac{1}{2e}$$

Since the integral is finite, the corresponding series **converges**.

(10) Let $f(x) = \frac{1}{x \ln x}$ and $a_j = f(j) = \frac{1}{j \ln j}$. (Fortunately, the series starts at $j = 2$; if it had been 1, we'd have division by zero since $\ln 1 = 0$.)

$$\int_{x=2}^{\infty} f(x)\, dx = \int_{x=2}^{\infty} \frac{dx}{x \ln x}$$

You can actually do this integral in your head. You might want to think about it for a moment. An indefinite integral is basically an anti-derivative, meaning that you're asking, "Which function can I take a derivative of and obtain the integrand as a result?" When you take a derivative of a logarithm, this puts the argument of the logarithm in the denominator of a fraction. So if the argument of the logarithm is another logarithm, the derivative puts a logarithm in the denominator, and then when you use the chain rule (see the solution to the previous exercise) to take a derivative of its argument, you also get $\frac{1}{x}$, which is exactly what we need. This is one way that you might reason out that the above integral is $\ln(\ln x)$. It's easy to check. Take a derivative of $\ln(\ln x)$ to see that you get $\frac{d}{dx}\ln(\ln x) = \frac{1}{\ln x}\frac{d}{dx}\ln x = \frac{1}{x \ln x}$. If you prefer to work this out with the chain rule, let $g = \ln x$ and $u = \ln g$ to get $\frac{du}{dx} = \frac{du}{dg}\frac{dg}{dx} = \frac{d}{dg}\ln g \frac{d}{dx}\ln x = \frac{1}{g}\frac{1}{x} = \frac{1}{\ln x}\frac{1}{x} = \frac{1}{x \ln x}$. (Or look up the integral in a table of integrals.)

$$\int_{x=2}^{\infty} \frac{dx}{x \ln x} = [\ln(\ln x)]_{x=1}^{\infty} = \infty$$

The integral diverges because the logarithm becomes infinite as $x \to \infty$, which means that the series also **<u>diverges</u>**.

Chapter 7 Answers

(1) Note that $(j+1)! = (j+1)j!$ For example, $7! = 7(6!)$. Also note that $[2(j+1)]! = (2j+2)! = (2j+2)(2j+1)(2j)!$. For example, $[2(6+1)]! = 14!$ agrees with $(2 \times 6 + 2)(2 \times 6 + 1)(2 \times 6)! = (14)(13)(12!)$. Since $a_j = \frac{j!}{(2j)!}$, it follows that $a_{j+1} = \frac{(j+1)!}{[2(j+1)]!} = \frac{(j+1)j!}{(2j+2)!} = \frac{(j+1)j!}{(2j+2)(2j+1)(2j)!}$.

$$L = \lim_{j\to\infty} \left|\frac{a_{j+1}}{a_j}\right| = \lim_{j\to\infty} \left|\frac{(j+1)j!}{(2j+2)(2j+1)(2j)!} \div \frac{j!}{(2j)!}\right|$$

$$L = \lim_{j\to\infty} \left|\frac{(j+1)j!}{(2j+2)(2j+1)(2j)!} \times \frac{(2j)!}{j!}\right| = \lim_{j\to\infty} \frac{(j+1)}{(2j+2)(2j+1)}$$

All of the factorials cancel out. Note that $2j + 2 = 2(j+1)$

$$L = \lim_{j\to\infty} \frac{(j+1)}{2(j+1)(2j+1)} = \lim_{j\to\infty} \frac{1}{2(2j+1)} = \lim_{j\to\infty} \frac{1}{4j+2} = 0$$

Since $L < 1$, the series **converges**.

(2) See the notes regarding factorials from the previous solution. Note that $[2(n+1)]! = (2n+2)! = (2n+2)(2n+1)(2n)!$ and $[3(n+1)]! = (3n+3)! = (3n+3)(3n+2)(3n+1)(3n)!$. If you're struggling to see this, try working out examples with numbers like we showed in the previous solution. Since $a_n = \frac{(2n)!n!}{(3n)!}$, it follows that $a_{n+1} = \frac{[2(n+1)]!(n+1)!}{[3(n+1)]!} = \frac{(2n+2)!(n+1)!}{(3n+3)!} = \frac{(2n+2)(2n+1)(2n)!(n+1)n!}{(3n+3)(3n+2)(3n+1)(3n)!}$.

$$L = \lim_{n\to\infty} \left|\frac{a_{n+1}}{a_n}\right| = \lim_{n\to\infty} \left|\frac{(2n+2)(2n+1)(2n)!\,(n+1)n!}{(3n+3)(3n+2)(3n+1)(3n)!} \div \frac{(2n)!\,n!}{(3n)!}\right|$$

$$L = \lim_{n\to\infty} \left|\frac{(2n+2)(2n+1)(2n)!\,(n+1)n!}{(3n+3)(3n+2)(3n+1)(3n)!} \times \frac{(3n)!}{(2n)!\,n!}\right|$$

All of the factorials cancel out. Note that $2n + 2 = 2(n+1)$ and $3n + 3 = 3(n+1)$.

$$L = \lim_{n\to\infty} \frac{(2n+2)(2n+1)(n+1)}{(3n+3)(3n+2)(3n+1)} = \lim_{n\to\infty} \frac{2(n+1)(2n+1)(n+1)}{3(n+1)(3n+2)(3n+1)}$$

$$L = \lim_{n\to\infty} \frac{2(2n+1)(n+1)}{3(3n+2)(3n+1)} = \lim_{n\to\infty} \frac{2(2n^2+3n+1)}{3(9n^2+9n+2)} = \lim_{n\to\infty} \frac{4n^2+6n+2}{27n^2+27n+6}$$

See the notes in Chapter 2 following Example 10. Divide the numerator and denominator each by n^2.

$$L = \frac{4 + \frac{6}{n} + \frac{2}{n^2}}{27 + \frac{27}{n} + \frac{6}{n^2}} = \frac{4+0+0}{27+0+0} = \frac{4}{27}$$

Since $L < 1$, the series **converges**.

(3) Since $a_k = \frac{3^k}{k!}$, it follows that $a_{k+1} = \frac{3^{k+1}}{(k+1)!} = \frac{3^{k+1}}{(k+1)k!}$.

$$L = \lim_{k\to\infty}\left|\frac{a_{k+1}}{a_k}\right| = \lim_{k\to\infty}\left|\frac{3^{k+1}}{(k+1)k!} \div \frac{3^k}{k!}\right| = \lim_{k\to\infty}\left|\frac{3^{k+1}}{(k+1)k!} \times \frac{k!}{3^k}\right|$$

$$L = \lim_{k\to\infty}\frac{3}{k+1} = 0$$

Since $L < 1$, the series **converges**.

(4) Note that $(m+1)^{m+1} = (m+1)^m(m+1)$ according to the rule $x^p x = x^{p+1}$.

Since $a_m = \frac{m^m}{m!}$, it follows that $a_{m+1} = \frac{(m+1)^{m+1}}{(m+1)!} = \frac{(m+1)^m(m+1)}{(m+1)m!}$.

$$L = \lim_{m\to\infty}\left|\frac{a_{m+1}}{a_m}\right| = \lim_{m\to\infty}\left|\frac{(m+1)^m(m+1)}{(m+1)m!} \div \frac{m^m}{m!}\right| = \lim_{m\to\infty}\left|\frac{(m+1)^m(m+1)}{(m+1)m!} \times \frac{m!}{m^m}\right|$$

$$L = \lim_{m\to\infty}\frac{(m+1)^m}{m^m} = \lim_{m\to\infty}\left(\frac{m+1}{m}\right)^m = \lim_{m\to\infty}\left(1+\frac{1}{m}\right)^m$$

We need to be careful here. It would be a **mistake** to look only at the $\frac{1}{m}$ and think that the limit equals one; it doesn't. Although the base decreases towards one, the exponent grows to infinity. You can check with your calculator that $\left(1+\frac{1}{1000}\right)^{1000} \approx 2.717$. As m grows even larger, your calculator can help convince you that $\left(1+\frac{1}{m}\right)^m$ doesn't converge to one. Recall from Example 15 and Exercise 40 in Chapter 2 that the way to deal with a limit with a variable exponent is to take a natural logarithm. Let $b_m = \left(1+\frac{1}{m}\right)^m$. Take the natural logarithm of both sides: $\ln b_m = \ln\left(1+\frac{1}{m}\right)^m$. Apply the rule $\ln p^q = q \ln p$ to get $\ln b_m = m \ln\left(1+\frac{1}{m}\right)$. This is an indeterminate form, since m grows to infinity while the logarithm approaches $\ln 1 = 0$. As with Exercise 40 in Chapter 2, the 'trick' here is to note that $m = \frac{1}{1/m}$ and write $\ln b_m = \frac{\ln\left(1+\frac{1}{m}\right)}{\frac{1}{m}}$. Now apply l'Hôpital's rule (recall Chapter 2).

$$\lim_{m\to\infty} \ln b_m = \lim_{m\to\infty} \frac{\ln\left(1+\frac{1}{m}\right)}{\frac{1}{m}} = \frac{\frac{d}{dx}\ln\left(1+\frac{1}{x}\right)\Big|_{x\to\infty}}{\frac{d}{dx}\frac{1}{x}\Big|_{x\to\infty}}$$

The derivative in the numerator is $\frac{1}{1+\frac{1}{x}}\left(-\frac{1}{x^2}\right)$, where the factor of $-\frac{1}{x^2}$ comes from the chain rule; after taking a derivative of the logarithm, then you take a derivative of its argument (which is $1+\frac{1}{x}$). The derivative in the denominator is $-\frac{1}{x^2}$. When we divide, the $-\frac{1}{x^2}$'s cancel out and we're left with:

$$\lim_{m\to\infty} \ln b_m = \lim_{x\to\infty} \frac{1}{1+\frac{1}{x}} = \frac{1}{1+0} = \frac{1}{1} = 1$$

Now we know that $\lim_{m\to\infty} \ln b_m = 1$, which means that $L = \lim_{m\to\infty} b_m = e$ (so that when we take the natural logarithm, we get $\ln e = 1$). If you happened to know that $\lim_{m\to\infty}\left(1+\frac{1}{m}\right)^m = e$, you could have saved a lot of work here. (Many math enthusiasts are aware that this limit equals e.) Since $L = e$ and since $e \approx 2.71828$ (we saw approximately this value earlier when we tried $m = 1000$), we found that $L > 1$, which means that the series **diverges**.

(5) Since $a_\ell = \frac{e^{-\ell}}{\sqrt{\ell}}$, it follows that $a_{\ell+1} = \frac{e^{-(\ell+1)}}{\sqrt{\ell+1}} = \frac{e^{-\ell-1}}{\sqrt{\ell+1}}$.

$$L = \lim_{\ell\to\infty}\left|\frac{a_{\ell+1}}{a_\ell}\right| = \lim_{\ell\to\infty}\left|\frac{e^{-\ell-1}}{\sqrt{\ell+1}} \div \frac{e^{-\ell}}{\sqrt{\ell}}\right| = \lim_{\ell\to\infty}\left|\frac{e^{-\ell-1}}{\sqrt{\ell+1}} \times \frac{\sqrt{\ell}}{e^{-\ell}}\right|$$

Note that $\frac{e^{-\ell-1}}{e^{-\ell}} = e^{-\ell-1-(-\ell)} = e^{-\ell-1+\ell} = e^{-1} = \frac{1}{e}$.

$$L = \lim_{\ell\to\infty}\left|\frac{\sqrt{\ell}}{e\sqrt{\ell+1}}\right| = \lim_{\ell\to\infty}\left(\frac{1}{e}\sqrt{\frac{\ell}{\ell+1}}\right) = \lim_{\ell\to\infty}\left(\frac{1}{e}\sqrt{\frac{1}{1+\frac{1}{\ell}}}\right) = \frac{1}{e}\sqrt{\frac{1}{1+0}} = \frac{1}{e}\sqrt{1} = \frac{1}{e}$$

Since $\frac{1}{e} \approx 0.367879$, we found that $L < 1$, which means that the series **converges**.

(6) Since $a_i = i!\,e^{-i}$, it follows that $a_{i+1} = (i+1)!\,e^{-(i+1)} = (i+1)i!\,e^{-i-1}$.

$$L = \lim_{i\to\infty}\left|\frac{a_{i+1}}{a_i}\right| = \lim_{i\to\infty}\left|\frac{(i+1)i!\,e^{-i-1}}{i!\,e^{-i}}\right| = \lim_{i\to\infty}\frac{(i+1)}{e} = \infty$$

Since $L > 1$, the series **diverges**. Recall from the previous solution that $\frac{e^{-i-1}}{e^{-i}} = \frac{1}{e}$.

(7) Since $a_k = \frac{(2k)!}{k!k!(4^k)}$, it follows that $a_{k+1} = \frac{[2(k+1)]!}{(k+1)!(k+1)!(4^{k+1})} = \frac{(2k+2)!}{(k+1)k!(k+1)k!(4^{k+1})} = \frac{(2k+2)(2k+1)(2k)!}{(k+1)k!(k+1)k!(4^{k+1})}$.

$$L = \lim_{k\to\infty}\left|\frac{a_{k+1}}{a_k}\right| = \lim_{k\to\infty}\left|\frac{(2k+2)(2k+1)(2k)!}{(k+1)k!\,(k+1)k!\,(4^{k+1})} \div \frac{(2k)!}{k!\,k!\,(4^k)}\right|$$

$$L = \lim_{k\to\infty}\left|\frac{(2k+2)(2k+1)(2k)!}{(k+1)k!\,(k+1)k!\,(4^{k+1})} \times \frac{k!\,k!\,(4^k)}{(2k)!}\right|$$

$$L = \lim_{k\to\infty}\frac{(2k+2)(2k+1)}{(k+1)(k+1)4} = \lim_{k\to\infty}\frac{2(k+1)(2k+1)}{4(k+1)(k+1)} = \lim_{k\to\infty}\frac{2k+1}{2(k+1)}$$

$$= \lim_{k\to\infty}\frac{2k+1}{2k+2} = \lim_{k\to\infty}\frac{2+\frac{1}{k}}{2+\frac{2}{k}} = \frac{2+0}{2+0} = 1$$

Since $L = 1$, the ratio test is **inconclusive**.

(8) We're given that $a_{k+1} = \frac{2k-1}{k+2}a_k$.

$$L = \lim_{k\to\infty}\left|\frac{a_{k+1}}{a_k}\right| = \lim_{k\to\infty}\left|\frac{2k-1}{k+2}\right| = \lim_{k\to\infty}\frac{2-\frac{1}{k}}{1+\frac{2}{k}} = \frac{2-0}{1+0} = \frac{2}{1} = 2$$

Since $L > 1$, the series **diverges**.

(9) We're given that $a_{n+1} = \frac{n}{n+2}a_n$.

$$L = \lim_{n\to\infty}\left|\frac{a_{n+1}}{a_n}\right| = \lim_{n\to\infty}\left|\frac{n}{n+2}\right| = \lim_{n\to\infty}\frac{1}{1+\frac{2}{n}} = \frac{1}{1+0} = \frac{1}{1} = 1$$

Since $L = 1$, the ratio test is **inconclusive**.

(10) We're given that $a_{k+1} = \frac{k+2}{3k}a_k$.

$$L = \lim_{k\to\infty}\left|\frac{a_{k+1}}{a_k}\right| = \lim_{k\to\infty}\left|\frac{k+2}{3k}\right| = \lim_{k\to\infty}\frac{1+\frac{2}{k}}{3} = \frac{1+0}{3} = \frac{1}{3}$$

Since $L < 1$, the series **converges**.

Chapter 8 Answers

(1) Here, $a_n = \left(\frac{6}{n}\right)^n$. Apply the rule $(x^n)^{1/n} = x^{n/n} = x^1 = x$.

$$L = \lim_{n\to\infty} |a_n|^{1/n} = \lim_{n\to\infty} \left|\left(\frac{6}{n}\right)^n\right|^{1/n} = \lim_{n\to\infty} \left(\frac{6}{n}\right)^{n/n} = \lim_{n\to\infty} \left(\frac{6}{n}\right)^1 = \lim_{n\to\infty} \frac{6}{n} = 0$$

Since $L < 1$, the series **converges**.

(2) Here, $a_m = \frac{m^8}{2^m}$. Apply the rule $(x^m)^{1/m} = x^{m/m} = x^1 = x$.

$$L = \lim_{m\to\infty} |a_m|^{1/m} = \lim_{m\to\infty} \left|\frac{m^8}{2^m}\right|^{1/m} = \lim_{m\to\infty} \frac{m^{8/m}}{2^{m/m}} = \lim_{m\to\infty} \frac{\left(m^{1/m}\right)^8}{2^1} = \frac{1^8}{2} = \frac{1}{2}$$

Since $L < 1$, the series **converges**. Note: $\lim_{m\to\infty} m^{1/m} = 1$ (see Chapter 2, Exercise 39).

(3) Here, $a_k = k^4 e^{-k} = \frac{k^4}{e^k}$. Apply the rule $(x^k)^{1/k} = x^{k/k} = x^1 = x$.

$$L = \lim_{k\to\infty} |a_k|^{1/k} = \lim_{k\to\infty} \left|\frac{k^4}{e^k}\right|^{1/k} = \lim_{k\to\infty} \frac{k^{4/k}}{e^{k/k}} = \lim_{k\to\infty} \frac{\left(k^{1/k}\right)^4}{e^1} = \frac{1^4}{e} = \frac{1}{e}$$

Since $L < 1$, the series **converges**. Note: $\frac{1}{e} \approx 0.367879$ and $\lim_{k\to\infty} k^{1/k} = 1$.

(4) Here, $a_j = \left(\frac{j}{j^2+3}\right)^j$. Apply the rule $\left(x^j\right)^{1/j} = x^{j/j} = x^1 = x$.

$$L = \lim_{j\to\infty} \left|a_j\right|^{1/j} = \lim_{j\to\infty} \left(\frac{j}{j^2+3}\right)^{j/j} = \lim_{j\to\infty} \left(\frac{j}{j^2+3}\right)^1 = \lim_{j\to\infty} \frac{j}{j^2+3} = \lim_{j\to\infty} \frac{\frac{1}{j}}{1+\frac{3}{j^2}} = \frac{0}{1+0} = 0$$

Since $L < 1$, the series **converges**. Notes: We divided the numerator and denominator each by j^2. See the notes following Example 10 in Chapter 2. (If you're thinking that $\frac{j}{j^2+3}$ should behave like $\frac{j}{j^2} = \frac{1}{j}$, which diverges, the reason this series doesn't diverge is that $\frac{j}{j^2+3}$ is raised to the power of j, which grows to infinity. The terms of this series reduce significantly faster than the harmonic series because of this exponent.)

(5) Here, $a_n = \left(1 - \frac{1}{2n}\right)^{2n}$. Apply the rule $(x^{2n})^{1/n} = x^{2n/n} = x^2$.

$$L = \lim_{n\to\infty} |a_n|^{1/n} = \lim_{n\to\infty} \left|\left(1 - \frac{1}{2n}\right)^{2n}\right|^{1/n} = \lim_{n\to\infty} \left(1 - \frac{1}{2n}\right)^{2n/n} = \lim_{n\to\infty} \left(1 - \frac{1}{2n}\right)^2$$

$$L = \lim_{n\to\infty}\left(\frac{2n}{2n} - \frac{1}{2n}\right)^2 = \lim_{n\to\infty}\left(\frac{2n-1}{2n}\right)^2 = \lim_{n\to\infty}\frac{4n^2-4n+1}{4n^2} = \lim_{n\to\infty}\frac{4-\frac{4}{n}+\frac{1}{n^2}}{4}$$

$$L = \frac{4-0+1}{4} = \frac{4}{4} = 1$$

Since $L = 1$, the root test is **inconclusive**.

(6) This solution is simpler if you shift the index (recall Chapter 3). Let $n = \ell + 2$. The shifted series will start at $n = 0 + 2 = 2$. Replace the exponent, $\ell + 2$, with n. Elsewhere, replace ℓ with $n - 2$ (since $n = \ell + 2$, it follows that $\ell = n - 2$).

$$\sum_{\ell=0}^{\infty}\left(\frac{8\ell+19}{6\ell+11}\right)^{\ell+2} = \sum_{n=2}^{\infty}\left[\frac{8(n-2)+19}{6(n-2)+11}\right]^n = \sum_{n=2}^{\infty}\left(\frac{8n-16+19}{6n-12+11}\right)^n = \sum_{n=2}^{\infty}\left(\frac{8n+3}{6n-1}\right)^n$$

Here, $a_n = \left(\frac{8n+3}{6n-1}\right)^n$. Apply the rule $(x^n)^{1/n} = x^{n/n} = x^1 = x$.

$$L = \lim_{n\to\infty}|a_n|^{1/n} = \lim_{n\to\infty}\left(\frac{8n+3}{6n-1}\right)^{n/n} = \lim_{n\to\infty}\left(\frac{8n+3}{6n-1}\right)^1 = \lim_{n\to\infty}\frac{8n+3}{6n-1}$$

$$= \lim_{n\to\infty}\frac{8+\frac{3}{n}}{6-\frac{1}{n}} = \frac{8+0}{6-0} = \frac{8}{6} = \frac{4}{3}$$

Since $L > 1$, the series **diverges**.

(7) Here, $a_i = \left(\frac{2}{3}\right)^i$. Apply the rule $\left(x^i\right)^{1/i} = x^{i/i} = x^1 = x$.

$$L = \lim_{i\to\infty}|a_i|^{1/i} = \lim_{i\to\infty}\left(\frac{2}{3}\right)^{i/i} = \lim_{i\to\infty}\left(\frac{2}{3}\right)^1 = \lim_{i\to\infty}\frac{2}{3} = \frac{2}{3}$$

Since $L < 1$, the series **converges**. Note: This is a geometric series with $r = \frac{2}{3}$.

(8) Here, $a_k = k\left(\frac{4}{3}\right)^k$. Apply the rule $(x^k)^{1/k} = x^{k/k} = x^1 = x$.

$$L = \lim_{k\to\infty}|a_k|^{1/k} = \lim_{k\to\infty}\left|k\left(\frac{4}{3}\right)^k\right|^{1/k} = \lim_{k\to\infty}k^{1/k}\left(\frac{4}{3}\right)^{k/k} = \lim_{k\to\infty}(1)\left(\frac{4}{3}\right)^1 = \frac{4}{3}$$

Since $L > 1$, the series **diverges**. Note: $\lim_{k\to\infty} k^{1/k} = 1$ (see Chapter 2, Exercise 39).

(9) First look at the odd denominators. Note that $\frac{1000}{27} = \frac{10^3}{3^3}$ and $\frac{10{,}000{,}000}{823{,}543} = \frac{10^7}{7^7}$.

These two terms suggest that the entire series can be made from $\frac{10^n}{n^n}$. Let's try it.

$$\sum_{n=1}^{\infty} \frac{10^n}{n^n} = \frac{10}{1} + \frac{100}{4} + \frac{1000}{27} + \frac{10{,}000}{256} + \frac{100{,}000}{3125} + \frac{1{,}000{,}000}{46{,}656} + \frac{10{,}000{,}000}{823{,}543} + \frac{100{,}000{,}000}{16{,}777{,}216} + \cdots$$

If you reduce the fractions, this series matches the given series. By looking only at the first 9 terms, one might be fooled into believing that the series will diverge, since all of the terms thus far are larger than one, but that's because the denominator is smaller than the numerator for the first 9 terms. After the 10th term, the denominator takes over. For example, once we reach $n = 20$, we get $\frac{10^{20}}{20^{20}} = \frac{1}{2^{20}} \approx 0.00000095$. We'll see that this series actually converges. The main idea is that n^n grows much faster for large values of n than 10^n does.

Here, $a_n = \frac{10^n}{n^n} = \left(\frac{10}{n}\right)^n$. Apply the rule $(x^n)^{1/n} = x^{n/n} = x^1 = x$.

$$L = \lim_{n\to\infty} |a_n|^{1/n} = \lim_{n\to\infty} \left(\frac{10}{n}\right)^{n/n} = \lim_{n\to\infty} \left(\frac{10}{n}\right)^1 = \lim_{n\to\infty} \frac{10}{n} = 0$$

Since $L < 1$, the series **converges**.

Chapter 9 Answers

(1) Note that $\sqrt{2k} = \sqrt{2}\sqrt{k} = 2^{1/2}k^{1/2}, 8 = 2^3$, and $\frac{8}{\sqrt{2}} = \frac{2^3}{2^{1/2}} = \frac{2^{6/2}}{2^{1/2}} = 2^{5/2}$.

$$\sum_{k=1}^{\infty} \frac{8}{\sqrt{2k}} = \sum_{k=1}^{\infty} \frac{2^3}{2^{1/2}k^{1/2}} = \sum_{k=1}^{\infty} \frac{2^{5/2}}{k^{1/2}}$$

This is a series of powers of the form $\sum \frac{a}{k^p}$ with $a = 2^{5/2}$ and $p = \frac{1}{2}$. This series is known to **diverge** when $p \leq 1$. (If the series happens to be a geometric series or a series of simple powers, you don't need to run a test; these are known series.)

(2) This is a geometric series with $r = \frac{7}{9}$. Since $|r| < 1$, this series **converges**. (If the series is a geometric series, you don't need to run a test. This series is well-known.) If you want to put it in standard form (like Chapter 3), simply shift the index. Let $k - 1 = n + 1$ such that $k = n + 2$. When $n = 2, k = 2 + 2 = 4$.

$$\sum_{n=2}^{\infty} \left(\frac{7}{9}\right)^{n+1} = \sum_{k=4}^{\infty} \left(\frac{7}{9}\right)^{k-1}$$

(3) We will use the limit comparison test with $c_\ell = \frac{1}{\ell^2}$. The terms are positive for $\ell > 1$.

$$\lim_{\ell\to\infty} \frac{a_\ell}{c_\ell} = \lim_{\ell\to\infty} \frac{\frac{\ell}{\ell^3 - 3}}{\frac{1}{\ell^2}} = \lim_{\ell\to\infty} \left(\frac{\ell}{\ell^3 - 3} \div \frac{1}{\ell^2}\right) = \lim_{\ell\to\infty} \left(\frac{\ell}{\ell^3 - 3} \times \frac{\ell^2}{1}\right) = \lim_{\ell\to\infty} \frac{\ell^3}{\ell^3 - 3}$$

$$= \lim_{\ell\to\infty} \frac{1}{1 - \frac{3}{\ell^3}} = \frac{1}{1 - 0} = \frac{1}{1} = 1$$

Since $\sum c_\ell = \sum \frac{1}{\ell^2}$ converges and $\lim_{\ell\to\infty} \frac{a_\ell}{c_\ell} = 1$ is finite, it follows that $\sum a_\ell = \sum \frac{\ell}{\ell^3 - 3}$ **converges**. (We just need this limit to be finite, unlike a ratio or root test.)

Note: For most of the problems, there is more than just one test that will work.

(4) We will use the ratio test. Since $a_i = \frac{9}{(2i)!}, a_{i+1} = \frac{9}{[2(i+1)]!} = \frac{9}{(2i+2)!} = \frac{9}{(2i+2)(2i+1)(2i)!}$.

$$L = \lim_{i\to\infty} \left|\frac{a_{i+1}}{a_i}\right| = \lim_{i\to\infty} \left|\frac{9}{(2i+2)(2i+1)(2i)!} \div \frac{9}{(2i)!}\right| = \lim_{i\to\infty} \left|\frac{9}{(2i+2)(2i+1)(2i)!} \times \frac{(2i)!}{9}\right|$$

$$L = \lim_{i\to\infty} \frac{1}{(2i+2)(2i+1)} = 0$$

Since $L < 1$, the series **converges**.

(5) We will use the integral test. Since the derivative of the denominator is $2x$, which is proportional to the numerator, this suggests make the substitution $u = x^2 + 9$ such that $du = 2x\,dx$. Note that $8x\,dx = 4\,du$. The new lower limit will be $0^2 + 9 = 9$.

$$\int_{x=0}^{\infty} \frac{8x}{x^2+9}\,dx = \int_{u=9}^{\infty} \frac{4}{u}\,du = 4[\ln u]_{u=9}^{\infty} = \infty$$

Since the integral diverges, the corresponding series **diverges**.

(6) We will use the root test with $a_j = j^{-j} = \frac{1}{j^j}$. Recall that $\left(x^j\right)^{1/j} = x^1 = x$.

$$L = \lim_{j\to\infty} \left|j^{-j}\right|^{1/j} = \lim_{j\to\infty} \left|\frac{1}{j^j}\right|^{1/j} = \lim_{j\to\infty} \frac{1}{(j^j)^{1/j}} = \lim_{j\to\infty} \frac{1}{j} = 0$$

Since $L < 1$, the series **converges**.

(7) We will use the simple divergence test. The terms are positive for $n > 1$.

$$\lim_{n\to\infty} \frac{3n^2-5}{2n^2+3n+4} = \lim_{n\to\infty} \frac{3-\frac{5}{n^2}}{2+\frac{3}{n}+\frac{4}{n^2}} = \frac{3-0}{2+0+0} = \frac{3}{2}$$

Since $\left\{\frac{3n^2-5}{2n^2+3n+4}\right\}$ doesn't converge to zero, this series **diverges**.

(8) We will use the root test with $a_k = k^5 e^{-k} = \frac{k^5}{e^k}$. Recall that $(x^k)^{1/k} = x^1 = x$. Note that $\lim_{k\to\infty} k^{1/k} = 1$ (see Chapter 2, Exercise 39).

$$L = \lim_{k\to\infty} |k^5 e^{-k}|^{1/k} = \lim_{k\to\infty} \left|\frac{k^5}{e^k}\right|^{1/k} = \lim_{k\to\infty} \frac{k^{5/k}}{(e^k)^{1/k}} = \lim_{k\to\infty} \frac{\left(k^{1/k}\right)^5}{e} = \frac{1}{e}$$

Since $L < 1$, the series **converges**. Note: $\frac{1}{e} \approx 0.367879$.

(9) Note that $\frac{\sqrt{i}}{i^2} = \frac{i^{1/2}}{i^2} = \frac{1}{i^{3/2}}$.

$$\sum_{i=1}^{\infty} \frac{\sqrt{i}}{i^2} = \sum_{i=1}^{\infty} \frac{i^{1/2}}{i^2} = \sum_{i=1}^{\infty} \frac{1}{i^{3/2}}$$

This is a series of powers of the form $\sum \frac{a}{i^p}$ with $a = 1$ and $p = \frac{3}{2}$. This series is known to **converge** when $p > 1$. (If the series happens to be a geometric series or a series of simple powers, you don't need to run a test; these are known series.)

(10) We will use the ratio test. Since $a_\ell = \frac{\ell^8}{6^\ell}$, $a_{\ell+1} = \frac{(\ell+1)^8}{6^{\ell+1}}$.

$$L = \lim_{\ell\to\infty}\left|\frac{a_{\ell+1}}{a_\ell}\right| = \lim_{\ell\to\infty}\left|\frac{(\ell+1)^8}{6^{\ell+1}} \div \frac{\ell^8}{6^\ell}\right| = \lim_{\ell\to\infty}\left|\frac{(\ell+1)^8}{6^{\ell+1}} \times \frac{6^\ell}{\ell^8}\right|$$

$$L = \lim_{\ell\to\infty}\left[\left(\frac{\ell+1}{\ell}\right)^8 \frac{1}{6}\right] = \frac{1}{6}\lim_{\ell\to\infty}\left(\frac{1+\frac{1}{\ell}}{1}\right)^8 = \frac{1}{6}\left(\frac{1+0}{1}\right)^8 = \frac{1}{6}\left(\frac{1}{1}\right)^8 = \frac{1}{6}(1)^8 = \frac{1}{6}$$

Since $L < 1$, the series **converges**. In one step, we divided the numerator and denominator each by ℓ (a strategy from Chapter 2).

(11) We will use the simple divergence test.

$$\lim_{m\to\infty}\left(1-\frac{1}{m^2}\right)^3 = (1-0)^3 = (1)^3 = 1$$

Since $\left\{\left(1-\frac{1}{m^2}\right)^3\right\}$ doesn't converge to zero, this series **diverges**.

Note: If the exponent had been m instead of 3, the limit would be more complicated. Recall Chapter 2, Example 15 and Exercises 38-40.

(12) See the note above. We will use the root test with $a_j = \left(\frac{8j-7}{7j+8}\right)^j$. Recall that $\left(x^j\right)^{1/j} = x^1 = x$.

$$L = \lim_{j\to\infty}\left|\left(\frac{8j-7}{7j+8}\right)^j\right|^{1/j} = \lim_{j\to\infty}\frac{8j-7}{7j+8} = \lim_{j\to\infty}\frac{8-\frac{7}{j}}{7+\frac{8}{j}} = \frac{8-0}{7+0} = \frac{8}{7}$$

Since $L > 1$, the series **diverges**.

(13) We will use the limit comparison test with $c_n = \frac{1}{2^n}$.

$$\lim_{n\to\infty}\frac{a_n}{c_n} = \lim_{\ell\to\infty}\frac{\frac{7}{2^n-n}}{\frac{1}{2^n}} = \lim_{\ell\to\infty}\left(\frac{7}{2^n-n} \div \frac{1}{2^n}\right) = \lim_{\ell\to\infty}\left(\frac{7}{2^n-n} \times \frac{2^n}{1}\right) = \lim_{\ell\to\infty}\frac{7(2^n)}{2^n-n}$$

$$= \lim_{\ell\to\infty}\frac{7}{1-\frac{n}{2^n}} = \frac{7}{1-0} = \frac{7}{1} = 7$$

Since $\sum c_n = \sum \frac{1}{2^n}$ converges and $\lim_{n\to\infty}\frac{a_n}{c_n} = 7$ is finite, it follows that $\sum a_n = \sum \frac{7}{2^n-n}$ **converges**. To see that $\frac{n}{2^n}$ approaches zero as $n \to \infty$, use l'Hôpital's rule (Chapter 2).

(14) We will use the integral test with **partial fractions** (Chapter 6).

$$\int_{x=2}^{\infty} \frac{6}{x(x+3)}\,dx$$

$$\frac{6}{x(x+3)} = \frac{p}{x} + \frac{q}{x+3}$$

Multiply both sides by $x(x+3)$.

$$6 = p(x+3) + qx = px + 3p + qx = (p+q)x + 3p$$

For this equation be true for any value of x, we need $p + q = 0$ and $3p = 6$. This gives $p = \frac{6}{3} = 2$ and $q = -p = -2$.

$$\int_{x=2}^{\infty} \frac{6}{x(x+3)}\,dx = \int_{x=2}^{\infty} \frac{2}{x}\,dx - \int_{x=2}^{\infty} \frac{2}{x+3}\,dx = 2[\ln x]_{x=2}^{\infty} - 2[\ln(x+3)]_{x=2}^{\infty}$$

Use the identity $\ln\left(\frac{t}{u}\right) = \ln t - \ln u$.

$$\int_{x=2}^{\infty} \frac{6}{x(x+3)}\,dx = 2\left[\ln\frac{x}{x+3}\right]_{x=2}^{\infty} = 2\lim_{x\to\infty} \ln\frac{x}{x+3} - 2\ln\frac{2}{2+3}$$

$$= 2\lim_{x\to\infty} \ln\frac{1}{1+\frac{3}{x}} - 2\ln\frac{2}{5} = 2\ln\frac{1}{1+0} - 2\ln\frac{2}{5} = 2\ln 1 - 2\ln\frac{2}{5} = 0 - 2\ln\frac{2}{5} = -2\ln\frac{2}{5}$$

Since the integral converges, the corresponding series **converges**. Notes: We divided the numerator and denominator each by x in order to evaluate the limit (recall Chapter 2). Since $\ln\frac{2}{5} \approx -0.916$, the integral equals ≈ 1.83 (which is positive).

(15) This is a geometric series with $r = \frac{5}{4}$. Since $|r| > 1$, this series **diverges**. (If the series is a geometric series, you don't need to run a test. This series is well-known.) If you want to put it in standard form (like Chapter 3), simply shift the index. Let $m - 2 = k - 1$ such that $k = m - 1$. When $m = 3$, $k = 3 - 1 = 2$.

$$\sum_{m=3}^{\infty} \left(\frac{5}{4}\right)^{m-2} = \sum_{k=2}^{\infty} \left(\frac{5}{4}\right)^{k-1}$$

(16) We will use the ratio test. Since $a_i = \frac{e^i}{i!}$, $a_{i+1} = \frac{e^{i+1}}{(i+1)!} = \frac{e^{i+1}}{(i+1)(i!)}$.

$$L = \lim_{i\to\infty}\left|\frac{a_{i+1}}{a_i}\right| = \lim_{i\to\infty}\left|\frac{e^{i+1}}{(i+1)(i!)} \div \frac{e^i}{i!}\right| = \lim_{i\to\infty}\left|\frac{e^{i+1}}{(i+1)(i!)} \times \frac{i!}{e^i}\right|$$

$$L = \lim_{i\to\infty} \frac{e}{i+1} = 0$$

Since $L < 1$, the series **converges**.

(17) We will use the root test with $a_\ell = \ell^6 \left(\frac{2}{3}\right)^\ell$. Recall that $\left(x^\ell\right)^{1/\ell} = x^1 = x$.

$$L = \lim_{\ell\to\infty} \left| \ell^6 \left(\frac{2}{3}\right)^\ell \right|^{1/\ell} = \lim_{\ell\to\infty} \left[\ell^{6/\ell} \left(\frac{2}{3}\right) \right] = \frac{2}{3} \lim_{\ell\to\infty} \left(\ell^{1/\ell}\right)^6 = \frac{2}{3}(1) = \frac{2}{3}$$

Since $L < 1$, the series **converges**. Note that $\lim_{\ell\to\infty} \ell^{1/\ell} = 1$ (see Chapter 2, Exercise 39).

(18) We will use the limit comparison test with $d_j = \frac{1}{j}$.

$$\lim_{j\to\infty} \frac{a_j}{d_j} = \lim_{j\to\infty} \frac{\frac{5j}{j^2 - \ln j}}{\frac{1}{j}} = \lim_{\ell\to\infty} \left(\frac{5j}{j^2 - \ln j} \div \frac{1}{j} \right) = \lim_{\ell\to\infty} \left(\frac{5j}{j^2 - \ln j} \times \frac{j}{1} \right) = \lim_{\ell\to\infty} \frac{5j^2}{j^2 - \ln j}$$

$$= \lim_{\ell\to\infty} \frac{5}{1 - \frac{\ln j}{j^2}} = \frac{5}{1-0} = \frac{5}{1} = 5$$

Since $\sum d_j = \sum \frac{1}{j}$ diverges and $\lim_{j\to\infty} \frac{a_j}{d_j} = 5$ is positive, it follows that $\sum a_j = \sum \frac{5j}{j^2 - \ln j}$ **diverges**. Note: To see that $\frac{\ln j}{j^2}$ approaches zero as $j \to \infty$, use l'Hôpital's rule (Chapter 2). The derivative of the numerator gives $\frac{1}{x}$, while a derivative of the denominator gives $2x$, and the division makes $\frac{1}{x} \div \frac{2x}{1} = \frac{1}{x} \times \frac{1}{2x} = \frac{1}{2x^2}$, which approaches zero as $x \to \infty$.

(19) We will use the ratio test. Since $a_k = \frac{k!}{k^k}$, $a_{k+1} = \frac{(k+1)!}{(k+1)^{k+1}} = \frac{(k+1)k!}{(k+1)^k(k+1)}$.

$$L = \lim_{k\to\infty} \left| \frac{a_{k+1}}{a_k} \right| = \lim_{k\to\infty} \left| \frac{(k+1)k!}{(k+1)^k(k+1)} \div \frac{k!}{k^k} \right| = \lim_{k\to\infty} \left| \frac{(k+1)k!}{(k+1)^k(k+1)} \times \frac{k^k}{k!} \right|$$

$$L = \lim_{k\to\infty} \frac{k^k}{(k+1)^k} = \lim_{k\to\infty} \left(\frac{k}{k+1} \right)^k$$

Be careful here. Although $\frac{k}{k+1}$ approaches one, the exponent is also growing to infinity. It would be incorrect to conclude that the limit approaches one. To see that the limit isn't approaching one, try using large values of k on your calculator. For example, $\left(\frac{100}{100+1}\right)^{100} = \left(\frac{100}{101}\right)^{100} \approx 0.3697$. You'll notice that this clearly doesn't converge to one.

Recall that when a limit involves the index in the exponent, you can deal with this by taking a natural logarithm (recall Chapter 2, Example 15 and Exercise 40). Let $b_k = \left(\frac{k}{k+1}\right)^k$. Take the natural logarithm of both sides to get $\ln b_k = \ln\left(\frac{k}{k+1}\right)^k$. Apply the rule $\ln p^q = q \ln p$ to get $\ln b_k = k \ln\left(\frac{k}{k+1}\right)$. This is an indeterminate form, since k grows to infinity while the logarithm approaches $\ln 1 = 0$. As with Exercise 40 in Chapter 2, the 'trick' here is to note that $k = \frac{1}{1/k}$ and write $\ln b_k = \frac{\ln\left(\frac{k}{k+1}\right)}{1/k}$. Now we can use l'Hôpital's rule.

$$\lim_{k\to\infty} \ln b_k = \lim_{k\to\infty} \frac{\ln\left(\frac{k}{k+1}\right)}{\frac{1}{k}} = \frac{\left.\frac{d}{dx}\ln\left(\frac{x}{x+1}\right)\right|_{x\to\infty}}{\left.\frac{d}{dx}\frac{1}{x}\right|_{x\to\infty}}$$

The derivative in the numerator is $\frac{x+1}{x}\frac{d}{dx}\frac{x}{x+1} = \frac{x+1}{x}\frac{(x+1)(1)-x(1)}{(x+1)^2} = \frac{x+1}{x}\frac{x+1-x}{(x+1)^2} = \frac{x+1-x}{x(x+1)} = \frac{1}{x(x+1)}$, where we applied the chain rule (taking a derivative of $\frac{x}{x+1}$ with respect to x after taking a derivative of the logarithm to get the reciprocal of its argument), and then used the quotient rule (to take the derivative of the quotient $\frac{x}{x+1}$). The derivative in the denominator is $-\frac{1}{x^2}$.

$$\lim_{k\to\infty} \ln b_k = \lim_{x\to\infty}\left[\frac{1}{x(x+1)} \div \frac{-1}{x^2}\right] = \lim_{x\to\infty}\left[\frac{-1}{x(x+1)} \times \frac{x^2}{1}\right] = \lim_{x\to\infty}\frac{-x^2}{x^2+x}$$

$$= \lim_{x\to\infty}\frac{-1}{1+\frac{1}{x}} = \frac{-1}{1+0} = -1$$

Now we know that $\lim_{k\to\infty} \ln b_k = -1$, which means that $L = \lim_{k\to\infty} b_k = \frac{1}{e}$. Why? If b_k approaches $\frac{1}{e} = e^{-1}$ as $k \to \infty$, then $\ln b_k$ approaches $\ln e^{-1} = -1$. So, we finally found that $L = \frac{1}{e}$. Since $L < 1$, the series **converges**. (Note: This agrees with the solution to Exercise 4 in Chapter 7. Here, $\sum a_k = \sum \frac{k!}{k^k}$ converges. From the reciprocal theorem we learned in Chapter 4, it follows that $\sum \frac{1}{a_k} = \sum \frac{k^k}{k!}$ diverges. But beware that the reciprocal theorem doesn't work the other way. Knowing that $\sum b_n$ diverges doesn't guarantee that $\sum \frac{1}{b_n}$ will converge.)

(20) We will use the limit comparison test with $c_n = \frac{1}{n^2}$. The terms are positive for $n > 1$.

$$\lim_{n\to\infty} \frac{a_n}{c_n} = \lim_{n\to\infty} \frac{\frac{n^4 - 3n^2 + 8}{n^6 + 9n^3 - 6}}{\frac{1}{n^2}} = \lim_{n\to\infty} \left(\frac{n^4 - 3n^2 + 8}{n^6 + 9n^3 - 6} \div \frac{1}{n^2} \right) = \lim_{n\to\infty} \left(\frac{n^4 - 3n^2 + 8}{n^6 + 9n^3 - 6} \times \frac{n^2}{1} \right)$$

$$= \lim_{n\to\infty} \frac{n^6 - 3n^4 + 8n^2}{n^6 + 9n^3 - 6} = \lim_{n\to\infty} \frac{1 - \frac{3}{n^2} + \frac{8}{n^4}}{1 + \frac{9}{n^3} - \frac{6}{n^6}} = \frac{1 - 0 + 0}{1 + 0 - 0} = \frac{1}{1} = 1$$

Since $\sum c_n = \sum \frac{1}{n^2}$ converges and $\lim_{n\to\infty} \frac{a_n}{c_n} = 1$ is finite, it follows that $\sum a_n = \sum \frac{n^4 - 3n^2 + 8}{n^6 + 9n^3 - 6}$ **converges**. (We just need this limit to be finite, unlike a ratio or root test.)

(21) We will use the root test with $a_j = \frac{j^j}{100^j}$. Recall that $(x^j)^{1/j} = x^1 = x$.

$$L = \lim_{j\to\infty} \left| \frac{j^j}{100^j} \right|^{1/j} = \lim_{j\to\infty} \frac{(j^j)^{1/j}}{(100^j)^{1/j}} = \lim_{j\to\infty} \frac{j}{100} = \infty$$

Since $L > 1$, the series **diverges**.

(22) We will use the simple divergence test. The corresponding sequence, $\left\{4 \sin^2\left(\frac{m\pi}{6}\right)\right\}$, doesn't converge to zero; rather, it diverges because it is a repeating sequence that never converges to any one value. Therefore, this series **diverges**.

(23) We will use the ratio test. Since $a_i = i^3 e^{-i} = \frac{i^3}{e^i}$, $a_{i+1} = \frac{(i+1)^3}{e^{i+1}}$.

$$L = \lim_{i\to\infty} \left| \frac{a_{i+1}}{a_i} \right| = \lim_{i\to\infty} \left| \frac{(i+1)^3}{e^{i+1}} \div \frac{i^3}{e^i} \right| = \lim_{i\to\infty} \left| \frac{(i+1)^3}{e^{i+1}} \times \frac{e^i}{i^3} \right|$$

$$L = \lim_{i\to\infty} \frac{(i+1)^3}{e(i^3)} = \frac{1}{e} \lim_{i\to\infty} \left(\frac{i+1}{i} \right)^3 = \frac{1}{e} \lim_{i\to\infty} \left(\frac{1 + \frac{1}{i}}{1} \right)^3 = \frac{1}{e} \left(\frac{1+0}{1} \right)^3 = \frac{1}{e}(1)^3 = \frac{1}{e}$$

Since $L < 1$, the series **converges**. Note: $\frac{1}{e} \approx 0.367879$.

(24) We will use the integral test. Note that $\sqrt{e^{-\ell}} = (e^{-\ell})^{1/2} = e^{-\ell/2}$ according to the rules $(e^p)^q = e^{pq}$ and $u^{1/2} = \sqrt{u}$.

$$\int_{x=1}^{\infty} \sqrt{e^{-x}}\, dx = \int_{u=1}^{\infty} e^{-x/2}\, du = \left[-2e^{-x/2}\right]_{x=1}^{\infty} = -\left[\frac{2}{e^{x/2}}\right]_{x=1}^{\infty}$$

$$\int_{x=1}^{\infty} \sqrt{e^{-x}}\,dx = -\lim_{x\to\infty}\frac{2}{e^{x/2}} - \left(-\frac{2}{e^{1/2}}\right) = 0 + \frac{2}{e^{1/2}} = \frac{2}{e^{1/2}}$$

Since the integral converges, the corresponding series **converges**.

(25) We will use the limit comparison test with $d_n = \frac{1}{n}$.

$$\lim_{n\to\infty}\frac{a_n}{d_n} = \lim_{n\to\infty}\frac{\frac{n}{n^2-2n+6}}{\frac{1}{n}} = \lim_{n\to\infty}\left(\frac{n}{n^2-2n+6} \div \frac{1}{n}\right) = \lim_{n\to\infty}\left(\frac{n}{n^2-2n+6} \times \frac{n}{1}\right)$$

$$= \lim_{n\to\infty}\frac{n^2}{n^2-2n+6} = \lim_{n\to\infty}\frac{1}{1-\frac{2}{n}+\frac{6}{n^2}} = \frac{1}{1-0+0} = \frac{1}{1} = 1$$

Since $\sum d_n = \sum \frac{1}{n}$ diverges and $\lim_{n\to\infty}\frac{a_n}{d_n} = 1$ is positive, it follows that $\sum a_n = \sum \frac{n}{n^2-2n+6}$ **diverges**. (We just need this limit to be positive, unlike a ratio or root test.)

(26) We will use the simple divergence test.

$$\lim_{k\to\infty}\sqrt{\frac{k-9}{9k+1}} = \lim_{k\to\infty}\sqrt{\frac{1-\frac{9}{k}}{9+\frac{1}{k}}} = \sqrt{\frac{1-0}{9+0}} = \sqrt{\frac{1}{9}} = \frac{1}{3}$$

Since $\left\{\sqrt{\frac{k-9}{9k+1}}\right\}$ doesn't converge to zero, this series **diverges**.

(27) We will use the limit comparison test with $c_j = \left(\frac{3}{4}\right)^j = \frac{3^j}{4^j}$, which is a geometric series that converges because $|r| = \frac{3}{4} < 1$.

$$\lim_{j\to\infty}\frac{a_j}{c_j} = \lim_{j\to\infty}\frac{\frac{3^j+j}{4^j}}{\frac{3^j}{4^j}} = \lim_{j\to\infty}\left(\frac{3^j+j}{4^j} \div \frac{3^j}{4^j}\right) = \lim_{j\to\infty}\left(\frac{3^j+j}{4^j} \times \frac{4^j}{3^j}\right) = \lim_{j\to\infty}\frac{3^j+j}{3^j}$$

$$= \lim_{j\to\infty}\left(1 + \frac{j}{3^j}\right) = 1 + 0 = 1$$

Since $\sum c_j = \sum \left(\frac{3}{4}\right)^j = \sum \frac{3^j}{4^j}$ converges and $\lim_{j\to\infty}\frac{a_j}{c_j} = 1$ is finite, it follows that $\sum a_j = \sum \frac{3^j+j}{4^j}$ **converges**. Note: To see that $\frac{j}{3^j}$ approaches zero as $j \to \infty$, use l'Hôpital's rule (Chapter 2). (We just need this limit to be finite, unlike a ratio or root test.)

(28) We will use the integral test. Let $u = x^2$ such that $du = 2x\,dx$ and $\frac{du}{2} = x\,dx$. The lower limit is unchanged since $1^2 = 1$.

$$\int_{x=1}^{\infty} xe^{-x^2}\,dx = \int_{u=1}^{\infty} e^{-u}\,du = [-e^{-u}]_{u=1}^{\infty} = -\left[\frac{1}{e^u}\right]_{u=1}^{\infty}$$

$$-\lim_{u\to\infty}\frac{1}{e^u} - \left(-\frac{1}{e^1}\right) = 0 + \frac{1}{e} = \frac{1}{e}$$

Since the integral converges, the corresponding series **converges**.

(29) We will use the limit comparison test with $c_\ell = \frac{1}{\ell^{3/2}} = \frac{1}{\ell\sqrt{\ell}}$. The terms of $\sum \frac{1}{\ell\sqrt{\ell}}$ are positive for $\ell > 0$.

$$\lim_{\ell\to\infty}\frac{a_\ell}{c_\ell} = \lim_{\ell\to\infty}\frac{\frac{6}{(\ell+2)\sqrt{\ell+4}}}{\frac{1}{\ell\sqrt{\ell}}} = \lim_{\ell\to\infty}\left(\frac{6}{(\ell+2)\sqrt{\ell+4}} \div \frac{1}{\ell\sqrt{\ell}}\right) = \lim_{\ell\to\infty}\left(\frac{6}{(\ell+2)\sqrt{\ell+4}} \times \frac{\ell\sqrt{\ell}}{1}\right)$$

$$= \lim_{\ell\to\infty}\frac{6\ell\sqrt{\ell}}{(\ell+2)\sqrt{\ell+4}} = \lim_{\ell\to\infty}\left(\frac{6\ell}{\ell+2}\sqrt{\frac{\ell}{\ell+4}}\right) = \lim_{\ell\to\infty}\left(\frac{6}{1+\frac{2}{\ell}}\sqrt{\frac{1}{1+\frac{4}{\ell}}}\right)$$

$$= \frac{6}{1+0}\sqrt{\frac{1}{1+0}} = \frac{6}{1}\sqrt{\frac{1}{1}} = 6\sqrt{1} = 6(1) = 6$$

Since $\sum c_\ell = \sum \frac{1}{\ell^{3/2}} = \sum \frac{1}{\ell\sqrt{\ell}}$ converges and $\lim_{\ell\to\infty}\frac{a_\ell}{c_\ell} = 6$ is finite, it follows that $\sum a_\ell = \sum \frac{6}{(\ell+2)\sqrt{\ell+4}}$ **converges**.

(30) (The terms are positive for $i > 1$.) We will use the root test with $a_i = \left(\frac{i^3+2i}{i^4-3i^2+1}\right)^i$. Recall that $\left(x^i\right)^{1/i} = x^1 = x$.

$$L = \lim_{i\to\infty}\left|\left(\frac{i^3+2i}{i^4-3i^2+1}\right)^i\right|^{1/i} = \lim_{i\to\infty}\frac{i^3+2i}{i^4-3i^2+1} = \lim_{i\to\infty}\frac{\frac{1}{i}+\frac{2}{i^3}}{1-\frac{3}{i^2}+\frac{1}{i^4}} = \frac{0+0}{1-0+0} = \frac{0}{1} = 0$$

Since $L < 1$, the series **converges**. Note: You need to be careful applying the simple divergence test because the exponent grows to infinity. For example, see the notes in the solution to Exercise 19. We're not saying that you can't use the simple divergence test, just that you need to deal with the exponent properly.

(31) We will use the ratio test. Since $a_n = \frac{(2n)!}{n!n!}$, $a_{n+1} = \frac{[2(n+1)]!}{(n+1)!(n+1)!} = \frac{(2n+2)!}{(n+1)(n!)(n+1)(n!)} = \frac{(2n+2)(2n+1)(2n)!}{(n+1)(n+1)n!n!}$.

$$L = \lim_{n\to\infty}\left|\frac{a_{n+1}}{a_n}\right| = \lim_{n\to\infty}\left|\frac{(2n+2)(2n+1)(2n)!}{(n+1)(n+1)n!\,n!} \div \frac{(2n)!}{n!\,n!}\right|$$

$$= \lim_{n\to\infty}\left|\frac{(2n+2)(2n+1)(2n)!}{(n+1)(n+1)n!\,n!} \times \frac{n!\,n!}{(2n)!}\right|$$

$$L = \lim_{n\to\infty}\frac{(2n+2)(2n+1)}{(n+1)(n+1)} = \lim_{n\to\infty}\frac{4n^2+6n+2}{n^2+2n+1} = \lim_{n\to\infty}\frac{4+\frac{6}{n}+\frac{2}{n^2}}{1+\frac{2}{n}+\frac{1}{n^2}} = \frac{4+0+0}{1+0+0} = 4$$

Since $L > 1$, the series **diverges**.

(32) We will use the root test with $a_j = \sqrt{j}\,e^{-j} = \frac{j^{1/2}}{e^j}$. Recall that $\left(x^j\right)^{1/j} = x^1 = x$.

$$L = \lim_{j\to\infty}\left|\frac{j^{1/2}}{e^j}\right|^{1/j} = \lim_{j\to\infty}\frac{\left(j^{1/2}\right)^{1/j}}{(e^j)^{1/j}} = \lim_{j\to\infty}\frac{j^{\frac{1}{2j}}}{e} = \lim_{j\to\infty}\frac{\left(j^{1/j}\right)^{1/2}}{e} = \lim_{j\to\infty}\frac{\sqrt{j^{1/j}}}{e}$$

$$\frac{\sqrt{1}}{e} = \frac{1}{e}$$

Since $L < 1$, the series **converges**. Note that $\lim_{j\to\infty} j^{1/j} = 1$ (see Chapter 2, Exercise 39).

Note: $\frac{1}{e} \approx 0.367879$.

(33) We will use the limit comparison test with $c_k = \frac{1}{7^k}$.

$$\lim_{k\to\infty}\frac{a_k}{c_k} = \lim_{k\to\infty}\frac{\frac{2k+4}{7^k(k-5)}}{\frac{1}{7^k}} = \lim_{k\to\infty}\left(\frac{2k+4}{7^k(k-5)} \div \frac{1}{7^k}\right) = \lim_{k\to\infty}\left(\frac{2k+4}{7^k(k-5)} \times \frac{7^k}{1}\right)$$

$$= \lim_{k\to\infty}\frac{2k+4}{k-5} = \lim_{k\to\infty}\frac{2+\frac{4}{k}}{1-\frac{5}{k}} = \frac{2+0}{1-0} = \frac{2}{1} = 2$$

Since $\sum c_k = \sum \frac{1}{7^k}$ converges and $\lim_{k\to\infty}\frac{a_k}{c_k} = 2$ is finite, it follows that $\sum a_k = \sum \frac{2k+4}{7^k(k-5)}$ **converges**.

(34) If you note that $\frac{3}{16} = \frac{9}{48}$ and $\frac{1}{672} = \frac{15}{10{,}080}$, the given series can be rewritten as:

$$\sum a_n = \frac{1}{2} + \frac{3}{2} + \frac{5}{4} + \frac{7}{12} + \frac{9}{48} + \frac{11}{240} + \frac{13}{1440} + \frac{15}{10{,}080} + \frac{17}{80{,}640} + \cdots$$

The numerators are the odd numbers, $2n + 1$. The denominators form the pattern

$$2, 2 \times 1 = 2, 2 \times 2 = 4, 4 \times 3 = 12, 12 \times 4 = 48, 48 \times 5 = 240, 240 \times 6 = 1440, \ldots$$

Note that factorials form a similar pattern: $1! = 0! \times 1$, $2! = 1! \times 2$, $3! = 2! \times 3$, $4! = 3! \times 4$, $5! = 4! \times 5$, ... Beginning with $0!$ (which equals one, not zero), the factorials form the pattern 1, 1, 2, 6, 24, 120, 720, 5040, 40,320, ... Note that the denominators are twice these values. We can make the denominators with the formula $n!\,(2)$. Using summation notation, this series can be written as

$$\sum_{n=0}^{\infty} \frac{2n+1}{n!\,(2)}$$

We will use the ratio test. Since $a_n = \frac{2n+1}{n!(2)}$, $a_{n+1} = \frac{2(n+1)+1}{(n+1)!(2)} = \frac{2n+2+1}{(n+1)n!(2)} = \frac{2n+3}{(n+1)n!(2)}$.

$$L = \lim_{n\to\infty} \left|\frac{a_{n+1}}{a_n}\right| = \lim_{n\to\infty} \left|\frac{2n+3}{(n+1)n!\,(2)} \div \frac{2n+1}{n!\,(2)}\right| = \lim_{n\to\infty} \left|\frac{2n+3}{(n+1)n!\,(2)} \times \frac{n!\,(2)}{2n+1}\right|$$

$$L = \lim_{n\to\infty} \frac{2n+3}{(n+1)(2n+1)} = \lim_{n\to\infty} \frac{2n+3}{2n^2+3n+1} = \lim_{n\to\infty} \frac{\frac{2}{n} + \frac{3}{n^2}}{2 + \frac{3}{n} + \frac{1}{n^2}} = \frac{0+0}{2+0+0} = \frac{0}{2} = 0$$

Since $L < 1$, the series **converges**. We divided the numerator and denominator each by n^2 (a strategy from Chapter 2).

(35) We'll present two solutions. First, look at the numerators: 2, 20, 5, 20, 10, 4, 5, 20, 10, ... This sequence is neither increasing nor decreasing. But every numerator does evenly divide into 20. Maybe the pattern of the fractions will make more sense if we rewrite every fraction with a numerator of 20. Let's try it.

$$\frac{20}{70} + \frac{20}{89} + \frac{20}{108} + \frac{20}{127} + \frac{20}{146} + \frac{20}{165} + \frac{20}{184} + \frac{20}{203} + \frac{20}{222} + \cdots$$

With this simple transformation, the pattern should now appear obvious. The denominators increase by 19. In summation notation, this series can be written as

$$\sum a_n = \sum_{n=0}^{\infty} \frac{20}{70 + 19n}$$

We will use the limit comparison test with $d_n = \frac{1}{n}$.

$$\lim_{n\to\infty}\frac{a_n}{d_n}=\lim_{n\to\infty}\frac{\frac{20}{70+19n}}{\frac{1}{n}}=\lim_{n\to\infty}\left(\frac{20}{70+19n}\div\frac{1}{n}\right)=\lim_{n\to\infty}\left(\frac{20}{70+19n}\times\frac{n}{1}\right)$$

$$=\lim_{n\to\infty}\frac{20n}{70+19n}=\lim_{n\to\infty}\frac{20}{\frac{70}{n}+19}=\frac{20}{0+19}=\frac{20}{19}$$

Since $\sum d_n=\sum\frac{1}{n}$ diverges and $\lim_{n\to\infty}\frac{a_n}{d_n}=\frac{20}{19}$ is positive, it follows that $\sum a_n=\sum\frac{20}{70+19n}$ **diverges**. (We just need this limit to be positive, unlike a ratio or root test.)

Alternate solution: Another way to solve this problem is to examine the reciprocals in decimal form. By looking at the reciprocals as decimals, you can write this series as

$$\sum a_k=\frac{1}{3.5}+\frac{1}{4.45}+\frac{1}{5.4}+\frac{1}{6.35}+\frac{1}{7.3}+\frac{1}{8.25}+\frac{1}{9.2}+\frac{1}{10.15}+\frac{1}{11.1}+\cdots$$

Compare this with the harmonic series starting at $k=4$.

$$\sum d_k=\sum_{k=4}^{\infty}\frac{1}{k}=\frac{1}{4}+\frac{1}{5}+\frac{1}{6}+\frac{1}{7}+\frac{1}{8}+\frac{1}{9}+\frac{1}{10}+\frac{1}{11}+\frac{1}{12}\cdots$$

Observe that $\frac{1}{3.5}>\frac{1}{4},\frac{1}{4.45}>\frac{1}{5},\frac{1}{5.4}>\frac{1}{6},\ldots,\frac{1}{11.1}>\frac{1}{12},\ldots$ (since a smaller denominator results in a larger fraction for the same numerator). The denominators of the given series grow by 0.95 each time. The given series can be expressed in summation notation as

$$\sum_{n=0}^{\infty}a_n=\frac{1}{3.5+0.95n}$$

Note that the harmonic series can be similarly expressed as

$$\sum_{n=0}^{\infty}d_n=\frac{1}{4+n}$$

4 is bigger than 3.5, such that $\frac{1}{4}<\frac{1}{3.5}$. Since n is bigger than $0.95n$, the denominator grows faster in $\frac{1}{4+n}$ than it does in $\frac{1}{3.5+0.95n}$, which makes $\frac{1}{3.5+0.95n}<\frac{1}{4+n}$ for any value of n in these series. Therefore, $a_n>d_n$. Every term is larger than every term of a diverging series. We will use the comparison test (Chapter 5). Since $\sum d_n=\sum\frac{1}{4+n}$ diverges (since the harmonic series diverges) and since $a_n>d_n$, it follows that $\sum a_n$ **diverges**.

Chapter 10 Answers

(1) We will compare $\sum|a_n|$ with $\sum c_n = \frac{1}{n^2}$. Since $0 \leq \left|\sin\left(\frac{n\pi}{6}\right)\right| \leq 1$, it follows that $|a_n| = \frac{\left|\sin\left(\frac{n\pi}{6}\right)\right|}{n^2} \leq \frac{1}{n^2} = c_n$. Since $|a_n| \leq c_n$ for all n and $\sum c_n$ is known to converge (recall Chapter 5), it follows that $\sum|a_n|$ **converges**. Since $\sum|a_n|$ converges, $\sum a_n$ also **converges**. This series converges **absolutely**. Note: Don't be fooled into thinking that $\left\{\frac{\sin\left(\frac{n\pi}{6}\right)}{n^2}\right\}$ diverges because the sine functions repeats; the presence of the $\frac{1}{n^2}$ causes the sequence to converge to zero even as the sine function cycles.

(2) Ignore the minus signs to see that $\sum|b_k|$ is a geometric series with $r = \frac{b_{k+1}}{b_k} = \frac{1}{3}$. Since $|r| \leq 1$, this series **converges**. Since $\sum|b_k|$ converges, $\sum b_k$ also **converges**. This series converges **absolutely**.

(3) Ignore the minus signs to see that $\sum|a_m| = \sum\frac{1}{m^2}$, which **converges** since $\sum\frac{1}{m^p}$ converges if $p > 1$ (recall Chapter 5). Since $\sum|a_m|$ converges, $\sum a_m$ also **converges**. This series converges **absolutely**.

(4) Ignore the minus signs to see that $\sum|b_j| = \sum\frac{1}{\sqrt{j}} = \sum\frac{1}{j^{1/2}}$, which **diverges** since $\sum\frac{1}{j^p}$ diverges if $p \leq 1$ (recall Chapter 5). When $\sum|b_j|$ diverges, **more information is needed** to determine whether or not $\sum b_j$ converges or diverges; **we can't conclude anything** about the convergence of $\sum b_j$ from this test of $\sum|b_j|$.

(5) Ignore the minus signs to see that $\sum|a_\ell| = \sum\frac{1}{5^\ell}$, which **converges** since $\sum\frac{1}{5^\ell}$ is a geometric series with $r = \frac{1}{5}$. Since $\sum|a_\ell|$ converges, $\sum a_\ell$ also **converges**. This series converges **absolutely**.

(6) Apply the simple divergence test. Since $\lim_{i\to\infty}\frac{2i^2-6}{i(4i+3)} = \lim_{i\to\infty}\frac{2i^2-6}{4i^2+3i} = \lim_{i\to\infty}\frac{2-\frac{6}{i^2}}{4+\frac{3}{i}} = \frac{2-0}{4+0} = \frac{2}{4} = \frac{1}{2}$, the sequence $\left\{(-1)^i\frac{2i^2-6}{i(4i+3)}\right\}$ doesn't converge to zero. Therefore, $\sum b_i$ **diverges**. It follows that $\sum|b_i|$ also **diverges**. (Contrast this with Exercise 4. Here we were able to test $\sum b_i$ directly, whereas in Exercise 4 we tested $\sum|b_j|$.)

(7) Ignore the minus signs to see that $\sum|a_k| = \sum\left(\frac{1}{4}\right)^{k+1}$ is a geometric series with $r = \frac{1}{4}$. For example, when $k = 4$, $\frac{|a_{k+1}|}{|a_k|} = \frac{|a_{4+1}|}{|a_4|} = \frac{|a_5|}{|a_4|} = \frac{(1/4)^{5+1}}{(1/4)^{4+1}} = \frac{(1/4)^6}{(1/4)^5} = \frac{1}{4}$. Since $|r| \leq 1$, this series **converges**. Since $\sum|a_k|$ converges, $\sum a_k$ also **converges**. This series converges **absolutely**. Note: You could test $\sum a_k = \sum\frac{1}{(-4)^{k+1}}$ directly, since $r = -\frac{1}{4}$ for this alternating geometric series and $|r| < 1$, but then you wouldn't be able to draw any conclusions about $\sum|a_k|$. (If $\sum|a_k|$ converges, we know that $\sum a_k$ converges, but knowing that $\sum a_k$ converges doesn't tell you if $\sum|a_k|$ converges. See the four bullet points in the chapter text that outline all four cases.)

(8) First, apply the root test (Chapter 8) to $\sum c_n = \sum e^{-n} = \sum\frac{1}{e^n}$. Note that $(e^n)^{1/n} = e^1 = e$. Since $L < 1$ (see below), this shows that $\sum c_n = \sum e^{-n} = \sum\frac{1}{e^n}$ converges. (Alternatively, note that $\sum c_n = \sum e^{-n} = \sum\frac{1}{e^n}$ is a geometric series. Since $r = e^{-1} = \frac{1}{e} \approx 0.367879$ is less than 1, we know that $\sum c_n$ converges.)

$$L = \lim_{n\to\infty}|c_n|^{1/n} = \lim_{n\to\infty}\left(\frac{1}{e^n}\right)^{1/n} = \lim_{n\to\infty}\frac{1}{(e^n)^{1/n}} = \lim_{n\to\infty}\frac{1}{e^1} = \frac{1}{e}$$

Compare $\sum|b_n| = \sum e^{-n}|\sin n|$ with $\sum c_n = \sum e^{-n} = \sum\frac{1}{e^n}$. Since $0 \leq |\sin n| \leq 1$, it follows that $|b_n| = e^{-n}|\sin n| \leq e^{-n} = c_n$. Since $|b_n| \leq c_n$ for all n and $\sum c_n$ is known to converge (recall Chapter 5), it follows that $\sum|b_n|$ **converges**. Since $\sum|b_n|$ converges, $\sum b_n$ also **converges**. This series converges **absolutely**.

(9) Apply the ratio test (Chapter 7) to $a_j = \frac{j^4}{(-2)^j}$.

$$L = \lim_{j\to\infty}\left|\frac{a_{j+1}}{a_j}\right| = \lim_{j\to\infty}\left|\frac{(j+1)^4}{(-2)^{j+1}} \div \frac{j^4}{(-2)^j}\right| = \lim_{j\to\infty}\left|\frac{(j+1)^4}{(-2)^{j+1}} \times \frac{(-2)^j}{j^4}\right|$$

$$= \lim_{j\to\infty}\left|\left(-\frac{1}{2}\right)\left(\frac{j+1}{j}\right)^4\right| = \lim_{j\to\infty}\frac{1}{2}\left(\frac{1+\frac{1}{j}}{1}\right)^4 = \frac{1}{2}\left(\frac{1+0}{1}\right)^4 = \frac{1}{2}(1)^4 = \frac{1}{2}$$

Since $L < 1$, $\sum a_j$ and $\sum|a_j|$ both **converge**. This series converges **absolutely**.

(10) Ignore the minus signs to see that $\sum|b_m| = \sum\frac{m!}{(2m)!}$. Apply the ratio test (Chapter 7) to $|b_m| = \frac{m!}{(2m)!}$.

$$L = \lim_{m\to\infty}\left|\frac{b_{m+1}}{b_m}\right| = \lim_{j\to\infty}\left|\frac{(m+1)!}{[2(m+1)]!} \div \frac{m!}{(2m)!}\right| = \lim_{m\to\infty}\left|\frac{(m+1)(m!)}{(2m+2)!} \times \frac{(2m)!}{m!}\right|$$

$$= \lim_{m\to\infty}\left|\frac{(m+1)(2m)!}{(2m+2)(2m+1)(2m)!}\right| = \lim_{m\to\infty}\left|\frac{(m+1)}{(2m+2)(2m+1)}\right|$$

$$= \lim_{m\to\infty}\frac{m+1}{4m^2+6m+2} = \lim_{m\to\infty}\frac{\frac{1}{m}+\frac{1}{m^2}}{4+\frac{6}{m}+\frac{2}{m^2}} = \frac{0+0}{4+0+0} = \frac{0}{4} = 0$$

Since $L < 1$, $\sum b_m$ and $\sum |b_m|$ both **<u>converge</u>**. This series converges **<u>absolutely</u>**.

(11) Ignore the minus signs to see that $\sum |a_m| = \sum \frac{1}{\sqrt{m^2+5m+3}}$. We will compare this with $\sum d_m = \frac{1}{m}$ using the limit comparison test (Chapter 5).

$$\lim_{m\to\infty}\frac{a_m}{d_m} = \lim_{m\to\infty}\left(\frac{1}{\sqrt{m^2+5m+3}} \div \frac{1}{m}\right) = \lim_{m\to\infty}\left(\frac{1}{\sqrt{m^2+5m+3}} \times m\right)$$

$$= \lim_{m\to\infty}\frac{m}{\sqrt{m^2+5m+3}} = \lim_{m\to\infty}\frac{\sqrt{m^2}}{\sqrt{m^2+5m+3}} = \lim_{m\to\infty}\sqrt{\frac{m^2}{m^2+5m+3}}$$

$$= \lim_{m\to\infty}\sqrt{\frac{1}{1+\frac{5}{m}+\frac{3}{m^2}}} = \sqrt{\frac{1}{1+0+0}} = \sqrt{\frac{1}{1}} = \sqrt{1} = 1$$

Since $\lim_{m\to\infty}\frac{a_m}{d_m}$ is positive and $\sum d_n$ is known to diverge (recall Chapter 5), it follows that $\sum |a_m|$ **<u>diverges</u>**. (Be careful here. If this had been a ratio test or a root test, a limit equal to 1 would have been inconclusive, but this was a limit comparison test. For a limit comparison test, when testing for divergence, we just need the limit to be positive.) When $\sum |a_m|$ diverges, **<u>more information is needed</u>** to determine whether or not $\sum a_m$ converges or diverges; **<u>we can't conclude anything</u>** about the convergence of $\sum a_m$ from this test of $\sum |a_m|$.

Chapter 11 Answers

(1) This is an alternating series. Each term changes sign. Examples: $a_1 = \frac{2}{(-1)^{1+5}} = \frac{2}{(-1)^6} = \frac{2}{1} = 2$, $a_2 = \frac{2}{(-1)^{2+5}} = \frac{2}{(-1)^7} = \frac{2}{-1} = -2$, $a_3 = \frac{2}{(-1)^{3+5}} = \frac{2}{(-1)^8} = \frac{2}{1} = 2$.

(2) This is an alternating series. Each term changes sign. Examples: $a_2 = (-4)^2 = 16$, $a_3 = (-4)^3 = -64$, $a_4 = (-4)^4 = 256$, $a_5 = (-4)^5 = -1024$.

(3) This series doesn't alternate because $2k + 1$ only makes odd numbers, like $2(0) + 1 = 1$, $2(1) + 1 = 3$, $2(2) + 1 = 5$, $2(3) + 1 = 7$, etc. Examples: $a_0 = \frac{(-1)^{2(0)+1}}{0+3} = \frac{(-1)^1}{3} = -\frac{1}{3}$, $a_1 = \frac{(-1)^{2(1)+1}}{1+3} = \frac{(-1)^3}{4} = -\frac{1}{4}$, $a_2 = \frac{(-1)^{2(2)+1}}{2+3} = \frac{(-1)^5}{5} = -\frac{1}{5}$.

(4) This series doesn't alternate because the exponents $n + 8$ and $n - 2$ are either both even or both odd. When the numerator is negative, so is the denominator, and the minus signs cancel out. Every term is positive. Examples: $a_2 = \frac{(-4)^{2+8}}{(-7)^{2-2}} = \frac{(-4)^{10}}{(-7)^0} = \frac{4^{10}}{1}$, $a_2 = \frac{(-4)^{3+8}}{(-7)^{3-2}} = \frac{(-4)^{11}}{(-7)^1} = \frac{-4^{11}}{-7} = \frac{-4^{11}}{-7} = \frac{4^{11}}{7}$.

(5) Every term doesn't change sign; sometimes two consecutive terms have the same sign. This doesn't meet the definition of an alternating series. Examples: $a_1 = \sin\left(\frac{\pi}{4}\right) = \frac{\sqrt{2}}{2}$, $a_2 = \sin\left(\frac{2\pi}{4}\right) = \sin\left(\frac{\pi}{2}\right) = 1$, $a_3 = \sin\left(\frac{3\pi}{4}\right) = \frac{\sqrt{2}}{2}$, $a_4 = \sin\left(\frac{4\pi}{4}\right) = \sin(\pi) = 0$, $a_5 = \sin\left(\frac{5\pi}{4}\right) = -\frac{\sqrt{2}}{2}$, $a_2 = \sin\left(\frac{6\pi}{4}\right) = \sin\left(\frac{3\pi}{2}\right) = -1$.

(6) This is an alternating series. Each term changes sign. Note that $\frac{\pi+3\ell\pi}{3} = \frac{\pi}{3} + \frac{3\ell\pi}{3} = \frac{\pi}{3} + \ell\pi$, so each term adds π radians to the angle, causing the cosine function to change sign. (Look at a graph of cosine to understand this.) Examples: $a_0 = \cos\left[\frac{\pi+3(0)\pi}{3}\right] = \cos\left(\frac{\pi}{3}\right) = \cos 60° = \frac{1}{2}$, $a_1 = \cos\left[\frac{\pi+3(1)\pi}{3}\right] = \cos\left(\frac{\pi+3\pi}{3}\right) = \cos\left(\frac{4\pi}{3}\right) = \cos 240° = -\frac{1}{2}$, $a_2 = \cos\left[\frac{\pi+3(2)\pi}{3}\right] = \cos\left(\frac{\pi+6\pi}{3}\right) = \cos\left(\frac{7\pi}{3}\right) = \cos 420° = \frac{1}{2}$, $a_3 = \cos\left[\frac{\pi+3(3)\pi}{3}\right] = \cos\left(\frac{\pi+9\pi}{3}\right) = \cos\left(\frac{10\pi}{3}\right) = \cos 600° = -\frac{1}{2}$. To convert from radians to degrees, multiply by $\frac{180}{\pi}$.

(7) Although the sequence "alternates" between $\frac{3}{2}$ and $\frac{1}{2}$, it doesn't meet the definition of an alternating series because **every term is positive**. (However, you can write it as

the sum of two series, where one alternates and one doesn't: $\sum \frac{(-1)^{j+1}+2}{2} = \sum \frac{(-1)^{j+1}}{2} + \sum \frac{2}{2}$. But since the combined series doesn't include any negative values, the combined series doesn't meet the definition of an alternating series.) Examples: $a_1 = \frac{(-1)^{1+1}+2}{2} = \frac{(-1)^2+2}{2} = \frac{1+2}{2} = \frac{3}{2}$, $a_2 = \frac{(-1)^{2+1}+2}{2} = \frac{(-1)^3+2}{2} = \frac{-1+2}{2} = \frac{1}{2}$, $a_3 = \frac{(-1)^{3+1}+2}{2} = \frac{(-1)^4+2}{2} = \frac{1+2}{2} = \frac{3}{2}$, $a_4 = \frac{(-1)^{4+1}+2}{2} = \frac{(-1)^5+2}{2} = \frac{-1+2}{2} = \frac{1}{2}$.

(8) This is an alternating series. Each term changes sign. Examples: $a_2 = (-1)^2 + \frac{1}{3} = 1 + \frac{1}{3} = \frac{4}{3}$, $a_3 = (-1)^3 + \frac{1}{3} = -1 + \frac{1}{3} = -\frac{2}{3}$, $a_4 = (-1)^4 + \frac{1}{3} = 1 + \frac{1}{3} = \frac{4}{3}$, $a_5 = (-1)^5 + \frac{1}{3} = -1 + \frac{1}{3} = -\frac{2}{3}$.

(9) This series is **absolutely convergent** because $\sum |a_m| = \sum \frac{1}{3^m}$ is convergent since it is a geometric series with $r = \frac{1}{3} < 1$.

(10) This series is **absolutely convergent** because $\sum |a_i| = \sum \frac{1}{i\sqrt{i}} = \sum \frac{1}{i^{3/2}}$ is convergent since $\sum \frac{1}{i^p}$ converges if $p > 1$ (Chapter 6).

(11) This series is **conditionally convergent**. It satisfies Leibniz's theorem because it is an alternating series, $\frac{1}{\sqrt{n}}$ is nonincreasing (for example, $\frac{1}{\sqrt{2}} \approx 0.707$ is larger than $\frac{1}{\sqrt{3}} \approx 0.577$), and $\lim_{n\to\infty} |a_n| = \lim_{n\to\infty} \frac{1}{\sqrt{n}} = 0$. The corresponding series of absolute values diverges: $\sum \frac{1}{\sqrt{n}} = \sum \frac{1}{n^{1/2}}$ since $\sum \frac{1}{n^p}$ diverges if $p \leq 1$ (Chapter 6).

(12) This series **diverges**. Apply the simple divergence test (Chapter 4).

$$\lim_{\ell\to\infty} |a_\ell| = \lim_{\ell\to\infty} \left|\frac{2^\ell}{\ell^2}\right| = \infty$$

Since $\{a_\ell\}$ doesn't converge to zero, this series diverges. Note that 2^ℓ grows much more rapidly than ℓ^2, which is why the above limit is infinite. If needed, apply l'Hôpital's rule (Chapter 2). Recall that $\frac{d}{dx} 2^x = 2^x \ln 2$. You will need to apply l'Hôpital's rule twice in a row since the first derivatives make $\frac{2^x \ln 2}{2x}$; the second time, the derivatives make $\frac{2^x \ln 2 \ln 2}{2}$.

(13) This series **diverges**. Apply the simple divergence test (Chapter 4).

$$\lim_{k\to\infty}|a_k| = \lim_{k\to\infty}\left|\frac{k^2-3}{2k^2+6k+5}\right| = \lim_{k\to\infty}\frac{1-\frac{3}{k^2}}{2+\frac{6}{k}+\frac{5}{k^2}} = \frac{1-0}{2+0+0} = \frac{1}{2}$$

Since $\{a_k\}$ doesn't converge to zero, this series diverges. (This wasn't a ratio test or a root test. In the simple divergence test, we just want to know if the limit is positive, in contrast to the ratio or root tests where the limit needs to be less than or greater than one.)

(14) This series is **conditionally convergent**. It satisfies Leibniz's theorem. First, it is an alternating series. Second, $\frac{\ln j}{j}$ is nonincreasing once $\ln j > j$, which corresponds to $j > e \approx 2.718$; since j is an integer, this means that $\frac{\ln j}{j}$ is nonincreasing once $j \geq 3$. One way to see this is to take a derivative of $\frac{\ln x}{x}$ with respect to x (using the quotient rule): $\frac{d}{dx}\frac{\ln x}{x} = \frac{x\left(\frac{1}{x}\right)-\ln x(1)}{x^2} = \frac{1-\ln x}{x^2}$. The numerator of the derivative, $1-\ln x$, is negative if $\ln x > 1$. Exponentiate both sides to get $x > e^1$, which means that $x > e \approx 2.718$. When the derivative is negative, the function decreases, so $\frac{\ln j}{j}$ is nonincreasing once $j \geq 3$. But the series begins at $j = 2$, and $\frac{\ln 2}{2} \approx 0.347$ isn't greater than $\frac{\ln 3}{3} \approx 0.366$. Is this a problem? No, it isn't a problem. We can write the series as

$$\sum_{j=2}^{\infty}\frac{\ln j}{j(-1)^{j-1}} = \frac{\ln 2}{2(-1)^{2-1}} + \sum_{j=3}^{\infty}\frac{\ln j}{j(-1)^{j-1}}$$

The series on the right meets the first condition of Leibniz's theorem; if the series on the right converges, so does the series on the left. (As we've often done, we're basically saying that if a series converges for $j > N$, it converges. Here, $N = 2$, so $j > 2$ or $j \geq 3$.) From $j = 3$ onward, $\frac{\ln j}{j}$ decreases. For example, $\frac{\ln 3}{3} \approx 0.366$ is greater than $\frac{\ln 4}{4} \approx 0.347$. The second condition of Leibniz's theorem is also satisfied: $\lim_{j\to\infty}|a_j| = \lim_{j\to\infty}\frac{\ln j}{j} = 0$. To see this, apply l'Hôpital's rule (Chapter 2) to $\frac{\ln x}{x}$. The first derivatives make $\frac{1/x}{1} = \frac{1}{x}$, which approaches zero as $x \to \infty$. This series converges conditionally because the corresponding series of absolute values diverges: $\sum \frac{\ln j}{j}$ diverges. To see that $\sum a_j = \sum \frac{\ln j}{j}$ diverges, apply the comparison test (Chapter 5). Compare it with $\sum d_j = \sum \frac{1}{j}$, which we know diverges (Chapter 6). Since $a_j = \frac{\ln j}{j}$ is greater than $d_j = \frac{1}{j}$ for all $j > 2$ (since $\ln j > 1$ if j is 3 or higher) and since $\sum d_j = \sum \frac{1}{j}$ diverges, it follows that $\sum a_j = \sum \frac{\ln j}{j}$ diverges.

(15) This series **diverges**. Apply the simple divergence test (Chapter 4).

$$\lim_{i\to\infty} |a_i| = \lim_{i\to\infty} \left|\frac{i!}{3^i}\right| = \infty$$

Since $\{a_i\}$ doesn't converge to zero, this series diverges. Note that $i!$ grows much more rapidly than 3^i, which is why the above limit is infinite. Recall from Chapter 2 that $\lim_{i\to\infty} \left|\frac{i!}{c^i}\right|$ diverges for any value of c (see Example 13 and the list of bullet points prior to the examples).

(16) This series **diverges**. Apply the simple divergence test (Chapter 4).

$$\lim_{n\to\infty} |a_n| = \lim_{n\to\infty} \left|\frac{2+\sqrt{n}}{\sqrt{2n+1}}\right| = \lim_{n\to\infty} \left|\frac{\frac{2}{\sqrt{n}}+1}{\sqrt{2+\frac{1}{n}}}\right| = \frac{0+1}{\sqrt{2+0}} = \frac{1}{\sqrt{2}}$$

Since $\{a_i\}$ doesn't converge to zero, this series diverges. (Again, with the simple divergence test, we just need the limit to be positive. Don't confuse this with the ratio or root tests, where a limit less than one results in convergence.)

Note: We divided by $\sqrt{n}$ in the numerator and denominator. In the denominator, this makes $\frac{\sqrt{2n+1}}{\sqrt{n}} = \sqrt{\frac{2n+1}{n}} = \sqrt{\frac{2n}{n} + \frac{1}{n}} = \sqrt{2 + \frac{1}{n}}$.

(17) First rewrite the series using summation notation. Observe that the denominators are perfect squares and that the number squared in the denominator is one less than the numerator.

$$\sum_{n=1}^{\infty} (-1)^{n+1} \left(\frac{n+1}{n^2}\right) = 2 - \frac{3}{4} + \frac{4}{9} - \frac{5}{16} + \frac{6}{25} - \frac{7}{36} + \frac{8}{49} - \frac{9}{64} + \frac{10}{81} - \frac{11}{100} + \cdots$$

This series is **conditionally convergent**. It satisfies Leibniz's theorem because it is an alternating series, $\frac{n+1}{n^2}$ is nonincreasing (for example, $\frac{3}{4} = 0.75$ is larger than $\frac{4}{9} \approx 0.444$ and $\frac{4}{9} \approx 0.444$ is larger than $\frac{5}{16} = 0.3125$), and $\lim_{n\to\infty} |a_n| = \lim_{n\to\infty} \frac{n+1}{n^2} = \lim_{n\to\infty} \frac{\frac{1}{n}+\frac{1}{n^2}}{1} = \frac{0}{1} = 0$ (where we divided by $\frac{1}{n^2}$ in both the numerator and denominator). The corresponding series of absolute values diverges: $\sum \frac{n+1}{n^2} = \sum \left(\frac{n}{n^2} + \frac{1}{n^2}\right) = \sum \left(\frac{1}{n} + \frac{1}{n^2}\right)$ since this is the sum of a diverging and converging series (recall from Chapter 6 that $\sum \frac{1}{n}$ diverges, whereas $\sum \frac{1}{n^2}$ converges).

(18) Every other term is the harmonic series $\left(1 + \frac{1}{2} + \frac{1}{3} + \frac{1}{4} + \cdots\right)$ and the remaining terms are $-\frac{1}{3} - \frac{1}{6} - \frac{1}{9} - \frac{1}{12} + \cdots$ (which are one-third of the harmonic series). Using summation notation, this series is:

$$\sum_{n=1}^{\infty} \left(\frac{1}{n} - \frac{1}{3n}\right) = 1 - \frac{1}{3} + \frac{1}{2} - \frac{1}{6} + \frac{1}{3} - \frac{1}{9} + \frac{1}{4} - \frac{1}{12} + \frac{1}{5} - \frac{1}{15} + \frac{1}{6} - \frac{1}{18} + \cdots$$

Since $\frac{1}{n} - \frac{1}{3n} = \frac{3}{3n} - \frac{1}{3n} = \frac{2}{3n}$, this series is $\sum \frac{2}{3n}$, which **diverges** because $\sum \frac{1}{n}$ diverges (and multiplying by 2/3 won't change that).

Chapter 12 Answers

(1) Apply the ratio test (Chapter 7).

$$L = \lim_{n\to\infty}\left|\frac{x^{n+1}}{n+1} \div \frac{x^n}{n}\right| = \lim_{n\to\infty}\left|\frac{x^{n+1}}{n+1} \times \frac{n}{x^n}\right| = \lim_{n\to\infty}\left|\frac{xn}{n+1}\right|$$

$$= \lim_{n\to\infty}\left|\frac{x}{1+\frac{1}{n}}\right| = \left|\frac{x}{1+0}\right| = \left|\frac{x}{1}\right| = |x|$$

The power series converges if $L = |x| < 1$, which equates to $-1 < x < 1$. The radius of convergence is one-half of this interval: $R = 1$. We also need to check the case $|x| = 1$, which equates to $x = -1$ or 1, since the ratio test is inconclusive for $L = 1$. When $x = -1$, the power series is $\sum \frac{(-1)^n}{n}$, which converges conditionally (Chapter 11), and when $x = 1$, the power series is $\sum \frac{1}{n}$, which diverges (Chapter 6). The interval of convergence is $-1 \leq x < 1$, with a $\leq$ sign to include -1.

(2) Apply the ratio test (Chapter 7).

$$L = \lim_{k\to\infty}\left|\frac{x^{k+1}}{(k+1)^2} \div \frac{x^k}{k^2}\right| = \lim_{k\to\infty}\left|\frac{x^{k+1}}{(k+1)^2} \times \frac{k^2}{x^k}\right| = \lim_{k\to\infty}\left|\frac{xk^2}{k^2+2k+1}\right|$$

$$= \lim_{k\to\infty}\left|\frac{x}{1+\frac{2}{k}+\frac{1}{k^2}}\right| = \left|\frac{x}{1+0+0}\right| = \left|\frac{x}{1}\right| = |x|$$

The power series converges if $L = |x| < 1$, which equates to $-1 < x < 1$. The radius of convergence is one-half of this interval: $R = 1$. We also need to check the case $|x| = 1$, which equates to $x = -1$ or 1, since the ratio test is inconclusive for $L = 1$. When $x = -1$, the power series is $\sum \frac{(-1)^k}{k^2}$, and when $x = 1$, the power series is $\sum \frac{1}{k^2}$. Both of these cases converge absolutely since $\sum \frac{1}{k^2}$ converges (recall Chapters 10-11). The interval of convergence is $-1 \leq x \leq 1$, with two $\leq$ signs.

(3) Apply the ratio test (Chapter 7).

$$L = \lim_{m\to\infty}\left|\frac{x^{m+1}}{(m+1)!} \div \frac{x^m}{m!}\right| = \lim_{m\to\infty}\left|\frac{x^{m+1}}{(m+1)m!} \times \frac{m!}{x^m}\right| = \lim_{m\to\infty}\left|\frac{x}{m+1}\right| = 0$$

The power series converges for all x, since $L = 0$ regardless of the value of x, such that the interval of convergence is $-\infty < x < \infty$ and the radius of convergence is $R = \infty$.

(4) Apply the ratio test (Chapter 7).

$$L = \lim_{j\to\infty}\left|\frac{x^{j+1}}{\sqrt{j+1}} \div \frac{x^j}{\sqrt{j}}\right| = \lim_{j\to\infty}\left|\frac{x^{j+1}}{\sqrt{j+1}} \times \frac{\sqrt{j}}{x^j}\right| = \lim_{j\to\infty}\left|x\sqrt{\frac{j}{j+1}}\right|$$

$$= \lim_{j\to\infty}\left|x\sqrt{\frac{1}{1+\frac{1}{j}}}\right| = \left|x\sqrt{\frac{1}{1+0}}\right| = \left|x\sqrt{\frac{1}{1}}\right| = \left|x\sqrt{1}\right| = |x|$$

The power series converges if $L = |x| < 1$, which equates to $-1 < x < 1$. The radius of convergence is one-half of this interval: $R = 1$. We also need to check the case $|x| = 1$, which equates to $x = -1$ or 1, since the ratio test is inconclusive for $L = 1$. When $x = -1$, the power series is $\sum \frac{(-1)^j}{\sqrt{j}}$, which converges conditionally (apply Leibniz's theorem from Chapter 11, noting that $\lim_{j\to\infty}\left|\frac{1}{\sqrt{j}}\right| = 0$), and when $x = 1$, the power series is $\sum \frac{1}{\sqrt{j}}$, which diverges (recall from Chapter 6 that $\sum \frac{1}{j^p}$ diverges if $p \leq 1$, and note that $\frac{1}{\sqrt{j}} = \frac{1}{j^{1/2}}$). The interval of convergence is $-1 \leq x < 1$, with a $\leq$ sign to include -1.

(5) Apply the root test (Chapter 8). Recall that $\left(x^i\right)^{1/i} = x^1 = x$.

$$L = \lim_{i\to\infty}\left|\frac{x^i}{5^i}\right|^{1/i} = \lim_{i\to\infty}\left|\frac{x}{5}\right| = \left|\frac{x}{5}\right|$$

The power series converges if $L = \left|\frac{x}{5}\right| < 1$, which equates to $-1 < \frac{x}{5} < 1$ or $-5 < x < 5$. The radius of convergence is one-half of this interval: $R = 5$. We also need to check the case $\left|\frac{x}{5}\right| = 1$, which equates to $x = -5$ or 5, since the root test is inconclusive for $L = 1$. When $x = -5$, the power series is $\sum \frac{(-5)^i}{5^i} = \sum\left(-\frac{5}{5}\right)^i = \sum(-1)^i$, and when $x = 5$, the power series is $\sum \frac{5^i}{5^i} = \sum 1$. Both of these cases diverge according to the simple divergence test (Chapter 4) since $\lim_{i\to\infty} a_i$ isn't zero. The complete interval is $-5 < x < 5$ (without using any $\leq$ signs).

(6) Apply the ratio test (Chapter 7).

$$L = \lim_{\ell\to\infty}\left|\frac{(\ell+1)^2 x^{\ell+1}}{\ell^2 x^\ell}\right| = \lim_{\ell\to\infty}\left|\frac{(\ell+1)^2}{\ell^2}x\right| = \lim_{\ell\to\infty}\left|x\left(\frac{\ell+1}{\ell}\right)^2\right|$$

$$= \lim_{\ell\to\infty}\left|x\left(\frac{1+\frac{1}{\ell}}{1}\right)^2\right| = \left|x\left(\frac{1+0}{1}\right)^2\right| = \left|x\left(\frac{1}{1}\right)^2\right| = |x(1)^2| = |x(1)| = |x|$$

The power series converges if $L = |x| < 1$, which equates to $-1 < x < 1$. The radius of convergence is one-half of this interval: $R = 1$. We also need to check the case $|x| = 1$, which equates to $x = -1$ or 1, since the ratio test is inconclusive for $L = 1$. When $x = -1$, the power series is $\sum \ell^2(-1)^\ell$, and when $x = 1$, the power series is $\sum \ell^2$. Both of these cases diverge according to the simple divergence test (Chapter 4) since $\lim\limits_{\ell\to\infty} a_\ell$ isn't zero. The complete interval is $-1 < x < 1$ (without using any $\leq$ signs).

(7) Apply the root test (Chapter 8). Recall that $(x^n)^{1/n} = x^1 = x$.

$$L = \lim_{n\to\infty}|x^n e^{-n}|^{1/n} = \lim_{n\to\infty}\left|\frac{x^n}{e^n}\right|^{1/n} = \lim_{n\to\infty}\left|\frac{x}{e}\right| = \left|\frac{x}{e}\right|$$

The power series converges if $L = \left|\frac{x}{e}\right| < 1$, which equates to $-1 < \frac{x}{e} < 1$ or $-e < x < e$. The radius of convergence is one-half of this interval: $R = e$. We also need to check the case $\left|\frac{x}{e}\right| = 1$, which equates to $x = -e$ or e, since the root test is inconclusive for $L = 1$. When $x = -e$, the power series is $\sum(-e)^n e^{-n} = \sum \frac{(-e)^n}{e^n} = \sum\left(-\frac{e}{e}\right)^n = \sum(-1)^n$, and when $x = e$, the power series is $\sum e^n e^{-n} = \sum e^{n-n} = \sum e^0 = \sum 1$. Both of these cases diverge according to the simple divergence test (Chapter 4) since $\lim\limits_{n\to\infty} a_n$ isn't zero. The complete interval is $-e < x < e$ (without using any $\leq$ signs).

(8) Apply the ratio test (Chapter 7).

$$L = \lim_{k\to\infty}\left|\frac{(x-4)^{k+1}}{k+4} \div \frac{(x-4)^k}{k+3}\right| = \lim_{k\to\infty}\left|\frac{(x-4)^{k+1}}{k+4} \times \frac{k+3}{(x-4)^k}\right| = \lim_{k\to\infty}\left|\frac{(x-4)(k+3)}{k+4}\right|$$

$$= \lim_{k\to\infty}\left|\frac{kx+3x-4k-12}{k+4}\right| = \lim_{k\to\infty}\left|\frac{x+\frac{3x}{k}-4-\frac{12}{k}}{1+\frac{4}{n}}\right| = \left|\frac{x+0-4-0}{1+0}\right| = \left|\frac{x-4}{1}\right| = |x-4|$$

The power series converges if $L = |x-4| < 1$, which equates to $-1 < x-4 < 1$ or $-3 < x < 5$. The radius of convergence is one-half of this interval: $R = 1$. We also need to check the case $|x-4| = 1$, which equates to $x = 3$ or 5, since the ratio test is inconclusive for $L = 1$. When $x = 3$, the power series is $\sum \frac{(-1)^k}{k+3}$, which converges conditionally (recall Leibniz's test from Chapter 11), and when $x = 5$, the power series is $\sum \frac{1}{k+3}$, which diverges (because $\sum \frac{1}{k}$ diverges; recall Chapter 5). The interval of convergence is $3 \leq x < 5$, with a $\leq$ sign to include the 3.

(9) Apply the root test (Chapter 8). Recall that $(x^m)^{1/m} = x^1 = x$.

$$L = \lim_{m\to\infty} \left|\frac{(x+6)^m}{4^m}\right|^{1/m} = \lim_{m\to\infty} \left|\frac{x+6}{4}\right| = \left|\frac{x+6}{4}\right|$$

The power series converges if $L = \left|\frac{x+6}{4}\right| < 1$, which equates to $-1 < \frac{x+6}{4} < 1$ or $-4 < x + 6 < 4$, which simplifies to $-10 < x < -2$. The radius of convergence is one-half of this interval: $R = 4$. We also need to check the case $\left|\frac{x+6}{4}\right| = 1$, which equates to $x = -10$ or -2, since the root test is inconclusive for $L = 1$. When $x = -10$, the power series is $\sum \frac{(-4)^m}{4^i} = \sum \left(-\frac{4}{4}\right)^m = \sum(-1)^m$, and when $x = -2$, the power series is $\sum \frac{4^m}{4^m} = \sum 1$. Both of these cases diverge according to the simple divergence test (Chapter 4) since $\lim\limits_{m\to\infty} a_m$ isn't zero. The complete interval is $-10 < x < -2$ (without using any $\leq$ signs).

(10) Apply the ratio test (Chapter 7).

$$L = \lim_{j\to\infty} \left|\frac{(2x-3)^{j+1}}{x^4} \div \frac{(2x-3)^j}{x^4}\right| = \lim_{j\to\infty} \left|\frac{(2x-3)^{j+1}}{x^4} \times \frac{x^4}{(2x-3)^j}\right| = \lim_{j\to\infty}|2x-3|$$
$$= |2x-3|$$

The power series converges if $L = |2x - 3| < 1$, which equates to $-1 < 2x - 3 < 1$ or $2 < 2x < 4$, which simplifies to $1 < x < 2$. The radius of convergence is one-half of this interval: $R = \frac{1}{2}$. We also need to check the case $|2x - 3| = 1$, which equates to $x = 1$ or 2, since the ratio test is inconclusive for $L = 1$. When $x = 1$, the power series is $\sum \frac{(-1)^j}{(-1)^4} = \sum \frac{(-1)^j}{1} = \sum(-1)^j$, and when $x = 2$, the power series is $\sum \frac{(1)^j}{(-1)^4} = \sum \frac{1}{1} = \sum 1$. Both of these cases diverge according to the simple divergence test (Chapter 4) since $\lim\limits_{j\to\infty} a_j$ isn't zero. The complete interval is $1 < x < 2$ (without using any $\leq$ signs).

(11) Follow Part A of Example 3.

$$\frac{d}{dx}\sum_{i=1}^{\infty}\frac{x^i}{i^2} = \sum_{i=1}^{\infty}\frac{ix^{i-1}}{i^2} = \sum_{i=1}^{\infty}\frac{x^{i-1}}{i}$$

(12) Follow Part B of Example 3.

$$\int\sum_{\ell=0}^{\infty}\frac{(x-4)^\ell}{\ell+1}dx = \sum_{\ell=0}^{\infty}\frac{(x-4)^{\ell+1}}{(\ell+1)^2} + c$$

(13) Follow Part A of Example 3.

$$\frac{d}{dx}\sum_{k=2}^{\infty}\frac{(x+3)^k}{5k} = \sum_{k=2}^{\infty}\frac{k(x+3)^{k-1}}{5k} = \sum_{k=2}^{\infty}\frac{(x+3)^{k-1}}{5}$$

(14) Before you follow Part B of Example 3, beware that this power series isn't in standard form because it has $3x + 2$ rather than the standard form of $x - a$. One way around this is to factor out the 3. (In the last step, we'll distribute the 3 back in.)

$$\int\sum_{n=1}^{\infty}\frac{(3x+2)^n}{\sqrt{n}}dx = \int\sum_{n=1}^{\infty}\frac{\left[3\left(x+\frac{2}{3}\right)\right]^n}{\sqrt{n}}dx = \int\sum_{n=1}^{\infty}\frac{3^n\left(x+\frac{2}{3}\right)^n}{\sqrt{n}}dx$$

$$= \sum_{n=1}^{\infty}\frac{3^n\left(x+\frac{2}{3}\right)^{n+1}}{(n+1)\sqrt{n}} + c = \sum_{n=1}^{\infty}\frac{3^{n+1}\left(x+\frac{2}{3}\right)^{n+1}}{3(n+1)\sqrt{n}} + c = \sum_{n=1}^{\infty}\frac{(3x+2)^{n+1}}{3(n+1)\sqrt{n}} + c$$

(15) We'll use the alternate method toward the end of Example 4.

$$\left(x + \frac{x^2}{2} + \frac{x^3}{3} + \frac{x^4}{4} + \frac{x^5}{5} + \cdots\right)\left(1 - \frac{x}{4} + \frac{x^2}{9} - \frac{x^3}{16} + \frac{x^4}{25} - \frac{x^5}{36} + \cdots\right)$$

$$= x - \frac{x^2}{4} + \frac{x^3}{9} - \frac{x^4}{16} + \frac{x^5}{25} + \frac{x^2}{2} - \frac{x^3}{8} + \frac{x^4}{18} - \frac{x^5}{32} + \frac{x^3}{3} - \frac{x^4}{12} + \frac{x^5}{27} + \frac{x^4}{4} - \frac{x^5}{16} + \frac{x^5}{5} + \cdots$$

$$= x + \left(-\frac{1}{4} + \frac{1}{2}\right)x^2 + \left(\frac{1}{9} - \frac{1}{8} + \frac{1}{3}\right)x^3 + \left(-\frac{1}{16} + \frac{1}{18} - \frac{1}{12} + \frac{1}{4}\right)x^4$$

$$+ \left(\frac{1}{25} - \frac{1}{32} + \frac{1}{27} - \frac{1}{16} + \frac{1}{5}\right)x^5 + \cdots$$

$$= x + \left(\frac{-1+2}{4}\right)x^2 + \left(\frac{8-9+24}{72}\right)x^3 + \left(\frac{-9+8-12+36}{144}\right)x^4$$

$$+ \left(\frac{864 - 675 + 800 - 1350 + 4320}{21{,}600}\right)x^5 + \cdots$$

$$= x + \frac{x^2}{4} + \frac{23}{72}x^3 + \frac{23}{144}x^4 + \frac{3959}{21{,}600}x^5 + \cdots$$

(16) We'll use the alternate method toward the end of Example 4.

$$\left(1 - x + \frac{x^2}{2} - \frac{x^3}{6} + \frac{x^4}{24} - \frac{x^5}{120} + \cdots\right)\left(x - \frac{x^2}{2} + \frac{x^3}{3} - \frac{x^4}{4} + \frac{x^5}{5} + \cdots\right)$$

$$= x - \frac{x^2}{2} + \frac{x^3}{3} - \frac{x^4}{4} + \frac{x^5}{5} - x^2 + \frac{x^3}{2} - \frac{x^4}{3} + \frac{x^5}{4} + \frac{x^3}{2} - \frac{x^4}{4} + \frac{x^5}{6} - \frac{x^4}{6} + \frac{x^5}{12} + \frac{x^5}{24} \cdots$$

$$= x + \left(-\frac{1}{2} - 1\right)x^2 + \left(\frac{1}{3} + \frac{1}{2} + \frac{1}{2}\right)x^3 + \left(-\frac{1}{4} - \frac{1}{3} - \frac{1}{4} - \frac{1}{6}\right)x^4$$

$$+ \left(\frac{1}{5} + \frac{1}{4} + \frac{1}{6} + \frac{1}{12} + \frac{1}{24}\right)x^5 + \cdots$$

$$= x + \left(\frac{-1-2}{2}\right)x^2 + \left(\frac{2+3+3}{6}\right)x^3 + \left(\frac{-6-8-6-4}{24}\right)x^4$$

$$+ \left(\frac{24+30+20+10+5}{120}\right)x^5 + \cdots$$

$$= x - \frac{3}{2}x^2 + \frac{8}{6}x^3 - \frac{24}{24}x^4 + \frac{89}{120}x^5 + \cdots = x - \frac{3}{2}x^2 + \frac{4}{3}x^3 - x^4 + \frac{89}{120}x^5 + \cdots$$

(17) If you know how to perform long division with polynomials, that strategy works (though it's easy to make an arithmetic mistake). However, here is a math secret: You don't actually need to know how to do long division with polynomials in order to divide polynomials. An alternative is to multiply as shown below:

$$\frac{x - \frac{x^3}{6} + \frac{x^5}{120} + \cdots}{1 - \frac{x^2}{2} + \frac{x^4}{24} + \cdots} = c_0 + c_1 x + c_2 x^2 + c_3 x^3 + c_4 x^4 + c_5 x^5 + \cdots$$

$$x - \frac{x^3}{6} + \frac{x^5}{120} + \cdots = \left(1 - \frac{x^2}{2} + \frac{x^4}{24} + \cdots\right)(c_0 + c_1 x + c_2 x^2 + c_3 x^3 + c_4 x^4 + c_5 x^5 + \cdots)$$

$$= c_0 + c_1 x + c_2 x^2 + c_3 x^3 + c_4 x^4 + c_5 x^5 - \frac{c_0 x^2}{2} - \frac{c_1 x^3}{2} - \frac{c_2 x^4}{2} - \frac{c_3 x^5}{2} + \frac{c_0 x^4}{24} + \frac{c_1 x^5}{24} + \cdots$$

$$= c_0 + c_1 x + \left(c_2 - \frac{c_0}{2}\right)x^2 + \left(c_3 - \frac{c_1}{2}\right)x^3 + \left(c_4 - \frac{c_2}{2} + \frac{c_0}{24}\right)x^4 + \left(c_5 - \frac{c_3}{2} + \frac{c_1}{24}\right)x^5$$

Compare $x - \frac{x^3}{6} + \frac{x^5}{120} + \cdots$ with the line above. Equate coefficients to see that:

$$c_0 = 0\,, c_1 = 1\,, c_2 - \frac{c_0}{2} = 0\,, c_3 - \frac{c_1}{2} = -\frac{1}{6}\,, c_4 - \frac{c_2}{2} + \frac{c_0}{24} = 0\,, c_5 - \frac{c_3}{2} + \frac{c_1}{24} = \frac{1}{120}$$

The equation $c_2 - \frac{c_0}{2} = 0$ leads to $c_2 = \frac{c_0}{2} = \frac{0}{2} = 0$. The equation $c_3 - \frac{c_1}{2} = -\frac{1}{6}$ leads to $c_3 - \frac{1}{2} = -\frac{1}{6}$, which leads to $c_3 = \frac{1}{2} - \frac{1}{6} = \frac{1}{3}$. The equation $c_4 - \frac{c_2}{2} + \frac{c_0}{24} = 0$ leads to

$c_4 - \frac{0}{2} + \frac{0}{24} = 0$, which means that $c_4 = 0$. The equation $c_5 - \frac{c_3}{2} + \frac{c_1}{24} = \frac{1}{120}$ leads to $c_5 - \frac{(1/3)}{2} + \frac{1}{24} = \frac{1}{120}$, which becomes $c_5 - \frac{1}{6} + \frac{1}{24} = \frac{1}{120}$ or $c_5 = \frac{1}{120} + \frac{1}{6} - \frac{1}{24}$, which gives $c_5 = \frac{1+20-5}{120} = \frac{16}{120} = \frac{2}{15}$. Plug in $c_0 = 0, c_1 = 1, c_2 = 0, c_3 = \frac{1}{3}, c_4 = 0$, and $c_5 = \frac{2}{15}$ to find the solution.

$$c_0 + c_1 x + c_2 x^2 + c_3 x^3 + c_4 x^4 + c_5 x^5 + \cdots = x + \frac{x^3}{3} + \frac{2x^5}{15} + \cdots$$

Check the answer: Plug in $x = 0.1$.

$$\frac{x - \frac{x^3}{6} + \frac{x^5}{120} + \cdots}{1 - \frac{x^2}{2} + \frac{x^4}{24} + \cdots} = \frac{0.1 - \frac{0.1^3}{6} + \frac{0.1^5}{120} + \cdots}{1 - \frac{0.1^2}{2} + \frac{0.1^4}{24} + \cdots}$$

$$\approx \frac{0.1 - 0.000166667 + 0.000000083 + \cdots}{1 - 0.005 + 0.000004167 + \cdots}$$

$$= \frac{0.099833416 + \cdots}{0.995004167 + \cdots} \approx 0.10033467 + \cdots$$

agrees with

$$x + \frac{x^3}{3} + \frac{2x^5}{15} = 0.1 + \frac{0.1^3}{3} + \frac{2(0.1)^5}{15}$$

$$\approx 0.1 + 0.000333333 + 0.000001333 \approx 0.10033467 + \cdots$$

Chapter 13 Answers

(1) Note: The odd derivatives are zero at $\frac{\pi}{2}$.

$$f(x) = \sin x \quad , \quad f'(x) = \cos x \quad , \quad f''(x) = -\sin x$$

$$f'''(x) = -\cos x \quad , \quad f^{(4)}(x) = \sin x \quad , \quad f^{(5)}(x) = \cos x$$

$$f\left(\frac{\pi}{2}\right) = \sin\frac{\pi}{2} = 1 \quad , \quad f'\left(\frac{\pi}{2}\right) = \cos\frac{\pi}{2} = 0 \quad , \quad f''\left(\frac{\pi}{2}\right) = -\sin\frac{\pi}{2} = -1$$

$$f'''\left(\frac{\pi}{2}\right) = -\cos\frac{\pi}{2} = 0 \quad , \quad f^{(4)}\left(\frac{\pi}{2}\right) = \sin\frac{\pi}{2} = 1 \quad , \quad f^{(5)}\left(\frac{\pi}{2}\right) = \cos\frac{\pi}{2} = 0$$

$$f(x) = f\left(\frac{\pi}{2}\right) + f'\left(\frac{\pi}{2}\right)\left(x - \frac{\pi}{2}\right) + \frac{f''\left(\frac{\pi}{2}\right)}{2}\left(x - \frac{\pi}{2}\right)^2 + \frac{f'''\left(\frac{\pi}{2}\right)}{6}\left(x - \frac{\pi}{2}\right)^3$$

$$+ \frac{f^{(4)}\left(\frac{\pi}{2}\right)}{24}\left(x - \frac{\pi}{2}\right)^4 + \frac{f^{(5)}\left(\frac{\pi}{2}\right)}{120}\left(x - \frac{\pi}{2}\right)^5 + \cdots$$

$$= 1 + 0 - \frac{1}{2}\left(x - \frac{\pi}{2}\right)^2 + 0 + \frac{1}{24}\left(x - \frac{\pi}{2}\right)^4 + 0 + \cdots$$

$$= 1 - \frac{1}{2}\left(x - \frac{\pi}{2}\right)^2 + \frac{1}{24}\left(x - \frac{\pi}{2}\right)^4 + \cdots$$

Careful: This differs from the common $\sin x = x - \frac{x^3}{3!} + \frac{x^5}{5!} + \cdots$ because here $a = \frac{\pi}{2}$ instead of the common $a = 0$. We'll explore the case $a = 0$ in Chapter 14.

Radius and interval of convergence: The nth term (starting with $n = 0$) is $\frac{(-1)^n}{(2n)!}\left(x - \frac{\pi}{2}\right)^{2n}$. Use the ratio test (Chapter 7).

$$L = \lim_{n\to\infty}\left|\frac{\left(x - \frac{\pi}{2}\right)^{2(n+1)}}{[2(n+1)]!} \div \frac{\left(x - \frac{\pi}{2}\right)^{2n}}{(2n)!}\right| = \lim_{n\to\infty}\left|\frac{\left(x - \frac{\pi}{2}\right)^{2n+2}}{(2n+2)!} \times \frac{(2n)!}{\left(x - \frac{\pi}{2}\right)^{2n}}\right|$$

$$L = \lim_{n\to\infty}\left|\frac{\left(x - \frac{\pi}{2}\right)^2}{(2n+2)(2n+1)}\right| = 0$$

Since this $L < 1$ for any (finite) value of x, the radius of convergence is $R = \infty$ and the interval of convergence is $-\infty < x < \infty$. (Note that the denominator becomes infinite as $n \to \infty$, while the numerator remains finite.)

Check the answer: Plug in $x = \frac{5\pi}{6}$ (corresponding to 150°). Note that $\frac{5\pi}{6} - \frac{\pi}{2} =$

$\frac{5\pi}{6} - \frac{3\pi}{6} = \frac{\pi}{3}$. We get $f\left(\frac{5\pi}{6}\right) = 1 - \frac{1}{2}\left(\frac{\pi}{3}\right)^2 + \frac{1}{24}\left(\frac{\pi}{3}\right)^4 + \cdots = 1 - 0.5483 + 0.0501 \approx$ 0.5018, which is close to the exact answer of $\sin\left(\frac{5\pi}{6}\right) = \frac{1}{2} = 0.5$.

(2) Note: All of the derivatives are identical since $\frac{d}{dx}e^x = e^x$.

$$f(x) = e^x \quad , \quad f'(x) = e^x \quad , \quad f''(x) = e^x$$

$$f'''(x) = e^x \quad , \quad f^{(4)}(x) = e^x \quad , \quad f^{(5)}(x) = e^x$$

$$f(1) = e^1 = e \quad , \quad f'(1) = e^1 = e \quad , \quad f''(1) = e^1 = e$$

$$f'''(1) = e^1 = e \quad , \quad f^{(4)}(1) = e^1 = e \quad , \quad f^{(5)}(1) = e^1 = e$$

$$f(x) = f(1) + f'(1)(x-1) + \frac{f''(1)}{2}(x-1)^2 + \frac{f'''(1)}{6}(x-1)^3 + \frac{f^{(4)}(1)}{24}(x-1)^4$$
$$+ \frac{f^{(5)}(1)}{120}(x-1)^5 + \cdots$$
$$= e + e(x-1) + \frac{e}{2}(x-1)^2 + \frac{e}{6}(x-1)^3 + \frac{e}{24}(x-1)^4 + \frac{e}{120}(x-1)^5 + \cdots$$

The nth term (starting with $n = 0$) is $\frac{e}{n!}(x-1)^n$. Use the ratio test (Chapter 7).

$$L = \lim_{n\to\infty}\left|\frac{e(x-1)^{n+1}}{(n+1)!} \div \frac{e(x-1)^n}{n!}\right| = \lim_{n\to\infty}\left|\frac{e(x-1)^{n+1}}{(n+1)n!} \times \frac{n!}{e(x-1)^n}\right| = \lim_{n\to\infty}\left|\frac{x-1}{n+1}\right| = 0$$

Since this $L < 1$ for any (finite) value of x, the radius of convergence is $R = \infty$ and the interval of convergence is $-\infty < x < \infty$.

Check the answer: Plug in $x = 1.1$. Note that $1.1 - 1 = 0.1$. Recall that $e \approx 2.71828$. We get $f(1) \approx 2.71828 + 2.71828(0.1) + \frac{2.71828}{2}(0.1)^2 + \frac{2.71828}{6}(0.1)^3 + \frac{2.71828}{24}(0.1)^4 + \frac{2.71828}{120}(0.1)^5 + \cdots \approx 2.71828 + 0.27183 + 0.01359 + 0.000453 + 0.00001 + 0.0000002 \approx 3.00416$, which is close to $e^{1.1} \approx 3.00417$.

(3) Apply the rules $x^{-p} = \frac{1}{x^p}$ and $\frac{d}{dx}x^{-p} = -px^{-p-1} = -\frac{p}{x^{p+1}}$.

$$f(x) = \ln x \quad , \quad f'(x) = x^{-1} = \frac{1}{x} \quad , \quad f''(x) = -x^{-2} = -\frac{1}{x^2}$$

$$f'''(x) = 2x^{-3} = \frac{2}{x^3} \quad , \quad f^{(4)}(x) = -6x^{-4} = -\frac{6}{x^4} \quad , \quad f^{(5)}(x) = 24x^{-5} = \frac{24}{x^5}$$

$$f(1) = \ln 1 = 0 \quad , \quad f'(1) = \frac{1}{1} = 1 \quad , \quad f''(1) = -\frac{1}{1^2} = -1$$

$$f'''(1) = \frac{2}{1^3} = 2 \quad , \quad f^{(4)}(1) = -\frac{6}{1^4} = -6 \quad , \quad f^{(5)}(1) = \frac{24}{1^5} = 24$$

$$f(x) = f(1) + f'(1)(x-1) + \frac{f''(1)}{2}(x-1)^2 + \frac{f'''(1)}{6}(x-1)^3 + \frac{f^{(4)}(1)}{24}(x-1)^4 + \frac{f^{(5)}(1)}{120}(x-1)^5 + \cdots$$

$$= 0 + 1(x-1) - \frac{1}{2}(x-1)^2 + \frac{2}{6}(x-1)^3 - \frac{6}{24}(x-1)^4 + \frac{24}{120}(x-1)^5 + \cdots$$

$$= (x-1) - \frac{(x-1)^2}{2} + \frac{(x-1)^3}{3} - \frac{(x-1)^4}{4} + \frac{(x-1)^5}{5} + \cdots$$

The nth term (starting with $n = 1$) is $\frac{(-1)^{n-1}}{n}(x-1)^n$. Use the ratio test (Chapter 7).

$$L = \lim_{n\to\infty}\left|\frac{(x-1)^{n+1}}{n+1} \div \frac{(x-1)^n}{n}\right| = \lim_{n\to\infty}\left|\frac{(x-1)^{n+1}}{n+1} \times \frac{n}{(x-1)^n}\right|$$

$$L = \lim_{n\to\infty}\left|(x-1)\frac{n}{n+1}\right| = \lim_{n\to\infty}\left|(x-1)\frac{1}{1+\frac{1}{n}}\right| = |x-1|(1) = |x-1|$$

In order for $L < 1$, we need $|x-1| < 1$, which means $-1 < x - 1 < 1$, which equates to $0 < x < 2$. The radius of convergence is one-half of this interval: $R = 1$. As we learned in Chapter 12, to establish the complete interval of convergence, we also need to check the endpoints. When $x = 0$, the series is $\sum \frac{(-1)^{n-1}}{n}(-1)^n = -\sum \frac{1}{n}$ because $(-1)^{n-1}$ and $(-1)^n$ have opposite signs for any given value of n; $-\sum \frac{1}{n}$ diverges because the harmonic series diverges (Chapter 6). When $x = 2$, the series is $\sum \frac{(-1)^{n-1}}{n}$ because $(2-1)^n = (1)^n = 1$; $\sum \frac{(-1)^{n-1}}{n}$ converges because the alternating harmonic series converges conditionally (Chapter 11). The **complete interval of convergence** is $0 < x \leq 2$ (with $\leq$ instead of $<$ before the 2).

Note: Compare this to the previous solutions. The previous solutions had factorials, where $\frac{n!}{(n+1)!} = \frac{1}{n+1}$, whereas this solution didn't have factorials, so it has $\frac{n}{n+1}$ instead. With the factorials, the first two problems had infinite radii of convergence, but in this problem the radius of convergence is finite.

Check the answer: Plug in $x = 1.1$. Note that $1.1 - 1 = 0.1$. We get $f(1.1) = 0.1 - \frac{(0.1)^2}{2} + \frac{(0.1)^3}{3} - \frac{(0.1)^4}{4} + \frac{(0.1)^5}{5} + \cdots = 0.1 - 0.005 + 0.000333 - 0.000025 + 0.000002 = 0.9531$, which matches $\ln(1.1) \approx 0.09531$.

(4) Recall that $\sqrt{x} = x^{1/2}$. This lets us use the rule $\frac{d}{dx}x^p = px^{p-1}$.

$$f(x) = x^{1/2} = \sqrt{x} \quad , \quad f'(x) = \frac{1}{2}x^{-1/2} = \frac{1}{2\sqrt{x}}$$

$$f''(x) = -\frac{1}{4}x^{-3/2} = -\frac{1}{4x\sqrt{x}} \quad , \quad f'''(x) = \frac{3}{8}x^{-5/2} = \frac{3}{8x^2\sqrt{x}}$$

$$f^{(4)}(x) = -\frac{15}{16}x^{-7/2} = -\frac{15}{16x^3\sqrt{x}} \quad , \quad f^{(5)}(x) = \frac{105}{32}x^{-9/2} = \frac{105}{32x^4\sqrt{x}}$$

$$f(9) = \sqrt{9} = 3 \quad , \quad f'(9) = \frac{1}{2\sqrt{9}} = \frac{1}{6}$$

$$f''(9) = -\frac{1}{4(9)\sqrt{9}} = -\frac{1}{108} \quad , \quad f'''(9) = \frac{3}{8(9)^2\sqrt{9}} = \frac{1}{648}$$

$$f^{(4)}(9) = -\frac{15}{16(9)^3\sqrt{9}} = -\frac{5}{11{,}664} \quad , \quad f^{(5)}(9) = \frac{105}{32x^4\sqrt{x}} = \frac{35}{209{,}952}$$

$$f(x) = f(9) + f'(9)(x-9) + \frac{f''(9)}{2}(x-9)^2 + \frac{f'''(9)}{6}(x-9)^3 + \frac{f^{(4)}(9)}{24}(x-9)^4$$

$$+ \frac{f^{(5)}(9)}{120}(x-9)^5 + \cdots$$

$$= 3 + \frac{(x-9)}{6} - \frac{(x-9)^2}{216} + \frac{(x-9)^3}{3888} - \frac{5(x-9)^4}{279{,}936} + \frac{7(x-9)^5}{5{,}038{,}848} + \cdots$$

To find the radius of convergence, first write a formula for the nth term. We'll do this in steps:

- Starting with the second derivative, the derivatives have the form $\frac{(-1)^{n+1}(2n-3)!!}{2^n x^{n-1}\sqrt{x}}$, where the two exclamation marks (!!) represent a **double factorial**, which means to multiply every other number. For example, when $n = 5$, this gives $[2(5) - 3]!! = (10 - 3)!! = 7!! = 7(5)(3)(1) = 105$. We'll use this double factorial notation as a way to multiply consecutive odd numbers.
- When $x = 9$, $x^{n-1}\sqrt{x} = 9^{n-1}\sqrt{9} = (3)9^{n-1} = (3)9^n 9^{-1} = \frac{(3)9^n}{9} = \frac{9^n}{3}$. When these derivatives are evaluated at $x = 9$, the above formula becomes $\frac{(-1)^{n+1}(2n-3)!!(3)}{2^n 9^n} = \frac{(-1)^{n+1}(2n-3)!!(3)}{18^n}$. For example, when $n = 4$, this formula gives $\frac{(-1)^{4+1}[2(4)-3]!!(3)}{18^4} = \frac{(-1)^5(5)(3)(1)(3)}{104{,}976} = -\frac{5}{11{,}664}$, which agrees with the fourth derivative.

- The Taylor series also includes $(x-9)^n$ as well as $\frac{1}{n!}$ in the denominator. The nth term (starting with $n = 2$) has the form $\frac{(-1)^{n+1}(2n-3)!!(3)}{18^n(n!)}(x-9)^n$. The series includes the terms for $n = 0$ and 1 in addition to the sum of terms of this form for $n \geq 2$:

$$3 + \frac{(x-9)}{6} + \sum_{n=2}^{\infty} \frac{(-1)^{n+1}(2n-3)!!\,(3)}{18^n(n!)}(x-9)^n$$

Use the ratio test (Chapter 7).

$$L = \lim_{n\to\infty} \left| \frac{[2(n+1)-3]!!\,(3)(x-9)^{n+1}}{18^{n+1}(n+1)!} \div \frac{(2n-3)!!\,(3)(x-9)^n}{18^n(n!)} \right|$$

$$L = \lim_{n\to\infty} \left| \frac{(2n-1)!!\,(3)(x-9)^{n+1}}{18^{n+1}(n+1)!} \times \frac{18^n(n!)}{(2n-3)!!\,(3)(x-9)^n} \right|$$

The (3)'s cancel out. Note that $\frac{(2n-1)!!}{(2n-3)!!} = \frac{2n-1}{1} = 2n-1$; as usual, when dividing consecutive terms of a factorial, only the largest term survives. Compare with $\frac{(n+1)!}{n!} = \frac{n+1}{1} = n+1$.

$$L = \lim_{n\to\infty} \left| \frac{(2n-1)(x-9)}{18(n+1)} \right| = \lim_{n\to\infty} \left| \frac{2nx - x - 18n + 9}{18n + 18} \right|$$

$$= \lim_{n\to\infty} \left| \frac{2x - \frac{x}{n} - 18 + \frac{9}{n}}{18 + \frac{18}{n}} \right| = \left| \frac{2x - 0 - 18 + 0}{18 + 0} \right| = \left| \frac{2x - 18}{18} \right| = \left| \frac{x}{9} - 1 \right|$$

In order for $L < 1$, we need $\left|\frac{x}{9} - 1\right| < 1$, which means $-1 < \frac{x}{9} - 1 < 1$, which equates to $0 < \frac{x}{9} < 2$, which simplifies to $0 < x < 18$. The radius of convergence is one-half of this interval: $R = 9$. (If you wish to establish the complete interval of convergence, you also need to check the endpoints. Whether or not the series converges at each endpoint is not as straightforward as most of the convergence tests in this book. The instructions didn't ask us to find the complete interval of convergence for this problem, so we'll leave it as an optional challenge exercise for the interested student.)

Check the answer: Plug in $x = 10$. Note that $10 - 9 = 1$. We get $f(10) = 3 + \frac{1}{6} -$

$\frac{1}{216}+\frac{1}{3888}-\frac{5}{279{,}936}+\frac{7}{5{,}038{,}848}+\cdots\approx 3+0.166667-0.004630+0.000257-0.000018+0.000001\approx 3.162277$, which is close to $\sqrt{10}\approx 3.162278$.

Chapter 14 Answers

(1) $f'(x) = f''(x) = f'''(x) = f^{(4)}(x) = f^{(5)}(x) = e^x$

$$f(0) = f'(0) = f''(0) = f'''(0) = f^{(4)}(0) = f^{(5)}(0) = e^0 = 1$$

$$f(x) = 1 + x + \frac{x^2}{2} + \frac{x^3}{6} + \frac{x^4}{24} + \frac{x^5}{120} + \cdots$$

The nth term (starting with $n = 0$) is $\frac{x^n}{n!}$. To find the radius of convergence, use the ratio test (Chapter 7).

$$L = \lim_{n\to\infty}\left|\frac{x^{n+1}}{(n+1)!} \div \frac{x^n}{n!}\right| = \lim_{n\to\infty}\left|\frac{x^{n+1}}{(n+1)!} \times \frac{n!}{x^n}\right| = \lim_{n\to\infty}\left|\frac{x}{n+1}\right| = 0$$

Since this $L < 1$ for any value of x (since the denominator, $n + 1$, grows infinite), the radius of convergence is $R = \infty$ and the interval of convergence is $-\infty < x < \infty$.

(2) $f'(x) = -\sin x$, $f''(x) = -\cos x$, $f'''(x) = \sin x$, $f^{(4)}(x) = \cos x$, $f^{(5)}(x) = -\sin x$

$$f(0) = 1 \,, f'(0) = 0 \,, f''(0) = -1 \,, f'''(0) = 0 \,, f^{(4)}(0) = 1 \,, f^{(5)}(0) = 0$$

$$f(x) = 1 + 0 - \frac{x^2}{2} + 0 + \frac{x^4}{24} + 0 + \cdots = 1 - \frac{x^2}{2} + \frac{x^4}{24} + \cdots$$

The nth term (starting with $n = 0$) is $\frac{(-1)^n x^{2n}}{(2n)!}$. To find the radius of convergence, use the ratio test (Chapter 7).

$$L = \lim_{n\to\infty}\left|\frac{x^{2(n+1)}}{[2(n+1)]!} \div \frac{x^{2n}}{(2n)!}\right| = \lim_{n\to\infty}\left|\frac{x^{2n+2}}{(2n+2)!} \times \frac{(2n)!}{x^{2n}}\right|$$

$$L = \lim_{n\to\infty}\left|\frac{x^2(2n)!}{(2n+2)(2n+1)(2n)!}\right| = \lim_{n\to\infty}\left|\frac{x^2}{(2n+2)(2n+1)}\right| = 0$$

Since this $L < 1$ for any value of x (since the denominator grows infinite), the radius of convergence is $R = \infty$ and the interval of convergence is $-\infty < x < \infty$.

(3) Note that this is the **hyperbolic** sine function, which is defined as $\sinh x = \frac{e^x - e^{-x}}{2}$. (It isn't the ordinary sine function from trig.) Observe that $\frac{d}{dx}\sinh x = \frac{d}{dx}\frac{e^x - e^{-x}}{2} = \frac{e^x + e^{-x}}{2} = \cosh x$ and $\frac{d}{dx}\cosh x = \frac{d}{dx}\frac{e^x + e^{-x}}{2} = \frac{e^x - e^{-x}}{2} = \sinh x$. It's important to realize that one of these derivatives is different from the ordinary trig derivatives; in particular, $\frac{d}{dx}\cosh x = \sinh x$ **doesn't have a minus sign**.

$f'(x) = \cosh x$, $f''(x) = \sinh x$, $f'''(x) = \cosh x$, $f^{(4)}(x) = \sinh x$, $f^{(5)}(x) = \cosh x$

$$f'(0) = f'''(0) = f^{(5)}(0) = \cosh 0 = \frac{e^0 + e^{-0}}{2} = \frac{1+1}{2} = \frac{2}{2} = 1$$

$$f(0) = f''(0) = f^{(4)}(0) = \sinh 0 = \frac{e^0 - e^{-0}}{2} = \frac{1-1}{2} = \frac{0}{2} = 0$$

$$f(x) = 0 + x + 0 + \frac{x^3}{6} + 0 + \frac{x^5}{120} + \cdots = x + \frac{x^3}{6} + \frac{x^5}{120} + \cdots$$

The Maclaurin series for $\sinh x$ is similar to $\sin x$, except that all terms have plus signs. The nth term (starting with $n = 1$) is $\frac{x^{2n+1}}{(2n+1)!}$. To find the radius of convergence, use the ratio test (Chapter 7).

$$L = \lim_{n\to\infty} \left| \frac{x^{2(n+1)+1}}{[2(n+1)+1]!} \div \frac{x^{2n+1}}{(2n+1)!} \right| = \lim_{n\to\infty} \left| \frac{x^{2n+3}}{(2n+3)!} \times \frac{(2n+1)!}{x^{2n+1}} \right|$$

$$L = \lim_{n\to\infty} \left| \frac{x^{3-1}(2n+1)!}{(2n+3)(2n+2)(2n+1)!} \right| = \lim_{n\to\infty} \left| \frac{x^2}{(2n+3)(2n+2)} \right| = 0$$

Since this $L < 1$ for any value of x (since the denominator grows infinite), the radius of convergence is $R = \infty$ and the interval of convergence is $-\infty < x < \infty$.

(4) Note that $\sqrt{u} = u^{1/2}$. Use the rule $\frac{d}{du} u^p = p u^{p-1}$.

$$f'(x) = \frac{1}{2}(x+4)^{-1/2} = \frac{1}{2(x+4)^{1/2}} = \frac{1}{2\sqrt{x+4}}$$

$$f''(x) = -\frac{1}{4}(x+4)^{-3/2} = -\frac{1}{4(x+4)^{3/2}} = -\frac{1}{4(x+4)\sqrt{x+4}}$$

$$f'''(x) = \frac{3}{8}(x+4)^{-5/2} = \frac{3}{8(x+4)^{5/2}} = \frac{3}{8(x+4)^2\sqrt{x+4}}$$

$$f^{(4)}(x) = -\frac{15}{16}(x+4)^{-7/2} = -\frac{15}{16(x+4)^{7/2}} = -\frac{15}{16(x+4)^3\sqrt{x+4}}$$

$$f^{(5)}(x) = \frac{105}{32}(x+4)^{-9/2} = \frac{105}{32(x+4)^{9/2}} = \frac{105}{32(x+4)^4\sqrt{x+4}}$$

Note that $u^{3/2} = u^1 u^{1/2} = u\sqrt{u}$, $u^{5/2} = u^2 u^{1/2} = u^2\sqrt{u}$, $u^{7/2} = u^3 u^{1/2} = u^3\sqrt{u}$, etc.

$$f(0) = \sqrt{0+4} = \sqrt{4} = 2$$

$$f'(0) = \frac{1}{2\sqrt{0+4}} = \frac{1}{2(2)} = \frac{1}{4}$$

$$f''(0) = -\frac{1}{4(0+4)\sqrt{0+4}} = -\frac{1}{4(4)(2)} = -\frac{1}{32}$$

$$f'''(0) = \frac{3}{8(0+4)^2\sqrt{0+4}} = \frac{3}{8(16)(2)} = \frac{3}{256}$$

$$f^{(4)}(0) = -\frac{15}{16(0+4)^3\sqrt{0+4}} = -\frac{15}{16(64)(2)} = -\frac{15}{2048}$$

$$f^{(5)}(0) = \frac{105}{32(0+4)^4\sqrt{0+4}} = \frac{105}{32(256)(2)} = \frac{105}{16{,}384}$$

$$f(x) = 2 + \frac{x}{4} - \frac{1}{32}\frac{x^2}{2} + \frac{3}{256}\frac{x^3}{6} - \frac{15}{2048}\frac{x^4}{24} + \frac{105}{16{,}384}\frac{x^5}{120} + \cdots$$

$$f(x) = 2 + \frac{x}{4} - \frac{x^2}{64} + \frac{x^3}{512} - \frac{5x^4}{16{,}384} + \frac{7x^5}{131{,}072} + \cdots$$

The nth term (starting with $n = 2$, that is skipping the first two terms) is $(-1)^{n+1}\frac{2}{8^n}\frac{x^n}{n!}(2n-3)!!$, where $(2n-3)!!$ is a **double factorial**, meaning to multiply every other odd number until reaching one. For example, $7!! = 7(5)(3)(1) = 105$. To find the radius of convergence, use the ratio test (Chapter 7).

$$L = \lim_{n\to\infty}\left|\frac{2}{8^{n+1}}\frac{x^{n+1}}{(n+1)!}[2(n+1)-3]!! \div \frac{2}{8^n}\frac{x^n}{n!}(2n-3)!!\right|$$

$$L = \lim_{n\to\infty}\left|\frac{2}{8^{n+1}}\frac{x^{n+1}}{(n+1)n!}(2n-1)!! \times \frac{8^n}{2}\frac{n!}{x^n}\frac{1}{(2n-3)!!}\right|$$

Note: Since $(2n-1)!!$ is a double factorial (see the definition of double factorial above), the first term after $2n-1$ is equal to $2n-3$; note that $2n-1$ and $2n-3$ are consecutive odd numbers.

$$L = \lim_{n\to\infty}\left|\frac{2}{2}\frac{8^n}{8^{n+1}}\frac{x^{n+1}}{x^n}\frac{n!}{(n+1)n!}\frac{(2n-1)(2n-3)!!}{(2n-3)!!}\right| = \lim_{n\to\infty}\left|\frac{x(2n-1)}{8(n+1)}\right|$$

$$L = \lim_{n\to\infty}\left|\frac{x\left(2-\frac{1}{n}\right)}{8\left(1+\frac{1}{n}\right)}\right| = \left|\frac{x(2-0)}{8(1+0)}\right| = \left|\frac{x}{4}\right|$$

Here, $L < 1$ if $\left|\frac{x}{4}\right| < 1$, which means that $|x| < 4$ or $-4 < x < 4$. The radius of convergence is one-half of this interval: $R = 4$. Normally, we would now check the endpoints, but this problem said we don't need to bother. (It's a bit tedious.)

Check the answer: Plug in $x = 0.1$ and compare with $f(0.1) = \sqrt{0.1+4} = \sqrt{4.1} \approx$ 2.02485.

$$f(0.1) = 2 + \frac{0.1}{4} - \frac{0.1^2}{64} + \frac{0.1^3}{512} - \frac{5(0.1^4)}{16{,}384} + \frac{7(0.1)^5}{131{,}072} + \cdots$$

$$\approx 2 + 0.025 - 0.000156 + 0.00000195 - 0.00000003 + 0.0000000001 + \cdots \approx 2.02485$$

(5) Multiply x^2 by the Maclaurin series for e^x.

$$f(x) = x^2 e^x = x^2\left(1 + x + \frac{x^2}{2} + \frac{x^3}{6} + \cdots\right) = x^2 + x^3 + \frac{x^4}{2} + \frac{x^5}{6} + \cdots$$

(6) Write the Maclaurin series for $\sin u$ and then use $u = x^2$

$$f(u) = 2\sin u = 2\left(u - \frac{u^3}{6} + \frac{u^5}{120} + \cdots\right) = 2u - \frac{u^3}{3} + \frac{u^5}{60} + \cdots$$

$$f(x) = 2\sin x^2 = 2x^2 - \frac{x^6}{3} + \frac{x^{10}}{60} + \cdots$$

(7) Note that $\frac{d}{dx}\ln(1+x) = \frac{1}{1+x}$, which means that $\ln(1+x) = \int \frac{dx}{1+x} + c$. This means that we can integrate the Maclaurin series for $\frac{1}{1+x}$ to obtain $\ln(1+x)$. (We'll set up a definite integral so as not to need to worry about a constant of integration.)

$$f(x) = \ln(1+x) = \int_{w=0}^{x} \frac{dw}{1+w} = \int_{w=0}^{x} (1 - w + w^2 - w^3 + w^4 + \cdots)\, dw$$

$$= x - \frac{x^2}{2} + \frac{x^3}{3} - \frac{x^4}{4} + \frac{x^5}{5} + \cdots$$

Note: Although this problem didn't ask for it, you could find that the interval of convergence is $-1 < x \leq 1$ by applying the ratio test to this series. (At the lower limit, $x = -1$, every term is negative, which gives a diverging harmonic series. At the upper limit, $x = 1$, the terms alternate sign, and the alternating harmonic series converges conditionally.)

(8) We'll write $f(u) = \ln(1+u)$ using the answer to Exercise 7 and then replace u with $-x$.

$$f(u) = \ln(1+u) = u - \frac{u^2}{2} + \frac{u^3}{3} - \frac{u^4}{4} + \frac{u^5}{5} + \cdots$$

$$f(x) = \ln(1-x) = -x - \frac{x^2}{2} - \frac{x^3}{3} - \frac{x^4}{4} - \frac{x^5}{5} + \cdots$$

Note: Although this problem didn't ask for it, you could find that the interval of convergence is $-1 \leq x < 1$ by applying the ratio test to this series. Contrast this with the interval of convergence for the previous exercise; this time the lower limit has the $\leq$ sign instead of the upper limit. (At the lower limit, $x = -1$, the terms alternate sign, and the alternating harmonic series converges conditionally. At the upper limit, $x = 1$, every term is negative, which gives a diverging harmonic series.)

(9) Contrast this with Exercise 6. Note that $\sin^2 u$ means to square the sine function, whereas $\sin u^2$ means to square the angle. In this problem, we need to square the function. We'll use the handy trig identity $\sin^2 u = \frac{1-\cos(2u)}{2}$. Begin by writing the Maclaurin series for ordinary cosine in terms of y.

$$\cos y = 1 - \frac{y^2}{2} + \frac{y^4}{24} - \frac{y^6}{720} + \cdots$$

Now replace y with $y = 2u$. (This still isn't x. Don't worry; we'll get there.)

$$\cos(2u) = 1 - \frac{(2u)^2}{2} + \frac{(2u)^4}{24} - \frac{(2u)^6}{720} + \cdots = 1 - \frac{4u^2}{2} + \frac{16u^4}{24} - \frac{64u^6}{720} + \cdots$$

$$= 1 - 2u^2 + \frac{2u^4}{3} - \frac{4u^6}{45} + \cdots$$

Now use the trig identity $\sin^2 u = \frac{1-\cos(2u)}{2}$.

$$\sin^2 u = \frac{1 - \cos(2u)}{2} = \frac{1}{2} - \frac{1}{2}\cos(2u) = \frac{1}{2} - \frac{1}{2}\left(1 - 2u^2 + \frac{2u^4}{3} - \frac{4u^6}{45} + \cdots\right)$$

Be careful distributing the minus sign.

$$\sin^2 u = \frac{1}{2} - \frac{1}{2} + u^2 - \frac{u^4}{3} + \frac{2u^6}{45} + \cdots = u^2 - \frac{u^4}{3} + \frac{2u^6}{45} + \cdots$$

Finally, replace u with $\frac{u}{2}$.

$$f(x) = \sin^2\left(\frac{x}{2}\right) = \frac{x^2}{4} - \frac{x^4}{3(16)} + \frac{2x^6}{45(64)} = \frac{x^2}{4} - \frac{x^4}{48} + \frac{x^6}{1440} + \cdots$$

Check the answer: Plug in $x = 0.1$ and compare with $f(0.1) = \sin^2\left(\frac{0.1}{2}\right) \approx \sin^2 0.05 \approx (0.049979169)^2 \approx 0.0024979$. Make sure that your calculator is in radians mode (not degrees), or convert 0.05 rad to 2.864788976° before using the sine function on your calculator.

$$f(0.1) = \frac{0.1^2}{4} - \frac{0.1^4}{48} + \frac{0.1^6}{1440} + \cdots \approx 0.0024979$$

(10) We will apply the series multiplication theorem (Chapter 12) following Example 3.

$$f(x) = e^x \cos x = \left(1 + x + \frac{x^2}{2} + \frac{x^3}{6} + \frac{x^4}{24} + \frac{x^5}{120} + \cdots\right)\left(1 - \frac{x^2}{2} + \frac{x^4}{24} + \cdots\right)$$

$$= 1 + x + \frac{x^2}{2} + \frac{x^3}{6} + \frac{x^4}{24} + \frac{x^5}{120} - \frac{x^2}{2} - \frac{x^3}{2} - \frac{x^4}{4} - \frac{x^5}{12} + \frac{x^4}{24} + \frac{x^5}{24} + \cdots$$

$$= 1 + x + \left(\frac{1}{2} - \frac{1}{2}\right)x^2 + \left(\frac{1}{6} - \frac{1}{2}\right)x^3 + \left(\frac{1}{24} - \frac{1}{4} + \frac{1}{24}\right)x^4 + \left(\frac{1}{120} - \frac{1}{12} + \frac{1}{24}\right)x^5 + \cdots$$

$$= 1 + x - \frac{x^3}{3} - \frac{x^4}{6} - \frac{x^5}{30} + \cdots$$

Check the answer: Plug in $x = 0.1$ and compare with $f(0.1) = e^{0.1} \cos 0.1 \approx 1.09965$. Make sure that your calculator is in radians mode (not degrees), or convert 0.1 rad to 5.729577951° before using the cosine function on your calculator.

$$f(0.1) = 1 + 0.1 - \frac{0.1^3}{3} - \frac{0.1^4}{6} - \frac{0.1^5}{30} + \cdots \approx 1.09965$$

(11) This looks like a division problem, but we'll actually be able to multiply. Begin by writing the **binomial series** (second page of the chapter) for $\frac{1}{\sqrt{1+u}}$.

$$\frac{1}{\sqrt{1+u}} = (1+u)^{-1/2} = 1 - \frac{u}{2} + \frac{(1)(3)u^2}{(2)(4)} + \cdots = 1 - \frac{u}{2} + \frac{3u^2}{8} + \cdots$$

Replace u with $u = x^2$.

$$\frac{1}{\sqrt{1+x^2}} = 1 - \frac{x^2}{2} + \frac{(1)(3)x^4}{(2)(4)} + \cdots = 1 - \frac{x^2}{2} + \frac{3x^4}{8} + \cdots$$

Multiply this series with the Maclaurin series for $\ln(1+x)$.

$$f(x) = \frac{\ln(1+x)}{\sqrt{1+x^2}} = \ln(1+x)\left(\frac{1}{\sqrt{1+x^2}}\right)$$

$$f(x) = \left(x - \frac{x^2}{2} + \frac{x^3}{3} - \frac{x^4}{4} + \frac{x^5}{5} + \cdots\right)\left(1 - \frac{x^2}{2} + \frac{3x^4}{8} + \cdots\right)$$

$$= x - \frac{x^2}{2} + \frac{x^3}{3} - \frac{x^4}{4} + \frac{x^5}{5} - \frac{x^3}{2} + \frac{x^4}{4} - \frac{x^5}{6} + \frac{3x^5}{8} + \cdots$$

$$= x - \frac{x^2}{2} + \left(\frac{1}{3} - \frac{1}{2}\right)x^3 + \left(-\frac{1}{4} + \frac{1}{4}\right)x^4 + \left(\frac{1}{5} - \frac{1}{6} + \frac{3}{8}\right)x^5 + \cdots$$

$$= x - \frac{x^2}{2} - \frac{x^3}{6} + \frac{49}{120}x^5 + \cdots$$

Since this binomial series is valid for $-1 < x \leq 1$, the combined series is valid for

$-1 < x < 1$ (to see if the endpoints are included, you need to check them manually). Check the answer: Plug in $x = 0.1$ and compare with $f(0.1) = \frac{\ln(1+0.1)}{\sqrt{1+0.1^2}} = \frac{\ln 1.1}{\sqrt{1.01}} \approx$ 0.094837.

$$f(0.1) = 0.1 - \frac{0.1^2}{2} - \frac{0.1^3}{6} + \frac{49}{120}0.1^5 + \cdots \approx 0.094837$$

Beware: These 'checks' aren't perfect. The higher-order terms like x^5 tend to be inherently much smaller, so that if your fractions are incorrect for the higher-order terms, the numerical answer might still be quite close. If your fractions or signs are incorrect for the first couple of terms, however, these checks will catch those mistakes.

(12) We'll divide series like we did in Example 4.

$$f(x) = \tan x = \frac{\sin x}{\cos x} = \frac{x - \frac{x^3}{6} + \frac{x^5}{120} + \cdots}{1 - \frac{x^2}{2} + \frac{x^4}{24} + \cdots}$$

If you remember how to do polynomial long division, you should put your skills to work. Since some students don't remember this, we'll present an alternative method. Since the numerator begins with x and the denominator begins with 1, the result of the division will begin with x.

$$\frac{x - \frac{x^3}{6} + \frac{x^5}{120} + \cdots}{1 - \frac{x^2}{2} + \frac{x^4}{24} + \cdots} = x + c_2x^2 + c_3x^3 + c_4x^4 + c_5x^5 + \cdots$$

Multiply both sides by $1 - \frac{x^2}{2} + \frac{x^4}{24} + \cdots$.

$$x - \frac{x^3}{6} + \frac{x^5}{120} + \cdots = \left(1 - \frac{x^2}{2} + \frac{x^4}{24} + \cdots\right)(x + c_2x^2 + c_3x^3 + c_4x^4 + c_5x^5 + \cdots)$$

$$x - \frac{x^3}{6} + \frac{x^5}{120} + \cdots = x - \frac{x^3}{2} + \frac{x^5}{24} + c_2x^2 - \frac{c_2x^4}{2} + c_3x^3 - \frac{c_3x^5}{2} + c_4x^4 + c_5x^5 + \cdots$$

$$x - \frac{x^3}{6} + \frac{x^5}{120} + \cdots = x + c_2x^2 + \left(c_3 - \frac{1}{2}\right)x^3 + \left(c_4 - \frac{c_2}{2}\right)x^4 + \left(\frac{1}{24} - \frac{c_3}{2} + c_5\right)x^5 + \cdots$$

Equate coefficients of the same power on both sides.

$$c_2 = 0 \quad , \quad c_3 - \frac{1}{2} = -\frac{1}{6} \quad , \quad c_4 - \frac{c_2}{2} = 0 \quad , \quad \frac{1}{24} - \frac{c_3}{2} + c_5 = \frac{1}{120}$$

All of the even coefficients are zero: $c_2 = c_4 = c_6 = \cdots = 0$. Solve for the odd coefficients in the other equations.

$$c_3 = \frac{1}{2} - \frac{1}{6} = \frac{3}{6} - \frac{1}{6} = \frac{2}{6} = \frac{1}{3}$$

$$c_5 = \frac{1}{120} - \frac{1}{24} + \frac{c_3}{2} = \frac{1}{120} - \frac{5}{120} + \frac{1}{3}\frac{1}{2} = -\frac{4}{120} + \frac{1}{6} = -\frac{1}{30} + \frac{5}{30} = \frac{4}{30} = \frac{2}{15}$$

$$f(x) = \tan x = x + 0x^2 + \frac{1}{3}x^3 + 0x^4 + \frac{2}{15}x^5 + \cdots$$

$$= x + \frac{x^3}{3} + \frac{2}{15}x^5 + \cdots$$

(13) We'll divide series like we did in the solution to the previous exercise.

$$f(x) = \frac{e^x}{\cos x} = \frac{1 + x + \frac{x^2}{2} + \frac{x^3}{6} + \frac{x^4}{24} + \frac{x^5}{120} + \cdots}{1 - \frac{x^2}{2} + \frac{x^4}{24} + \cdots} = 1 + c_1x + c_2x^2 + c_3x^3 + c_4x^4 + c_5x^5 + \cdots$$

$$1 + x + \frac{x^2}{2} + \frac{x^3}{6} + \frac{x^4}{24} + \frac{x^5}{120} + \cdots$$

$$= \left(1 - \frac{x^2}{2} + \frac{x^4}{24} + \cdots\right)(1 + c_1x + c_2x^2 + c_3x^3 + c_4x^4 + c_5x^5 + \cdots)$$

$$= 1 - \frac{x^2}{2} + \frac{x^4}{24} + c_1x - c_1\frac{x^3}{2} + c_1\frac{x^5}{24} + c_2x^2 - \frac{c_2x^4}{2} + c_3x^3 - \frac{c_3x^5}{2} + c_4x^4 + c_5x^5 + \cdots$$

$$= 1 + c_1x + \left(c_2 - \frac{1}{2}\right)x^2 + \left(c_3 - \frac{c_1}{2}\right)x^3 + \left(\frac{1}{24} - \frac{c_2}{2} + c_4\right)x^4 + \left(\frac{c_1}{24} - \frac{c_3}{2} + c_5\right)x^5 + \cdots$$

$$c_1 = 1 \quad , \quad c_2 - \frac{1}{2} = \frac{1}{2} \quad , \quad c_3 - \frac{c_1}{2} = \frac{1}{6}$$

$$\frac{1}{24} - \frac{c_2}{2} + c_4 = \frac{1}{24} \quad , \quad \frac{c_1}{24} - \frac{c_3}{2} + c_5 = \frac{1}{120}$$

$$c_1 = 1 \quad , \quad c_2 = \frac{1}{2} + \frac{1}{2} = 1 \quad , \quad c_3 = \frac{c_1}{2} + \frac{1}{6} = \frac{1}{2} + \frac{1}{6} = \frac{3}{6} + \frac{1}{6} = \frac{4}{6} = \frac{2}{3}$$

$$c_4 = \frac{1}{24} - \frac{1}{24} + \frac{c_2}{2} = \frac{c_2}{2} = \frac{1}{2}$$

$$c_5 = \frac{1}{120} - \frac{c_1}{24} + \frac{c_3}{2} = \frac{1}{120} - \frac{1}{24} + \frac{2}{3}\frac{1}{2} = \frac{1}{120} - \frac{5}{120} + \frac{1}{3} = -\frac{4}{120} + \frac{1}{3} = -\frac{1}{30} + \frac{10}{30} = \frac{9}{30} = \frac{3}{10}$$

$$f(x) = \frac{e^x}{\cos x} = 1 + x + x^2 + \frac{2}{3}x^3 + \frac{x^4}{2} + \frac{3x^5}{10} + \cdots$$

Check the answer: Plug in $x = 0.1$ and compare with $f(0.1) = \frac{e^{0.1}}{\cos 0.1} \approx 1.11072$.

Make sure that your calculator is in radians mode (not degrees), or convert 0.1 rad to 5.729577951° before using the cosine function on your calculator.

$$f(0.1) = 1 + 0.1 + 0.1^2 + \frac{2}{3}0.1^3 + \frac{0.1^4}{2} + \frac{3(0.1^5)}{10} + \cdots \approx 1.11072$$

(14) Begin with the binomial series (from the beginning of the chapter) for $\frac{1}{1+u}$.

$$\frac{1}{1+u} = (1+u)^{-1} = 1 - u + u^2 - u^3 + u^4 + \cdots$$

Make the substitution $u = 8x^3$.

$$f(x) = \frac{1}{8x^3+1} = 1 - 8x^3 + 64x^6 + \cdots$$

Check the answer: Plug in $x = 0.1$ and compare with $f(0.1) = \frac{1}{8(0.1)^3+1} \approx 0.992063$.

$$f(0.1) = 1 - 8(0.1)^3 + 64(0.1)^6 + \cdots \approx 0.992064$$

(15) We can make this look like one of the binomial series (from the beginning of the chapter) if we factor out the 3.

$$f(x) = \sqrt{12x^2+3} = \sqrt{3(4x^2+1)} = \sqrt{3}\sqrt{4x^2+1}$$

We'll write the binomial series for $\sqrt{u+1}$.

$$\sqrt{u+1} = \sqrt{1+u} = (1+u)^{1/2} = 1 + \frac{u}{2} - \frac{u^2}{(2)(4)} + \cdots = 1 + \frac{u}{2} - \frac{u^2}{8} + \cdots$$

Replace u with $u = 4x^2$.

$$\sqrt{4x^2+1} = 1 + \frac{4x^2}{2} - \frac{16x^4}{8} + \cdots = 1 + 2x^2 - 2x^4 + \cdots$$

Multiply this series by $\sqrt{3}$.

$$f(x) = \sqrt{12x^2+3} = \sqrt{3}\sqrt{4x^2+1} = \sqrt{3} + 2x^2\sqrt{3} - 2x^4\sqrt{3} + \cdots$$

Check the answer: Plug in $x = 0.1$ and compare with $f(0.1) = \sqrt{12(0.1)^2+3} \approx$ 1.766352.

$$f(0.1) = \sqrt{3} + 2(0.1)^2\sqrt{3} - 2(0.1)^4\sqrt{3} + \cdots \approx 1.766345$$

(16) If you're familiar with your derivatives of inverse trig functions, you should know that $\frac{d}{dx}\sin^{-1}x = \frac{1}{\sqrt{1-x^2}}$, which means that $\int \frac{dx}{\sqrt{1-x^2}} = \sin^{-1}x + c$. We'll obtain the Maclaurin series for inverse sine from the following definite integral.

$$f(x) = \sin^{-1}x = \int_{w=0}^{x} \frac{dw}{\sqrt{1-w^2}}$$

First we need the binomial series for $\frac{1}{\sqrt{1-w^2}}$. We'll begin with the series for $\frac{1}{\sqrt{1+u}}$.

$$\frac{1}{\sqrt{1+u}} = 1 - \frac{u}{2} + \frac{(1)(3)}{(2)(4)}u^2 + \cdots = 1 - \frac{u}{2} + \frac{3}{8}u^2 + \cdots$$

Replace u with $u = -w^2$. With this minus sign, every term will now be positive.

$$\frac{1}{\sqrt{1-w^2}} = 1 + \frac{w^2}{2} + \frac{3}{8}w^4 + \cdots$$

Now integrate term by term.

$$f(x) = \sin^{-1} x = \int_{w=0}^{x} \frac{dw}{\sqrt{1-w^2}} = \int_{w=0}^{x} \left(1 + \frac{w^2}{2} + \frac{3}{8}w^4 + \cdots\right) dw = x + \frac{x^3}{6} + \frac{3x^5}{40} + \cdots$$

(17) Although e^{-w^2} doesn't have an algebraic anti-derivative, you can integrate its Maclaurin series. Recall the Maclaurin series for e^{-w^2} from Example 5.

$$e^{-w^2} = 1 - w^2 + \frac{w^4}{2} + \cdots$$

$$f(x) = \int_{w=0}^{x} e^{-w^2}\, dw = \int_{w=0}^{x} \left(1 - w^2 + \frac{w^4}{2} + \cdots\right) dx = x - \frac{x^3}{3} + \frac{x^5}{10} + \cdots$$

(18) Begin with the binomial series expansion for $\frac{1}{\sqrt{1+x}}$.

$$\frac{1}{\sqrt{1+x}} = (1+x)^{-1/2} = 1 - \frac{x}{2} + \frac{(1)(3)}{(2)(4)}x^2 - \frac{(1)(3)(5)}{(2)(4)(6)}x^3 + \frac{(1)(3)(5)(7)}{(2)(4)(6)(8)}x^4$$

$$+ \frac{(1)(3)(5)(7)(9)}{(2)(4)(6)(8)(10)}x^5 + \frac{(1)(3)(5)(7)(9)(11)}{(2)(4)(6)(8)(10)(12)}x^6 + \cdots$$

$$= 1 - \frac{x}{2} + \frac{3}{8}x^2 - \frac{5}{16}x^3 + \frac{35}{128}x^4 - \frac{63}{256}x^5 + \frac{231}{1024}x^6 + \cdots$$

$$f(x) = \frac{d}{dx}\frac{1}{\sqrt{1+x}} = \frac{d}{dx}\left(1 - \frac{x}{2} + \frac{3}{8}x^2 - \frac{5}{16}x^3 + \frac{35}{128}x^4 - \frac{63}{256}x^5 + \frac{231}{1024}x^6 + \cdots\right)$$

$$f(x) = -\frac{1}{2} + \frac{3}{4}x - \frac{15}{16}x^2 + \frac{35}{32}x^3 - \frac{315}{256}x^4 + \frac{693}{512}x^5 + \cdots$$

Check the answer: Plug in $x = 0.1$.

$$f(0.1) = -\frac{1}{2} + \frac{3}{4}(0.1) - \frac{15}{16}(0.1)^2 + \frac{35}{32}(0.1)^3 - \frac{315}{256}(0.1)^4 + \frac{693}{512}(0.1)^5 + \cdots$$

$$f(0.1) \approx -0.433390762$$

The derivative is $\frac{d}{dx}\frac{1}{\sqrt{1+x}} = \frac{d}{dx}(1+x)^{-1/2} = -\frac{1}{2}(1+x)^{-3/2}$. At $x = 0.1$, the derivative equals $-\frac{1}{2}(1+.1)^{-3/2} = -\frac{1}{2}(1.1)^{-3/2} \approx -0.433392086$.

Chapter 15 Answers

(1) You get this series by setting $x = 1$ in the Maclaurin series for $\tan^{-1} x$ (which has an interval of convergence of $-1 \leq x \leq 1$). Therefore, this series converges to $\tan^{-1}(1) = \frac{\pi}{4}$ in radians. (If your calculator is in degrees mode, you'll need to convert $45°$ to $\frac{\pi}{4}$ radians.) Note: Although this series (times 4) could be used to approximate π to many decimal places, it converges relatively slowly since x isn't close to zero. There are many series that converge to a fraction of π (or a power of π, like π^2). If you want to approximate π numerically, you want to choose one that converges rapidly.

(2) This is a geometric series (Chapter 4). Before using the formula from Chapter 4, you need to put the series in standard form. We need $\frac{1}{2^{n-1}}$ instead of $\frac{1}{2^n}$. To do this, note that $\frac{3}{2^{n-1}} = \frac{3(2)}{2^{n-1}(2)} = \frac{6}{2^{n-1+1}} = \frac{6}{2^n}$. So all we need to do is replace $\frac{6}{2^n}$ with $\frac{3}{2^{n-1}}$.

$$\sum_{n=1}^{\infty} \frac{6}{2^n} = \sum_{n=1}^{\infty} \frac{3}{2^{n-1}} = \sum_{n=1}^{\infty} 3\left(\frac{1}{2}\right)^{n-1}$$

Now use the formula for an infinite geometric series with $a = 3$ and $r = \frac{1}{2}$.

$$\frac{a}{1-r} = \frac{3}{1-\frac{1}{2}} = \frac{3}{1/2} = 3 \div \frac{1}{2} = 3 \times \frac{2}{1} = 6$$

Check the answer: Add up the beginning terms and compare this sum with 6.

$$\frac{6}{2} + \frac{6}{4} + \frac{6}{8} + \frac{6}{16} + \frac{6}{32} + \frac{6}{64} + \frac{6}{128} + \frac{6}{256} + \frac{6}{512}$$
$$\approx 3 + 1.5 + 0.75 + 0.375 + 0.1875 + 0.0938 + 0.0469 + 0.0234 + 0.0117 = 5.99$$

(3) These denominators are factorials: $2! = 2$, $3! = 6$, $4! = 24$, $5! = 120$, $6! = 720$, and $7! = 5040$. Which series do we know which has these factorials? The Maclaurin series for e^{-x} has these factorials along with alternating minus signs. Start by writing the Maclaurin series for e^{-x} with $x = 1$.

$$e^{-1} = \frac{1}{e} = 1 - 1 + \frac{1}{2} - \frac{1}{6} + \frac{1}{24} - \frac{1}{120} + \frac{1}{720} - \frac{1}{5040} + \cdots$$

The two 1's cancel out and we get the desired series.

$$e^{-1} = \frac{1}{e} = \frac{1}{2} - \frac{1}{6} + \frac{1}{24} - \frac{1}{120} + \frac{1}{720} - \frac{1}{5040} + \cdots$$

This series converges to $e^{-1} = \frac{1}{e} \approx 0.367879441$.

(4) Try writing out the first handful of terms. Note that $3^0 = 1$ and $0! = 1$.

$$\sum_{k=0}^{\infty} \frac{3^{2k}}{(2k)!} = 1 + \frac{3^2}{2!} + \frac{3^4}{4!} + \frac{3^6}{6!} + \frac{3^8}{8!} + \frac{3^{10}}{10!} + \cdots$$

This is the Maclaurin series for **hyperbolic cosine** with $x = 3$. (Unlike the Maclaurin series for the ordinary cosine function from trig, which has alternating minus signs, the hyperbolic cosine function has all plus signs.) Recall the formula for hyperbolic cosine from Chapter 14. Plug in $x = 3$.

$$\cosh 3 = \frac{e^3 + e^{-3}}{2} \approx \frac{20.08553692 + 0.049787068}{2} \approx 10.06766199$$

Check: If you add the 6 terms shown above, you should get approximately 10.0665.

(5) Each denominator is the product of two consecutive integers.

$$\frac{1}{2} + \frac{1}{6} + \frac{1}{12} + \frac{1}{20} + \frac{1}{30} + \frac{1}{42} + \frac{1}{56} + \frac{1}{72} + \frac{1}{90} + \cdots$$

$$= \frac{1}{1(2)} + \frac{1}{2(3)} + \frac{1}{3(4)} + \frac{1}{4(5)} + \frac{1}{5(6)} + \frac{1}{6(7)} + \frac{1}{7(8)} + \frac{1}{8(9)} + \frac{1}{9(10)} + \cdots$$

In summation notation, this series is

$$\sum_{n=1}^{\infty} \frac{1}{n(n+1)}$$

Use the method of **partial fractions** (Chapter 4). First write

$$\frac{1}{n(n+1)} = \frac{x}{n} + \frac{y}{n+1}$$

Multiply by $n(n+1)$ on both sides.

$$1 = x(n+1) + ny = xn + x + ny = n(x+y) + x$$

Both sides of this equation will be equal for all values of n if $x = 1$ and $y = -x = -1$. This means that we may rewrite the sum as

$$\sum_{n=1}^{\infty} \frac{1}{n(n+1)} = \sum_{n=1}^{\infty} \left(\frac{1}{n} - \frac{1}{n+1}\right) = \frac{1}{1} - \frac{1}{2} + \frac{1}{2} - \frac{1}{3} + \frac{1}{3} - \frac{1}{4} + \frac{1}{4} - \frac{1}{5} + \frac{1}{5} - \frac{1}{6} + \cdots$$

Every term cancels out except for the first and the last, and the last term approaches zero as n grows infinite. Therefore, the sum equals 1.

Check the answer: Add some terms from the given series.

$$\frac{1}{2}+\frac{1}{6}+\frac{1}{12}+\frac{1}{20}+\frac{1}{30}+\frac{1}{42}+\frac{1}{56}+\frac{1}{72}+\frac{1}{90}+$$
$$= 0.5 + 0.1667 + 0.0833 + 0.05 + 0.0333 + 0.0238 + 0.0179 + 0.0139 + 0.0111 = 0.9$$

Since this series converges somewhat slowly, you will need to add several more terms before it more noticeably converges to 1.

(6) Write out the first several terms of the series.

$$\sum_{i=1}^{\infty}\frac{i}{2^{i-1}}=\frac{1}{1}+\frac{2}{2}+\frac{3}{4}+\frac{4}{8}+\frac{5}{16}+\frac{6}{32}+\frac{7}{64}+\frac{8}{128}+\frac{9}{256}+\frac{10}{512}+\cdots$$

There are two things going on in this series. First, the numerators are 1, 2, 3, 4, 5, ... Second, the denominators are powers of 2. Can you think of a Maclaurin series from Chapter 14 that can get these two features? Since Maclaurin series involve powers of x and the given series has powers of 2 in the denominator, this suggests that this series was made using $x = \frac{1}{2}$. Consider the series below.

$$1 + 2x + 3x^2 + 4x^3 + 5x^4 + 6x^5 + 7x^6 + 8x^7 + 9x^8 + 10x^9 + \cdots$$

If you plug $x = \frac{1}{2}$ into this series, you get the given series. Now which Maclaurin series from Chapter 14 resembles this series? It looks like the series for $\frac{1}{(1+x)^2}$, except that it's missing the minus signs. We can insert those minus signs in the right place by replacing x with $-x$. First, we'll write the **binomial** series $\frac{1}{(1+u)^2}$.

$$\frac{1}{(1+u)^2} = (1+u)^{-2} = 1 - 2u + 3u^2 - 4u^3 + 5u^4 - 6u^5 + 7u^6 - 8u^7 + 9u^8 + \cdots$$

Now make the substitution $u = -x$.

$$\frac{1}{(1-x)^2} = (1-x)^{-2} = 1 + 2x + 3x^2 + 4x^3 + 5x^4 + 6x^5 + 7x^6 + 8x^7 + 9x^8 + \cdots$$

This shows that the series adds up to $\frac{1}{\left(1-\frac{1}{2}\right)^2} = \frac{1}{1/4} = 4$.

Check the answer: Add some terms from the given series.

$$\frac{1}{1}+\frac{2}{2}+\frac{3}{4}+\frac{4}{8}+\frac{5}{16}+\frac{6}{32}+\frac{7}{64}+\frac{8}{128}+\frac{9}{256}+\frac{10}{512}+\frac{11}{1024}+\frac{12}{2048}+\frac{13}{4096}+\frac{14}{8192}$$
$$\approx 1 + 1 + 0.75 + 0.5 + 0.3125 + 0.1875 + 0.1094 + 0.0625 + 0.0352 + 0.0195$$
$$+ 0.0107 + 0.0059 + 0.0032 + 0.0017 \approx 3.9981$$

(7) There are two parts to this series. One part goes 2, 6, 12, 20, 30, 42, ..., while the other part has powers of three: $3^2 = 9, 3^3 = 27, 3^4 = 81, 3^5 = 243, 3^6 = 729$, and $3^7 = 2187$. If this series came from a Maclaurin series, the powers of x could have made the powers of three. Try replacing the powers of 3 with powers of x. More precisely, make the substitution $\frac{1}{3} = x$ to get the series below.

$$\frac{x^2}{2} - \frac{x^3}{6} + \frac{x^4}{12} - \frac{x^5}{20} + \frac{x^6}{30} - \frac{x^7}{42} + \cdots$$

Does this look like any Maclaurin series from Chapter 14? Not quite. The best match seems to be a logarithm. See below. Can you see any relationship between the series below and the series above? There is a relationship. Hint: It involves calculus.

$$\ln(1+x) = x - \frac{x^2}{2} + \frac{x^3}{3} - \frac{x^4}{4} + \frac{x^5}{5} - \frac{x^6}{6} + \cdots$$

Try taking a derivative of the top series with respect to x.

$$\frac{d}{dx}\left(\frac{x^2}{2} - \frac{x^3}{6} + \frac{x^4}{12} - \frac{x^5}{20} + \frac{x^6}{30} - \frac{x^7}{42} + \cdots\right) = x - \frac{x^2}{2} + \frac{x^3}{3} - \frac{x^4}{4} + \frac{x^5}{5} - \frac{x^6}{6} + \cdots$$

This shows that the given series is the integral of $\ln(1+x)$. More precisely, it is the definite integral below, since we used $x = \frac{1}{3}$ to get the series above.

$$\int_{x=0}^{1/3} \ln(1+x)\, dx$$

You might recall from calculus that $\int \ln u = u\ln(u) - u$ (or you could look this up in a table of integrals). Check: $\frac{d}{du}(u\ln u - u) = \ln u + \frac{u}{u} - 1 = \ln u + 1 - 1 = \ln u$.

Here, we need $u = x + 1$.

$$\int_{x=0}^{1/3} \ln(1+x)\, dx = [(1+x)\ln(1+x) - (1+x)]_{x=0}^{1/3}$$

$$= \frac{4}{3}\ln\frac{4}{3} - \frac{4}{3} - (1\ln 1 - 1) = \frac{4}{3}\ln\frac{4}{3} - \frac{4}{3} - (0-1) = \frac{4}{3}\ln\frac{4}{3} - \frac{4}{3} + 1$$

$$= \frac{4}{3}\ln\frac{4}{3} - \frac{1}{3} \approx 0.050242763$$

Check the answer: Add some terms from the given series.

$$\frac{1}{2(9)} - \frac{1}{6(27)} + \frac{1}{12(81)} - \frac{1}{20(243)} + \frac{1}{30(729)} + \cdots$$

$$\approx 0.05556 - 0.00617 + 0.00103 - 0.00021 + 0.00005 = 0.5026$$

(This approximation is actually bigger than the actual value, but that's okay. This is an alternating series. The next term is negative.)

(8) Write out the first terms of the series.

$$\sum_{j=1}^{\infty} \frac{(-1)^{j+1}}{j(2^j)} = \frac{1}{1(2)} - \frac{1}{2(2^2)} + \frac{1}{3(2^3)} - \frac{1}{4(2^4)} + \frac{1}{5(2^5)} - \frac{1}{6(2^6)} + \cdots$$

Make the substitution $x = \frac{1}{2}$.

$$x - \frac{x^2}{2} + \frac{x^3}{3} - \frac{x^4}{4} + \frac{x^5}{5} - \frac{x^6}{6} + \cdots$$

This is the Maclaurin series for $\ln(1+x)$ with $x = \frac{1}{2}$. This series converges to $\ln\left(1 + \frac{1}{2}\right) = \ln\left(\frac{3}{2}\right) = \ln 1.5 \approx 0.405465108$.

Check the answer: Add some terms from the given series.

$$\frac{1}{1(2)} - \frac{1}{2(2^2)} + \frac{1}{3(2^3)} - \frac{1}{4(2^4)} + \frac{1}{5(2^5)} - \frac{1}{6(2^6)} + \frac{1}{7(2^7)} - \frac{1}{8(2^8)}$$

$$\approx 0.5 - 0.125 + 0.0417 - 0.0156 + 0.0063 - 0.0026 + 0.0011 - 0.0005 = 0.4054$$

(9) This is the Maclaurin series for tangent with $x = 1$ radian. The series converges to $\tan 1 \approx \tan 57.296° \approx 1.5574$. (You'll need to work out several more terms using the strategy from Exercise 12 of Chapter 14 before the sum appears to converge to 1.5574.)

(10) Write out the first terms of the series.

$$\sum_{m=3}^{\infty} \frac{1}{(2m)!} = \frac{1}{6!} + \frac{1}{8!} + \frac{1}{10!} + \frac{1}{12!} + \frac{1}{14!} + \frac{1}{16!} + \cdots$$

This is similar to the Maclaurin series for **hyperbolic cosine**. (Unlike the Maclaurin series for the ordinary cosine function from trig, which has alternating minus signs, the hyperbolic cosine function has all plus signs.)

$$\cosh x = 1 + \frac{x^2}{2!} + \frac{x^4}{4!} + \frac{x^6}{6!} + \frac{x^8}{8!} + \frac{x^{10}}{10!} + \frac{x^{12}}{12!} + \frac{x^{14}}{14!} + \frac{x^{16}}{16!} + \cdots$$

The only differences are that we need to plug in $x = 1$ and the given sum skips the first three terms. Comparing the two series, we can solve for the given series as follows:

$$\sum_{m=3}^{\infty} \frac{1}{(2m)!} = \cosh 1 - 1 - \frac{1^2}{2!} - \frac{1^4}{4!}$$

Recall from Chapter 14 the formula for hyperbolic cosine.

$$\sum_{m=3}^{\infty} \frac{1}{(2m)!} = \frac{e^1 + e^{-1}}{2} - 1 - \frac{1}{2} - \frac{1}{24} \approx \frac{2.718282 + 0.367879}{2} - 1.5 - 0.041667$$

$$\approx 1.543081 - 1.5 - 0.041667 = 0.001414$$

Check the answer: Add some terms from the given series. This series converges rather rapidly compared to other series we've examined, so we don't need many terms to recognize the convergence.

$$\frac{1}{6!} + \frac{1}{8!} + \frac{1}{10!} \approx 0.00138889 + 0.0000248 + 0.00000028 = 0.001414$$

(11) This is the binomial series for $\sqrt{1+x}$ with $x = 1$ (recall Chapter 14). This series converges to $\sqrt{2} \approx 1.414213562$.

(12) Write out the first terms of the series. Note that $2^0 = 1$ and $0! = 1$.

$$\sum_{\ell=0}^{\infty} \frac{2^\ell}{\ell!} = \frac{2^0}{0!} + \frac{2^1}{1!} + \frac{2^2}{2!} + \frac{2^3}{3!} + \frac{2^4}{4!} + \frac{2^5}{5!} + \frac{2^6}{6!} + \cdots$$

Compare with the Maclaurin series for e^x.

$$e^x = 1 + x + \frac{x^2}{2!} + \frac{x^3}{3!} + \frac{x^4}{4!} + \frac{x^5}{5!} + \frac{x^6}{6!} + \cdots$$

If you let $x = 2$, you get the given series. The given series adds up to $e^2 \approx 7.3891$.

Check the answer: Add up several terms.

$$\frac{2^0}{0!} + \frac{2^1}{1!} + \frac{2^2}{2!} + \frac{2^3}{3!} + \frac{2^4}{4!} + \frac{2^5}{5!} + \frac{2^6}{6!} + \frac{2^7}{7!} + \frac{2^8}{8!} + \frac{2^9}{9!} + \frac{2^{10}}{10!} + \cdots$$

$$\approx 1 + 2 + 2 + 1.3333 + 0.6667 + 0.2667 + 0.0889 + 0.0254 + 0.0063 + 0.0014 + 0.0003 = 7.389$$

(13) This is a **geometric series**: $a_{n+1} = -a_n\sqrt{3}$.

$$9 - 3\sqrt{3} + 3 - \sqrt{3} + 1 - \frac{\sqrt{3}}{3} + \frac{1}{3} + \cdots = \sum_{n=1}^{\infty} \frac{9}{\left(-\sqrt{3}\right)^{n-1}} = \sum_{n=1}^{\infty} 9\left(-\frac{1}{\sqrt{3}}\right)^{n-1}$$

For example, when $n = 0$, we get $\frac{9}{1} = 9$, when $n = 1$, we get $\frac{9}{-\sqrt{3}} = -\frac{9}{\sqrt{3}}\frac{\sqrt{3}}{\sqrt{3}} = -\frac{9\sqrt{3}}{3} = -3\sqrt{3}$, when $n = 2$, we get $\frac{9}{\left(-\sqrt{3}\right)^2} = \frac{9}{3} = 3$, and when $n = 3$, we get $\frac{9}{\left(-\sqrt{3}\right)^3} = \frac{9}{-3\sqrt{3}} =$

$-\frac{3}{\sqrt{3}} = -\frac{3}{\sqrt{3}}\frac{\sqrt{3}}{\sqrt{3}} = -\frac{3\sqrt{3}}{3} = -\sqrt{3}$. Use the formula for an infinite geometric series from Chapter 4. Here, $a = 9$ and $r = -\frac{1}{\sqrt{3}}$. Note that r is **negative** because this series alternates.

$$\frac{a}{1-r} = \frac{9}{1-\left(-\frac{1}{\sqrt{3}}\right)} \approx \frac{9}{1+0.577350269} \approx 5.7058$$

Check the answer: Add some terms from the given series.

$$9 - 3\sqrt{3} + 3 - \sqrt{3} + 1 - \frac{\sqrt{3}}{3} + \frac{1}{3} - \frac{1}{3\sqrt{3}} + \frac{1}{9} - \frac{1}{9\sqrt{3}} + \frac{1}{27} - \frac{1}{27\sqrt{3}} + \frac{1}{81} - \frac{1}{81\sqrt{3}}$$
$$\approx 9 - 5.1962 + 3 - 1.7321 + 1 - 0.5774 + 0.3333 - 0.1925 + 0.1111 - 0.0642$$
$$+ 0.037 - 0.0214 + 0.0123 - 0.0071 = 5.7028$$

Be careful using your calculator. For example, for $\frac{1}{3\sqrt{3}}$, if you enter $1 \div 3\sqrt{3}$, most calculators will give an incorrect answer (thinking that you want to divide by 3 but multiply by $\sqrt{3}$). Use parentheses to avoid this problem: $1 \div (3\sqrt{3}) \approx 0.1925$.

(14) Use the method of **partial fractions** (Chapter 4).

$$\frac{2k+1}{k^2(k+1)^2} = \frac{x}{k^2} + \frac{y}{(k+1)^2}$$

Multiply both sides by $k^2(k+1)^2$.

$$2k + 1 = x(k+1)^2 + k^2y = x(k^2 + 2k + 1) + k^2y$$
$$2k + 1 = k^2x + 2kx + x + k^2y = k^2(x+y) + (2k+1)x$$

Both sides of this equation will be equal for any value of k if $x + y = 0$ (meaning that $y = -x$) and if $x = 1$. Thus, $x = 1$ and $y = -1$.

$$\sum_{k=4}^{\infty} \frac{2k+1}{k^2(k+1)^2} = \sum_{k=4}^{\infty} \left[\frac{1}{k^2} - \frac{1}{(k+1)^2}\right]$$

Write this series out longhand.

$$\sum_{k=4}^{\infty} \frac{2k+1}{k^2(k+1)^2} = \frac{1}{4^2} - \frac{1}{5^2} + \frac{1}{5^2} - \frac{1}{6^2} + \frac{1}{6^2} - \frac{1}{7^2} + \frac{1}{7^2} - \frac{1}{8^2} + \frac{1}{8^2} - \frac{1}{9^2} + \cdots$$

Every term cancels out except for the first and the last, and the last term approaches zero as n grows infinite. Therefore, the sum equals $\frac{1}{4^2} = \frac{1}{16} = 0.0625$.

Check the answer: Add some terms from the given series.

$$\sum_{k=4}^{\infty} \frac{2k+1}{k^2(k+1)^2} = \frac{9}{16(25)} + \frac{11}{25(36)} + \frac{13}{36(49)} + \frac{15}{49(64)} + \frac{17}{64(81)} + \frac{19}{81(100)} + \frac{21}{100(121)} + \cdots$$

$$\approx 0.0225 + 0.0122 + 0.0074 + 0.0048 + 0.0033 + 0.0023 + 0.0017 = 0.0542$$

You'll need to add several more terms to see that this series converges close to 0.0625 since it converges relatively slowly.

(15) Each denominator is the product of two integers.

$$\frac{1}{3} + \frac{1}{8} + \frac{1}{15} + \frac{1}{24} + \frac{1}{35} + \frac{1}{48} + \frac{1}{63} + \frac{1}{80} + \cdots$$

$$= \frac{1}{1(3)} + \frac{1}{2(4)} + \frac{1}{3(5)} + \frac{1}{4(6)} + \frac{1}{5(7)} + \frac{1}{6(8)} + \frac{1}{7(9)} + \frac{1}{8(10)} + \cdots$$

Using summation notation, this series is

$$\sum_{n=1}^{\infty} \frac{1}{n(n+2)}$$

Use the method of **partial fractions** (Chapter 4).

$$\frac{1}{n(n+2)} = \frac{x}{n} + \frac{y}{n+2}$$

Multiply both sides by $n(n+2)$.

$$1 = x(n+2) + ny = nx + 2x + ny = n(x+y) + 2x$$

Both sides of this equation will be equal for any value of n if $x = \frac{1}{2}$ and $y = -x = -\frac{1}{2}$.

We'll factor the $\frac{1}{2}$ out in front of the sum; this $\frac{1}{2}$ multiplies the entire series.

$$\sum_{n=1}^{\infty} \frac{1}{n(n+2)} = \frac{1}{2}\sum_{n=1}^{\infty}\left(\frac{1}{n} - \frac{1}{n+2}\right)$$

Write this series out longhand.

$$\sum_{n=1}^{\infty} \frac{1}{n(n+2)} = \frac{1}{2}\left(\frac{1}{1} - \frac{1}{3} + \frac{1}{2} - \frac{1}{4} + \frac{1}{3} - \frac{1}{5} + \frac{1}{4} - \frac{1}{6} + \frac{1}{5} - \frac{1}{7} + \frac{1}{6} - \frac{1}{8} + \frac{1}{7} - \frac{1}{9} + \cdots\right)$$

Every term in parentheses cancels out except for $\frac{1}{1} + \frac{1}{2}$ and two terms near the end, and the terms near the end approach zero as n grows infinite. Therefore, the sum equals $\frac{1}{2}\left(1 + \frac{1}{2}\right) = \frac{1.5}{2} = 0.75$.

Check the answer: Add some terms from the given series.

$$\frac{1}{3}+\frac{1}{8}+\frac{1}{15}+\frac{1}{24}+\frac{1}{35}+\frac{1}{48}+\frac{1}{63}+\frac{1}{80}+\frac{1}{99}+\frac{1}{120}+\frac{1}{143}+\frac{1}{168}+\frac{1}{195}+\frac{1}{224}+\frac{1}{255}$$
$$\approx 0.3333+0.125+0.0667+0.0417+0.0286+0.0208+0.0159+0.0125+0.0101$$
$$+0.0083+0.007+0.006+0.0051+0.0045+0.0039=0.6894$$

You'll need to add several more terms to see that this series converges close to 0.75 since it converges relatively slowly.

(16) Rewrite the series in terms of powers of 2 and 3 as follows.

$$\frac{1}{2^1}+\frac{1}{3^1}-\frac{1}{3}\left(\frac{1}{2^3}+\frac{1}{3^3}\right)+\frac{1}{5}\left(\frac{1}{2^5}+\frac{1}{3^5}\right)-\frac{1}{7}\left(\frac{1}{2^7}+\frac{1}{3^7}\right)+\frac{1}{9}\left(\frac{1}{2^9}+\frac{1}{3^9}\right)+\cdots$$

Observe that we can split this into the sum of two series.

$$\frac{1}{2^1}-\frac{1}{3}\frac{1}{2^3}+\frac{1}{5}\frac{1}{2^5}-\frac{1}{7}\frac{1}{2^7}+\frac{1}{9}\frac{1}{2^9}+\cdots+\frac{1}{3^1}-\frac{1}{3}\frac{1}{3^3}+\frac{1}{5}\frac{1}{3^5}-\frac{1}{7}\frac{1}{3^7}+\frac{1}{9}\frac{1}{3^9}+\cdots$$

The first series is the Maclaurin series for $\tan^{-1} x$ with $x=\frac{1}{2}$, while the second series is the Maclaurin series for $\tan^{-1} x$ with $x=\frac{1}{3}$. The combined series converges to $\tan^{-1}\frac{1}{2}+\tan^{-1}\frac{1}{3}=\frac{\pi}{4}$ radians (equivalent to 45°), which is approximately equal to 0.785398163 radians. This combined series provides a better approximation for calculating several digits of π on a computer than the series in Exercise 1 because $\frac{1}{2}$ and $\frac{1}{3}$ are closer to zero than 1 is; because $\frac{1}{2}$ and $\frac{1}{3}$ are closer to zero, those series converge faster.

Note: You can find other combinations of inverse tangents with smaller arguments that converge even faster, such as $\frac{\pi}{4}=4\tan^{-1}\frac{1}{5}-\tan^{-1}\frac{1}{239}$ (discovered by John Machin in the early 1700's).

(17) We'll present two different solutions. First, recall from the text prior to Example 1 that

$$\sum_{n=1}^{\infty}\frac{1}{n^2}=\frac{1}{1}+\frac{1}{2^2}+\frac{1}{3^2}+\frac{1}{4^2}+\frac{1}{5^2}+\frac{1}{6^2}+\frac{1}{7^2}+\frac{1}{8^2}+\frac{1}{9^2}+\frac{1}{10^2}+\cdots=\frac{\pi^2}{6}$$

Also, recall from Example 6 that

$$\sum_{n=0}^{\infty}\frac{1}{(2n+1)^2}=\frac{1}{1^2}+\frac{1}{3^2}+\frac{1}{5^2}+\frac{1}{7^2}+\frac{1}{9^2}+\cdots=\frac{\pi^2}{8}$$

If we subtract these two series, all of the terms with odd denominators will cancel out and we'll just be left with terms with even denominators.

$$\frac{1}{2^2}+\frac{1}{4^2}+\frac{1}{6^2}+\frac{1}{8^2}+\frac{1}{10^2}+\cdots=\frac{\pi^2}{6}-\frac{\pi^2}{8}=\frac{4\pi^2}{28}-\frac{3\pi^2}{24}=\frac{\pi^2}{24}\approx 0.411233517$$

Alternate solution: Rewrite the series using summation notation. Note that $2n$ makes an even number.

$$\sum_{n=1}^{\infty}\frac{1}{(2n)^2}=\frac{1}{4}+\frac{1}{16}+\frac{1}{36}+\frac{1}{64}+\frac{1}{100}+\frac{1}{144}+\frac{1}{196}+\cdots$$

Since $\frac{1}{(2n)^2}=\frac{1}{2^2n^2}=\frac{1}{4n^2}$, this sum is one-quarter of the sum for $\sum\frac{1}{n^2}$.

$$\sum_{n=1}^{\infty}\frac{1}{(2n)^2}=\sum_{n=1}^{\infty}\frac{1}{2^2n^2}=\sum_{n=1}^{\infty}\frac{1}{4n^2}=\frac{1}{4}\sum_{n=1}^{\infty}\frac{1}{n^2}=\frac{1}{4}\frac{\pi^2}{6}=\frac{\pi^2}{24}\approx 0.411233517$$

(18) Write the series out longhand.

$$\sum_{n=1}^{\infty}\frac{(-1)^{n+1}}{n^2}=\frac{1}{1^2}-\frac{1}{2^2}+\frac{1}{3^2}-\frac{1}{4^2}+\frac{1}{5^2}-\frac{1}{6^2}+\frac{1}{7^2}-\frac{1}{8^2}+\frac{1}{9^2}-\frac{1}{10^2}+\cdots$$

Separate this into two separate series as follows.

$$\sum_{n=1}^{\infty}\frac{(-1)^{n+1}}{n^2}=\frac{1}{1^2}+\frac{1}{3^2}+\frac{1}{5^2}+\frac{1}{7^2}+\frac{1}{9^2}+\cdots-\frac{1}{2^2}-\frac{1}{4^2}-\frac{1}{6^2}-\frac{1}{8^2}-\frac{1}{10^2}+\cdots$$

As we saw in the solution to Exercise 17, the first series has a sum of $\frac{\pi^2}{8}$ and the second series has a sum of $-\frac{\pi^2}{24}$ (since every term is negative). The total sum is

$$\frac{\pi^2}{8}-\frac{\pi^2}{24}=\frac{3\pi^2}{24}-\frac{\pi^2}{24}=\frac{2\pi^2}{24}=\frac{\pi^2}{12}$$

(19) (A) Plug $x=\frac{1}{2}$ into the Maclaurin series for e^{-x}.

$$e^{-x}=1-x+\frac{x^2}{2}-\frac{x^3}{6}+\frac{x^4}{24}-\frac{x^5}{120}+\cdots$$

$$e^{-1/2}=1-\frac{1}{2}+\frac{1}{2}\left(\frac{1}{4}\right)-\frac{1}{6}\left(\frac{1}{8}\right)+\frac{1}{24}\left(\frac{1}{16}\right)-\frac{1}{120}\left(\frac{1}{32}\right)+\cdots$$

(B) Use your calculator to add the first six terms.

$$S_6\approx 1-0.5+0.125-0.02083+0.0026-0.00026=0.60651$$

(C) Let $n=0$ correspond to the term $f(a)$, as in Example 2. There are 6 terms, but the last term corresponds to $n = 5$ because we begin from zero. The remainder is based on the next term, $n + 1 = 5 + 1 = 6$, which involves the sixth derivative.

$$f^{(5+1)}(w) = \frac{d^{5+1}}{dw^{5+1}} e^{-w} = \frac{d^6}{dw^6} e^{-w} = e^{-w}$$

Note that $\frac{d}{dw} e^{-w} = -e^{w}$ and $\frac{d^2}{dw^2} e^{-w} = e^{-w}$; the odd derivatives are negative and the even derivatives are positive. For the remainder theorem, $a \leq w \leq x$. In this case, $a = 0$ (for a Maclaurin series) and we're evaluating the function at $x = \frac{1}{2}$, such that $0 \leq w \leq \frac{1}{2}$. Over this interval, the sixth derivative, $f^6(w) = e^{-w}$, has a maximum of 1 when $w = 0$. If we choose $q = r = 1$, for example, we can be sure that $|f^{(6)}(w)| \leq qr^6$ since $qr^6 = 1(1^6) = 1$ matches the maximum value of e^{-w}. Use the remainder estimation theorem, setting x equal to $\frac{1}{2}$ and $n = 5$ (where, again, $n = 5$ even though there are 6 terms because we start from zero).

$$|R_5(x)| \leq 1\frac{1^{5+1}|x - 0|^{5+1}}{(5 + 1)!} = 1\frac{|1/2|^6}{6!} = \frac{1}{64(720)} \approx 0.000022$$

Note that we would have obtained this same remainder if we had used the alternating series estimation theorem, since the first unused term is $\frac{1}{2^6}\frac{1}{6!} = \frac{1}{64}\frac{1}{720}$.

(D) Add the remainder from Part C to the finite sum of Part B: $0.60651 + 0.00002 = 0.60653$.

(E) According to a calculator, $e^{-1/2} = e^{-0.5} \approx 0.60653$. This agrees with our approximation to the 5th decimal place. (By keeping additional terms in the finite sum, you could get an approximation to a further decimal place.)

(20) (A) If you realize that $(k^2 + 1)^2 = k^4 + 2k^2 + 1$, you can rewrite this sum as

$$\sum_{k=1}^{\infty} \frac{k}{(k^2 + 1)^2}$$

Work out the first ten terms.

$$S_{10} = \frac{1}{2^2} + \frac{2}{5^2} + \frac{3}{10^2} + \frac{4}{17^2} + \frac{5}{26^2} + \frac{6}{37^2} + \frac{7}{50^2} + \frac{8}{65^2} + \frac{9}{82^2} + \frac{10}{101^2}$$

$$S_{10} \approx 0.25 + 0.08 + 0.03 + 0.01384 + 0.0074 + 0.00438 + 0.0028 + 0.00189 + 0.00134 + 0.00098 = 0.39263$$

(B) As noted in Part A, $\frac{k}{k^4+2k^2+1} = \frac{k}{(k^2+1)^2}$. Integrate the function $f(x) = \frac{x}{(x^2+1)^2}$. One integral will begin from 10 and the other will begin from $10 + 1 = 11$ because the finite sum went up to $k = 10$. Make the substitution $u = x^2 + 1$ such that $du = 2x\,dx$. Replace $(x^2 + 1)^2$ with u^2 and replace $x\,dx$ with $\frac{du}{2}$. When x is 10, note that $u = 10^2 + 1 = 101$, so the new integral begins at 101.

$$\int_{x=10}^{\infty} \frac{x}{(x^2+1)^2} dx = \int_{u=101}^{\infty} \frac{du}{2u^2} = \frac{1}{2} \int_{u=101}^{\infty} u^{-2}\,du = \frac{1}{2}\left[\frac{u^{-1}}{-1}\right]_{u=101}^{\infty} = -\frac{1}{2}\left[\frac{1}{u}\right]_{u=101}^{\infty}$$

$$= -0 - \left(-\frac{1}{2}\right)\left(\frac{1}{101}\right) = \frac{1}{202} \approx 0.00495$$

For the integral starting at $x = 11$, simply change u's initial value from 101 to $11^2 + 1 = 121 + 1 = 122$ in the last line.

$$\int_{x=11}^{\infty} \frac{x}{(x^2+1)^2} dx = -\frac{1}{2}\left[\frac{1}{u}\right]_{u=122}^{\infty} = -0 - \left(-\frac{1}{2}\right)\left(\frac{1}{122}\right) = \frac{1}{244} \approx 0.0041$$

Observe that $\frac{x}{(x^2+1)^2}$ decreases over the interval $10 \leq x < \infty$. The remainder of the series (that is, the terms starting from $n = 11$) lies between 0.00495 and 0.0041. The average of these values gives an estimate of the remainder of 0.00453.

(C) Add the sum of the first 10 terms from Part A to our estimate of the remainder from Part B to estimate the complete (infinite) sum: $0.39263 + 0.00453 \approx 0.397$. (Since the remainder lies between 0.0041 and 0.0049, the 4th decimal place has a lot of uncertainty, which is why we reported our answer to 3 decimal places.)

(21) (A) $a_1 = 1$, $a_2 = a_{1+1} = \frac{1^2}{2(1)^2+3} a_1 = \frac{1}{5}(1) = \frac{1}{5} = 0.2$

$$a_3 = a_{2+1} = \frac{2^2}{2(2)^2+3} a_2 = \frac{4}{11}\left(\frac{1}{5}\right) = \frac{4}{55} \approx 0.07273$$

$$a_4 = a_{3+1} = \frac{3^2}{2(3)^2+3} a_3 = \frac{9}{21}\left(\frac{4}{55}\right) = \frac{12}{385} \approx 0.03117$$

$$a_5 = a_{4+1} = \frac{4^2}{2(4)^2+3} a_4 = \frac{16}{35}\left(\frac{12}{385}\right) = \frac{192}{13{,}475} \approx 0.01425$$

$$a_6 = a_{5+1} = \frac{5^2}{2(5)^2+3} a_5 = \frac{25}{53}(0.01425) \approx 0.00672$$

$$a_7 = a_{6+1} = \frac{6^2}{2(6)^2+3} a_6 = \frac{36}{75}(0.00672) \approx 0.00323$$

$$a_8 = a_{7+1} = \frac{7^2}{2(7)^2+3} a_7 = \frac{49}{101}(0.00323) \approx 0.00157$$

$$S_8 \approx 1 + 0.2 + 0.07273 + 0.03117 + 0.01425 + 0.00672 + 0.00323 + 0.00157$$
$$= 1.32967$$

(B) The ratio of any two consecutive terms is $\frac{a_{n+1}}{a_n} = \frac{n^2}{2n^2+3}$. We want to find two constants r_1 and r_2 such that $r_1 \le \frac{a_{n+1}}{a_n} \le r_2 < 1$ for $n \ge 8$ (since in this problem, there are $N = 8$ terms in the finite series approximation), which means $r_1 \le \frac{n^2}{2n^2+3} \le r_2 < 1$.

- Observe that $\left\{\frac{n^2}{2n^2+3}\right\}$ is an increasing sequence: $\frac{1}{5}, \frac{4}{11}, \frac{9}{21}, \frac{16}{35}, \ldots$ For $n \ge 8$, we can be sure that $\frac{n^2}{2n^2+3}$ is no smaller than $\frac{8^2}{2(8)^2+3} = \frac{64}{131}$. We may thus choose $r_1 = \frac{64}{131}$.
- Observe that $\left\{\frac{n^2}{2n^2+3}\right\}$ approaches the limiting value of $\frac{1}{2}$ (recall Chapter 2) as n grows infinite. Thus, $\frac{n^2}{2n^2+3}$ is no greater than $\frac{1}{2}$. We may thus choose $r_2 = \frac{1}{2}$.

Every term of this series is positive.

$$\frac{r_1 a_n}{1 - r_1} \le R_n \le \frac{r_2 a_n}{1 - r_2}$$

Use $n = 8$, $a_8 \approx 0.00157$, $r_1 = \frac{64}{131} \approx 0.48855$, and $r_2 = \frac{1}{2} = 0.5$.

$$\frac{(0.48855)(0.00157)}{1 - 0.48855} \le R_8 \le \frac{(0.5)(0.00157)}{1 - 0.5}$$
$$0.0015 \le R_8 \le 0.00157$$

The remainder of the series (that is, the terms starting from $n = 9$) lies between 0.0015 and 0.00157. The average of these values gives an estimate of the remainder of 0.00154.

(C) Add the sum of the first eight terms from Part A to our estimate of the remainder from Part B to estimate the complete (infinite) sum: $1.3297 + 0.0015 = 1.3312$.

(22) (A) Use the summation formula.

$$S_6 = -\frac{1}{1\sqrt{1}} + \frac{1}{2\sqrt{2}} - \frac{1}{3\sqrt{3}} + \frac{1}{4\sqrt{4}} - \frac{1}{5\sqrt{5}} + \frac{1}{6\sqrt{6}}$$
$$\approx -1 + 0.35355 - 0.19245 + 0.125 - 0.08944 + 0.06804 = -0.7353$$

Calculator note: If you enter $1/6\sqrt{6}$ on a calculator, most calculators will give you an incorrect answer. You probably need to enter $1/(6\sqrt{6})$ instead.

Note: This series doesn't converge as rapidly as some of the other series that we've seen. If you really want a good approximation, many more terms would be desired.

(B) This alternating series meets all of the conditions of Leibniz's theorem (Chapter 11): the terms alternate sign, the absolute values of the terms are nonincreasing, and the sequence of absolute values converges to zero. According to the alternating series estimation theorem, the error is less than the absolute value of the first unused term: $|a_7| = \frac{1}{7\sqrt{7}} \approx 0.054$. The remainder, R_7, which is the infinite sum minus the finite approximation, has the same sign as the 7th term, which is negative: $(-1)^7 = -1$. Since the infinite sum, S, minus the finite approximation, -0.7353, is negative, this means that the infinite sum is more negative than the finite approximation: $S < -0.7353$. Since the error is 0.054, the infinite sum must lie in the range $-0.7893 < S < -0.7353$.

(23) (A) Start with this equation from the text (prior to Example 1).

$$\left(1-\frac{x}{\pi}\right)\left(1+\frac{x}{\pi}\right)\left(1-\frac{x}{2\pi}\right)\left(1+\frac{x}{2\pi}\right)\left(1-\frac{x}{3\pi}\right)\left(1+\frac{x}{3\pi}\right)\cdots = 1-\frac{x^2}{6}+\frac{x^4}{120}+\cdots$$

If we multiply the infinite product out on the left-hand side, the nonzero terms proportional to x^4 will involve two pairs of $\frac{x}{n\pi}$'s, like $\left(-\frac{x}{2\pi}\right)\left(\frac{x}{2\pi}\right)\left(-\frac{x}{3\pi}\right)\left(\frac{x}{3\pi}\right) = \frac{x^4}{2^2 3^2 \pi^4}$. Terms of the form $\left(\frac{x}{2\pi}\right)\left(-\frac{x}{3\pi}\right)\left(\frac{x}{3\pi}\right)\left(-\frac{x}{5\pi}\right)$, which don't have two matching pairs, will cancel out. For example, in this case, $\left(-\frac{x}{2\pi}\right)\left(-\frac{x}{3\pi}\right)\left(\frac{x}{3\pi}\right)\left(-\frac{x}{5\pi}\right) + \left(-\frac{x}{2\pi}\right)\left(-\frac{x}{3\pi}\right)\left(\frac{x}{3\pi}\right)\left(\frac{x}{5\pi}\right) + \left(\frac{x}{2\pi}\right)\left(-\frac{x}{3\pi}\right)\left(\frac{x}{3\pi}\right)\left(-\frac{x}{5\pi}\right) + \left(\frac{x}{2\pi}\right)\left(-\frac{x}{3\pi}\right)\left(\frac{x}{3\pi}\right)\left(\frac{x}{5\pi}\right) = 0$. The terms proportional to x^4 will add up to $\frac{x^4}{120}$ from the right-hand side.

$$\frac{x^4}{1^2 2^2 \pi^4}+\frac{x^4}{1^2 3^2 \pi^4}+\frac{x^4}{1^2 4^2 \pi^4}+\frac{x^4}{1^2 5^2 \pi^4}+\cdots+\frac{x^4}{2^2 3^2 \pi^4}+\frac{x^4}{2^2 4^2 \pi^4}+\frac{x^4}{2^2 5^2 \pi^4}+\cdots$$
$$+\frac{x^4}{3^2 4^2 \pi^4}+\frac{x^4}{3^2 5^2 \pi^4}+\cdots+\frac{x^4}{4^2 5^2 \pi^4}+\cdots=\frac{x^4}{120}$$

The left-hand side can be written with the following double sum. Here, $n > m$.

$$\sum_{m=1}^{\infty}\sum_{n=m+1}^{\infty}\frac{x^4}{m^2 n^2 \pi^4} = \frac{x^4}{120}$$

Divide by x^4 on both sides and multiply by π^4.

$$\sum_{m=1}^{\infty}\sum_{n=m+1}^{\infty}\frac{1}{m^2 n^2} = \frac{\pi^4}{120}$$

(B) Apply the series multiplication theorem (Chapter 12).

$$\left(\frac{1}{1}+\frac{1}{4}+\frac{1}{9}+\frac{1}{16}+\frac{1}{25}+\cdots\right)\left(\frac{1}{1}+\frac{1}{4}+\frac{1}{9}+\frac{1}{16}+\frac{1}{25}+\cdots\right)=\frac{\pi^4}{36}$$

This is similar to the previous double sum, except that this includes cases where $n = m$ and where $n < m$.

$$\sum_{m=1}^{\infty}\sum_{n=1}^{\infty}\frac{1}{m^2n^2}=\frac{\pi^4}{36}$$

(C) We can split the above double sum into three double sums as follows. The first sum is when $n < m$, the second sum is when $n = m$, and the last sum is when $n > m$.

$$\sum_{n=1}^{\infty}\sum_{m=n+1}^{\infty}\frac{1}{m^2n^2}+\sum_{n=1}^{\infty}\frac{1}{n^4}+\sum_{m=1}^{\infty}\sum_{n=m+1}^{\infty}\frac{1}{m^2n^2}=\frac{\pi^4}{36}$$

By symmetry, the first and last double sums are identical.

$$\sum_{n=1}^{\infty}\frac{1}{n^4}+2\sum_{m=1}^{\infty}\sum_{n=m+1}^{\infty}\frac{1}{m^2n^2}=\frac{\pi^4}{36}$$

In Part A, we found that this double sum adds up to $\frac{\pi^4}{120}$.

$$\sum_{n=1}^{\infty}\frac{1}{n^4}+2\left(\frac{\pi^4}{120}\right)=\frac{\pi^4}{36}$$

$$\sum_{n=1}^{\infty}\frac{1}{n^4}+\frac{\pi^4}{60}=\frac{\pi^4}{36}$$

$$\sum_{n=1}^{\infty}\frac{1}{n^4}=\frac{\pi^4}{36}-\frac{\pi^4}{60}=\frac{5\pi^4}{180}-\frac{3\pi^4}{180}=\frac{2\pi^4}{180}=\frac{\pi^4}{90}$$

Check the answer: $\frac{1}{1^4}+\frac{1}{2^4}+\frac{1}{3^4}+\frac{1}{4^4}+\frac{1}{5^4}+\frac{1}{6^4}+\frac{1}{7^4}+\frac{1}{8^4}\approx 1.0818$ agrees with $\frac{\pi^4}{90}\approx$ 1.0823.

Index

WAS THIS BOOK HELPFUL?

Much effort and thought were put into this book, such as:

- Introducing the main ideas at the beginning of each chapter. The goal was to be concise while also covering the pertinent information.
- Solving examples step by step to serve as a helpful guide.
- Including not just answers, but even the solutions to most of the problems.
- Striving to make abstract or difficult concepts clear to all students, while also trying to inspire and challenge advanced students; it's a difficult balancing act.

If you appreciate the effort that went into making this book possible, there is a simple way that you could show it:

Please take a moment to post an honest review.

For example, you can review this book at Amazon.com or Goodreads.com.

Even a short review can be helpful and will be much appreciated. If you are not sure what to write, following are a few ideas, though it is best to describe what is important to you.

- Was it helpful to have notes/solutions in addition to the answers at the back of the book?
- Were the examples useful?
- Were you able to understand the ideas at the beginning of the chapter?
- Did this book offer good practice for you?
- Would you recommend this book to others? If so, why?

Do you believe that you found a mistake? Please email the author, Chris McMullen, at greekphysics@yahoo.com to ask about it. One of two things will happen:

- You might discover that it wasn't a mistake after all and learn why.
- You might be right, in which case the author will be grateful and future readers will benefit from the correction. Everyone is human.

ABOUT THE AUTHOR

Dr. Chris McMullen has over 20 years of experience teaching university physics in California, Oklahoma, Pennsylvania, and Louisiana. Dr. McMullen is also an author of math and science workbooks. Whether in the classroom or as a writer, Dr. McMullen loves sharing knowledge and the art of motivating and engaging students.

The author earned his Ph.D. in phenomenological high-energy physics (particle physics) from Oklahoma State University in 2002. Originally from California, Chris McMullen earned his Master's degree from California State University, Northridge, where his thesis was in the field of electron spin resonance.

As a physics teacher, Dr. McMullen observed that many students lack fluency in essential math skills. In an effort to help students of all ages and levels become fluent in mathematics, he published a series of math workbooks on fractions, long division, word problems, algebra, geometry, trigonometry, logarithms, calculus, probability, differential equations, complex numbers, and more. Dr. McMullen has also published a variety of science books, including astronomy, chemistry, and physics workbooks.

Author, Chris McMullen, Ph.D.

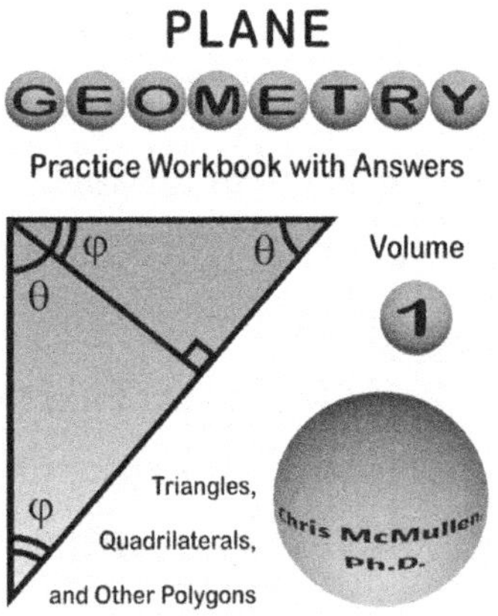
PLANE
GEOMETRY
Practice Workbook with Answers
Volume
1
Triangles,
Quadrilaterals,
and Other Polygons
Chris McMullen
Ph.D.

PLANE
GEOMETRY
Practice Workbook with Answers
Volume
2
Circles, Chords,
Secants, and Tangents
Chris McMullen
Ph.D.

GEOMETRY
PROOFS
ESSENTIAL PRACTICE PROBLEMS
WORKBOOK WITH FULL SOLUTIONS
CHRIS MCMULLEN, PH.D.

Intermediate
ALGEBRA
Skills Practice Workbook
with Answers
Chris McMullen, Ph.D.

101 Involved
ALGEBRA
Problems with Answers
Chris McMullen, Ph.D.

COMPLEX
NUMBERS
√3 + 4i
5 + 10i
4 − 3i
cosh(π/6 i)
ln i
Chris McMullen, Ph.D.

Chris McMullen, Ph.D.
log3 81 = ?
log4 32 = ?
ln e = ?
e0 = ?
ln(ex) = ?
Essential Skills Practice Workbook with Answers
2a = ?
cosh(x − y) = ?
LOGARITHMS
and
EXPONENTIALS

TRIGONOMETRY
Essentials Practice Workbook with Answers
Master Basic Trig Skills
cos(α + β) =
cos α cos β − sin α sin β
Improve Your Math Fluency Series
Chris McMullen, Ph.D.

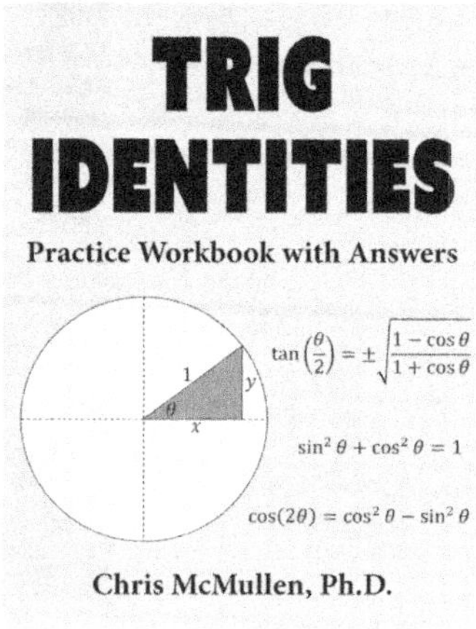
TRIG
IDENTITIES
Practice Workbook with Answers
tan(θ/2) = ±√((1 − cos θ)/(1 + cos θ))
sin² θ + cos² θ = 1
cos(2θ) = cos² θ − sin² θ
Chris McMullen, Ph.D.

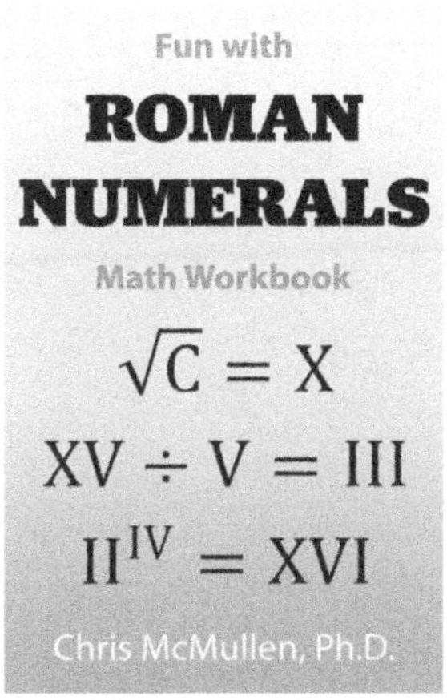
Fun with
ROMAN
NUMERALS
Math Workbook
$\sqrt{\text{C}} = \text{X}$
$\text{XV} \div \text{V} = \text{III}$
$\text{II}^{\text{IV}} = \text{XVI}$
Chris McMullen, Ph.D.

300+
MATHEMATICAL
PATTERN
PUZZLES
NUMBER PATTERN RECOGNITION AND REASONING
CHRIS MCMULLEN, PH.D.

PYRAMID MATH
PUZZLE CHALLENGE
CAN YOU FILL IN THE MISSING NUMBERS?
CHRIS MCMULLEN, PH.D.

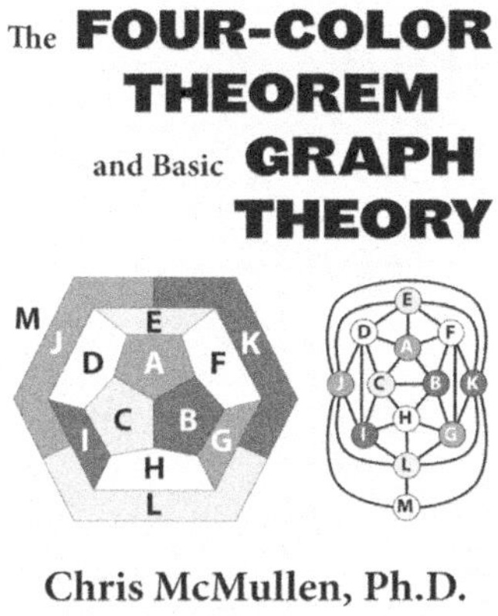
The FOUR-COLOR
THEOREM
and Basic GRAPH
THEORY
Chris McMullen, Ph.D.

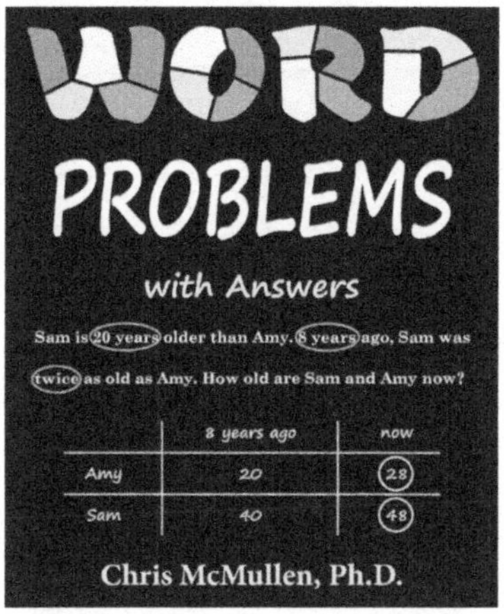
WORD
PROBLEMS
with Answers
Sam is 20 years older than Amy. 8 years ago, Sam was
twice as old as Amy. How old are Sam and Amy now?
Chris McMullen, Ph.D.

ESSENTIAL
PROBABILITY
PRACTICE WORKBOOK
WITH ANSWERS
CHRIS MCMULLEN, PH.D.

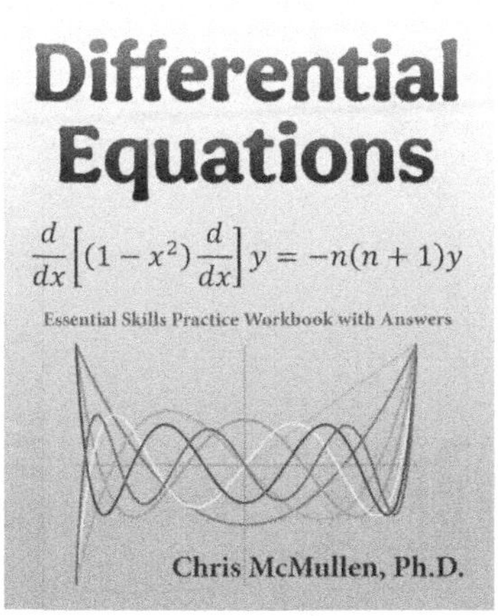
Differential
Equations
$\frac{d}{dx}\left[(1-x^2)\frac{d}{dx}\right]y = -n(n+1)y$
Essential Skills Practice Workbook with Answers
Chris McMullen, Ph.D.

Essential
CALCULUS
Skills Practice Workbook
with Full Solutions
$\frac{d}{dx}\tan(5x)$
$\int \sqrt{1-x^2}\,dx$
Chris McMullen, Ph.D.

CALCULUS
with Multiple Variables
Essential Skills Workbook
$\frac{\partial^2}{\partial x \partial y} 3\ln(x+y)$
Includes Vector Calculus
and Full Solutions
$\int_{x=0}^{2}\int_{y=4}^{9} x\sqrt{y}\,dx\,dy$
Chris McMullen, Ph.D.

Made in the USA
Las Vegas, NV
31 July 2025

25632410R00190